Evolutionary Biology – Concepts, Biodiversity, Macroevolution and Genome Evolution

Pierre Pontarotti
Editor

Evolutionary Biology – Concepts, Biodiversity, Macroevolution and Genome Evolution

Springer

Editor
Dr. Pierre Pontarotti
UMR 6632
Université d'Aix-Marseille/CNRS
Laboratoire Evolution Biologique et Modélisation
3 Place Victor Hugo
case 19
13331 Marseille Cedex 03
France
Pierre.Pontarotti@univ-provence.fr

ISBN 978-3-642-20762-4 e-ISBN 978-3-642-20763-1
DOI 10.1007/978-3-642-20763-1
Springer Heidelberg Dordrecht London New York

Library of Congress Control Number: 2011932535

Cover design: deblik, Berlin, Germany

Printed on acid-free paper

Springer is part of Springer Science+Business Media (www.springer.com)

Preface

For the 14th time, the Evolutionary Biology Meeting at Marseilles (EBM) took place in the two-thousand-year old city. The aim of this congress is to allow scientists involved in the research in evolutionary biology and in the application of evolutionary biology concepts, from mathematics to epistemology, to meet, exchange, and start interdisciplinary collaborations.

The organizers see the Marseille's meeting as a scientific exchange platform and a booster for the use of evolutionary based approaches not only in biology, but also in other scientific areas.

The outputs of the meeting are proved by numerous collaborations initiated during the meeting and achieved by several peer-reviewed articles published over the past years.

This year, more than 100 presentations – talks, flash presentations and posters – were selected by the committee. Those works really reflected the epistemological positioning of the meeting. We have chosen one fifth of the most representative ones to make this book.

The book will give the reader an overview of the state of the art in the evolutionary biology field. As during the congress, this area of science will be considered from an epistemological point of view. This work is the fourth that we have published further to the EBM. We want to underline that the four books are complementary and should be understood as tomes.

The 19 selected articles are organized according to the following categories:

1. Concepts in evolution
2. Evolution and biodiversity
3. Macroevolution
4. Genome evolution

We would like to thank the scientists who contributed to this book, the meeting participants as well as our sponsors: the University of Provence, the CNRS, the GDR BIM, the Conseil Général 13, the Municipality of Marseille and Genoscreen. We also wish to thank the Springer's Editions staff for its competences and help.

Finally, we want to give thanks to the members of the Association pour l'Etude de l'Evolution Biologique (AEEB) and the members of the Evolutionary Biology Modeling Laboratory.

We sincerely want to thank the AEEB coordinator, Axelle Pontarotti, for the remarkable coordination of the meeting. The scientific outputs of the meeting – initiated international collaborations, scientific exchanges – are due, of course, to the quality of the participants, but also to the Marseilles way of hosting of which Axelle is an outstanding ambassador.

We wish to thank also our new coordinator, Marie-Hélène Rome, for her help with the book.

Marseilles, France
May 2011

Gaëlle Pontarotti
Association pour l'Etude de l'Evolution Biologique (AEEB)
Pierre Pontarotti
Directeur de recherche CNRS
President of the AEEB

Contents

Contributors

Ornella Affinito Stazione Zoologica Anton Dohrn, Villa Comunale, 80121 Naples, Italy, ornella.affinito@szn.it

Juan R. Alvarez-Dominguez Department of Molecular Biology, Princeton University, Princeton, NJ, USA

Olga Amosova Department of Molecular Biology, Princeton University, Princeton, NJ, USA

Tina Begum Bioinformatics Centre, Bose Institute, Kolkata, India

Alexander Belyayev Laboratory of Plant Molecular Cytogenetics, Institute of Evolution, 250 University of Haifa, Mt. Carmel, Haifa, Israel, belyayev@research.haifa.ac.il

Eugene V. Berezikov Institute of Cytology and Genetics, Siberian Branch of the Russian Academy of Sciences, Novosibirsk, Russia; Hubrecht Institute, RNAAS, Utrecht, The Netherlands

Andrey V. Bocharnikov Department of Natural Sciences, Novosibirsk State University, Novosibirsk, Russia

Didier Bouchon Université de Poitiers, UMR CNRS 6556 Ecologie Evolution Symbiose, Poitiers, France

Nicolas Cerveau Université de Poitiers, UMR CNRS 6556 Ecologie Evolution Symbiose, Poitiers, France

Sandip Chakraborty Bioinformatics Centre, Bose Institute, Kolkata, India

Richard Cordaux Université de Poitiers, UMR CNRS 6556 Ecologie Evolution Symbiose, Poitiers, France, richard.cordaux@univ-poitiers.fr

Jacques Dainat UMR6632, Evolutionary Biology and Modeling, Université de Provence, Marseille, France

Etienne G.J. Danchin INRA, CNRS, Université de Nice-Sophia Antipolis, UMR 1301, 400 route des Chappes, B.P. 167, F-06903 Sophia-Antipolis Cedex, France, etienne.danchin@sophia.inra.fr

Elodie Darbo UMR6632, Evolutionary Biology and Modeling, Université de Provence, Marseille, France

Karine Van Doninck Unit of Research in Organism Biology (URBO), University of Namur (FUNDP), Namur, Belgium

Ronald E. Ellis Department of Molecular Biology, The UMDNJ School of Osteopathic Medicine, B303 Science Center, Stratford, NJ, USA, ron.ellis@umdnj.edu

Mark C. Field Department of Pathology, University of Cambridge, Cambridge, United Kingdom

Jean-François Flot Unit of Research in Organism Biology (URBO), University of Namur (FUNDP), Namur, Belgium

Jacques R. Fresco Department of Molecular Biology, Princeton University, Princeton, NJ, USA, jrfresco@princeton.edu

Toni Gabaldón Comparative genomics group, Bioinformatics and Genomics Programme, Centre for Genomic Regulation (CRG), UPF, Barcelona, Spain

Tapash Chandra Ghosh Bioinformatics Centre, Bose Institute, Kolkata, India, tapash@boseinst.ernet.in

Damian Glumcher Department of Molecular Biology, Princeton University, Princeton, NJ, USA

Philippe Gouret UMR6632, Evolutionary Biology and Modeling, Université de Provence, Marseille, France, philippe.gouret@univ-provence.fr

Konstantin V. Gunbin Institute of Cytology and Genetics, Siberian Branch of the Russian Academy of Sciences, Novosibirsk, Russia

Yiqing Guo Department of Molecular Biology, The UMDNJ School of Osteopathic Medicine, Stratford, NJ, USA

John Harting Department of Biological Sciences, Texas Tech University, Lubbock, TX, USA, john.harting@ttu.edu

Bratati Kahali Bioinformatics Centre, Bose Institute, Kolkata, India

Rod W. King CSIRO, Plant Industry, Canberra, ACT 2601, Australia, rod.king@csiro.au

Margie Kinnersley Division of Biological Sciences, University of Montana, Missoula, MT, USA

Lila V. Koumandou Department of Pathology, University of Cambridge, Tennis Court Road, Cambridge CB2 1QP, UK; Biomedical Research Foundation, Academy of Athens, Athens, Greece, koumandou@cantab.net

Sébastien Leclercq Université de Poitiers, UMR CNRS 6556 Ecologie Evolution Symbiose, Poitiers, France

Anthony Levasseur INRA, UMR1163 de Biotechnologie des Champignons Filamenteux, IFR86-BAIM, Universités de Provence et de la Méditerranée, ESIL, 163 avenue de Luminy, CP 925, 13288 Marseille Cedex 09, France; Universités Aix-Marseille 1 et 2, UMR1163, 163 avenue de Luminy, CP 925, 13288 Marseille Cedex 09, France

Eladio Liñán Área y Museo de Paleontología, Depto. de Ciencias de la Tierra, Fac. de Ciencias, Universidad de Zaragoza, Zaragoza, Spain, linan@unizar.es

Rudi Loesel Unit of Developmental Biology and Morphology of Animals, Institute for Biology II (Zoology), RWTH Aachen University, Aachen, Germany, loesel@bio2.rwth-aachen.de

Dorra Louati UMR6632, Evolutionary Biology and Modeling, Université de Provence, Marseille, France; (LAMSIN-IRD) ENIT, Ecole Nationale d'Ingénieurs de Tunis BP 37, Le Belvèdere 1002-Tunis, Tunisia

Marina Marcet-Houben Comparative genomics group, Bioinformatics and Genomics Programme, Centre for Genomic Regulation (CRG), UPF, Barcelona, Spain, mmarcet@crg.es

Elliott H. Margulies Genome Informatics Section, Genome Technology Branch, National Human Genome Research Institute, National Institutes of Health, Bethesda, MD, USA, elliott@nhgri.nih.gov

Michel Laurin Département Histoire de la Terre, UMR 7207, CNRS/MNHN/UPMC, Centre de Recherches sur la Paleodiversité et les Paléoenvironments, Muséum national d'Histoire naturelle, Bâtiment de Géologie, Case Postale 48, 43 rue Buffon, F-75231 Paris Cedex 05, France, michel.laurin@upmc.fr

Julien Paganini UMR6632, Evolutionary Biology and Modeling, Université de Provence, Marseille, France

Anthony Papadopoulos Department of Biological Sciences, Texas Tech University, Lubbock, TX, USA, anthony.papadopoulos@ttu.edu

Stephen C. J. Parker Program in Bioinformatics, Boston University, Boston, MA, USA; Genome Informatics Section, Genome Technology Branch, National Human Genome Research Institute, National Institutes of Health, Bethesda, MD, USA, stephen.parker@nih.gov

Laetitia Perfus-Barbeoch INRA, CNRS, Université de Nice-Sophia Antipolis, UMR 1301, 400 route des Chappes, B.P. 167, F-06903 Sophia-Antipolis Cedex, France

Soumita Podder Bioinformatics Centre, Bose Institute, Kolkata, India

Pierre Pontarotti UMR6632, Evolutionary Biology and Modeling, Université de Provence, Marseille, France, Pierre.Pontarotti@univ-provence.fr

Gabriele Procaccini Stazione Zoologica Anton Dohrn, Villa Comunale, 80121 Naples, Italy, gabriele.procaccini@szn.it

Olga Raskina Laboratory of Plant Molecular Cytogenetics, Institute of Evolution, University of Haifa, Mt. Carmel, Haifa, Israel

Sean H. Rice Department of Biological Sciences, Texas Tech University, Lubbock, TX, USA, sean.h.rice@ttu.edu

Frank R. Rosenzweig Division of Biological Sciences, University of Montana, Missoula, MT, USA, Frank.Rosenzweig@mso.umt.edu

Kamalika Sen Bioinformatics Centre, Bose Institute, Kolkata, India

Gavin Sherlock Department of Genetics, Stanford University School of Medicine, Stanford, CA, USA

Paolo Sordino Stazione Zoologica Anton Dohrn, Villa Comunale, 80121 Naples, Italy, paolo.sordino@szn.it

Helena Storchova Institute of Experimental Botany v.v.i., Academy of Sciences of the Czech Republic, Rozvojová 223, 165 00 Prague 6, Lysolaje, Czech Republic, storchova@ueb.cas.cz

Rafael Torres Department of Molecular Biology, Princeton University, Princeton, NJ, USA

Francesco Toscano Stazione Zoologica Anton Dohrn, Villa Comunale, 80121 Naples, Italy, francesco.toscano@szn.it

Thomas D. Tullius Department of Chemistry, Boston University, Boston, MA, USA; Program in Bioinformatics, Boston University, Boston, MA, USA, tullius@bu.edu

José Antonio Gámez Vintaned Área de Paleontología, Depto. de Geología, Fac. de Biológicas, Universitat de València, Burjassot, Valencia, Spain, j.antonio.gamez@uv.es; gamez@unizar.es

Oleg V. Vishnevsky Institute of Cytology and Genetics, Siberian Branch of the Russian Academy of Sciences, 630090, Prospekt Lavrentyeva 10, Novosibirsk, Russia; Chair of Information Biology, Department of Natural Sciences, Novosibirsk State University, Novosibirsk, Russia, oleg@bionet.nsc.ru

Peter Wei Department of Molecular Biology, Princeton University, Princeton, NJ, USA

Jared W. Wenger Department of Genetics, Stanford University School of Medicine, Stanford, CA, USA

Andrey Yu. Zhuravlev Área y Museo de Paleontología, Depto. de Ciencias de la Tierra, Fac. de Ciencias, Universidad de Zaragoza, Zaragoza, Spain; Geological Institute, Russian Academy of Sciences, Moscow, Russia, ayzhur@mail.ru

Part I
Concepts

Chapter 1
Site-Specific Self-Catalyzed DNA Depurination, the Basis of a Spontaneous Mutagenic Mechanism of Wide Evolutionary Significance

Jacques R. Fresco, Olga Amosova, Peter Wei, Juan R. Alvarez-Dominguez, Damian Glumcher, and Rafael Torres

Abstract This chapter focuses on the nature of site-specific self-catalyzed DNA depurination as a spontaneous mechanism inherent in the chemical structure and dynamics of DNA that has contributed to evolutionary change. It describes the essential molecular features of the mechanism, the short consensus sequence elements that form the catalytic intermediate, the basics of the reactions that lead to the creation of apurinic sites, and the means by which those sites give rise to substitution and short deletion mutations. The consensus sequences are widely distributed in double-stranded genomes across the phyla at high frequency that increases up the phylogenetic tree. In the human genome, they constitute $>2 \times 10^6$ potential mutagenic sites, non-randomly scattered among very many genes, some containing multiple sites. Examples are presented of genes in which the mutations coincide with their self-depurination consensus sequences, the most striking being those in the β-globin gene that are responsible for six anemias and two β-thalassemias. Those of the olfactory receptor genes and the hypervariable regions of the immunoglobulin genes are shown to have utilized the mechanism to evolve their high degree of diversity and/or to develop their contemporaneous diversity for their present function.

1.1 Introduction

Spontaneous mutations are among the primary engines of evolutionary change. Until now, the major mode of their occurrence has been thought to be a consequence of errors in DNA replication, resulting in substitution and frameshift mutations. In reality, such mutational errors are not due to the enzymatic process having gone awry. Rather, substitution mutations are the consequences of intrinsic

---

J.R. Fresco • O. Amosova • P. Wei • J.R. Alvarez-Dominguez • D. Glumcher • R. Torres
Department of Molecular Biology, Princeton University, Princeton, NJ 08544, USA
e-mail: jrfresco@princeton.edu

P. Pontarotti (ed.), *Evolutionary Biology – Concepts, Biodiversity, Macroevolution and Genome Evolution*, DOI 10.1007/978-3-642-20763-1_1,

chemical equilibria associated with the tautomers of the bases and the steric isomers of the nucleotide precursors of DNA replication (Topal and Fresco 1976). Frame-shift mutations, on the other hand, arise as a consequence of template or growing strand DNA misfoldings resulting in sequences that misdirect the polymerase, which is possible because of the flexibility of the nucleic acid backbone (Fresco and Alberts 1960).

Recently, we discovered a very different source of spontaneous mutations that is due to a self-catalyzed, site-specific DNA depurination mechanism inherent in a DNA sequence element 14–16 base pairs long (Amosova et al. 2006). The mechanism is mediated by this stem-loop-forming sequence that is widely present in all the more than 100 double-stranded genomes we have examined from Archaea to Homo sapiens, occurring within many lower forms with a frequency of every ~5,000 base pairs for self-depurination of G-residues, and gradually rising up the phylogenetic tree to every ~3,000 base pairs for self-depurination of those pairs in man, i.e., $\sim 1.25 \times 10^6$ such sites.

Self-depurination has the potential to occur wherever the consensus sequences specific for self-catalysis of depurination of G or of A residues (see below and Fig. 1.1) are present, and so give rise to apurinic sites in the DNA backbone. Those sites can, in turn, result in point mutations as a consequence of their highly error-prone repair (Boiteux and Guillet 2004; Chakravarti et al. 2000; Korolev 2005; Simonelli et al. 2005). On the one hand, such repair can cause a substitution mutation. Alternatively, if the apurinic site gives rise to a backbone break by either

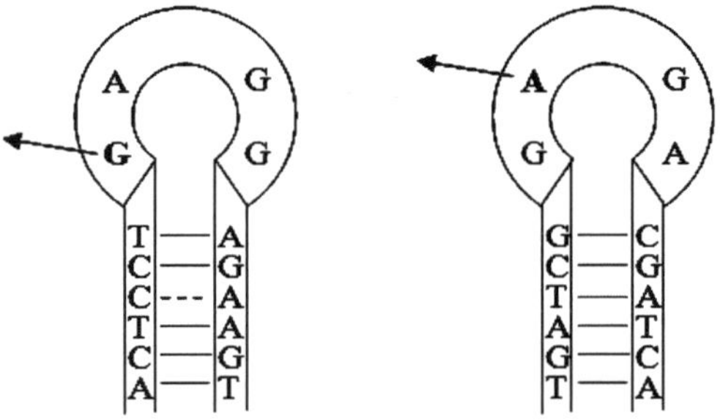

Fig. 1.1 Schematic diagrams of the stem-loop structure of the catalytic intermediates for site-specific self-catalyzed depurination in DNA of the indicated G-residue (*left*) and A-residue (*right*). In each case, the self-depurinating stem-loop is one side of a cruciform extruded from duplex DNA. The complementary inverted repeat sequence forms the other stem-loop of the cruciform that is, however, incapable of self-depurination. Certain sequence elements of the self-depurinating stem-loops are required for self-depurination to occur: the residues shown for each tetra-loop; particular complementary base pairs at the base of the loop, T•A or G•C on the left, and T•A, G•C or A•T on the right. These should be followed by any sequence of four or more additional complementary base pairs that can generally tolerate one mismatched pair (as shown on the left) or a single extrahelical residue. Thus, the specificity for self-depurination lies in the loop sequence and first base pair. The helical stem otherwise functions to maintain some strained loop structure required for the self-catalysis, and by its stability, to extend the lifetime of the intermediate and so enhance the kinetics of the self-catalysis

the action of the widely occurring enzyme apurinic endonuclease (Korolev 2005) or by way of the well-known spontaneous β-elimination reaction (Lhomme et al. 1999; Sugiyama et al. 1994), the strand break site results in a frayed end accessible to exonuclease attack, particularly if the first base pair after the break is A•T, and so leads to a short deletion mutation.

In this report, some essential features of the underlying self-depurination mechanism are first described. We then proceed to issues of biological relevance, including indications that the mechanism has played a role in the evolution of some biological phenomena. Because the consensus sequence for self-depurination of G-residues was discovered several years ago, and that for A-residues only very recently, our data is much more extensive for the former. Nevertheless, it will become apparent that both mechanisms are very similar, and appear to have played comparable mutagenic roles in several biological processes, including those of molecular evolution.

1.2 Essential Features of DNA Self-Catalyzed Depurination

Self-catalyzed depurination is a remarkably site-specific reaction that does not involve the direct participation of either a protein enzyme or any multivalent cation or cofactor, and can occur under essentially physiological conditions (Amosova et al. 2006). As such, it represents the first natural deoxyribozyme activity discovered. It is mediated by formation of two very similar stem-loop-forming consensus sequences, which we have found to be present at rather high frequency in every double-stranded DNA genome searched, from the lowest to the highest form.

Figure 1.1 shows schematically the stem-loop structure of two sequences that are highly site-specific for self-catalyzed depurination of a G-residue (left) and an A-residue (right). As the figure indicates, the self-depurination mechanism removes the 5′G-residue of the loop in the former case, and the A-residue toward the 5′ end in the latter one. In either case, the initial product of the catalytic event is an apurinic site in the loop sequence. This chain backbone site is thereby labilized, carrying a potential for intracellular backbone cleavage either by the enzyme apurinic endonuclease, or else as a result of its susceptibility to spontaneous backbone cleavage by a β-elimination reaction that can occur at slightly alkaline intracellular pH. Such an apurinic site, susceptible to error-prone repair, is potentially highly mutagenic and can give rise to a substitution or a short deletion. It is this resultant mutagenic potential that confers on self-depurination a role in evolution, all the more so because the self-catalytic depurination rate we have measured in vitro (Amosova et al. 2006) occurs 10^4–$10^5\times$ faster than the background spontaneous depurination rate that has been estimated in vivo (Lindahl and Nyberg 1972).

DNA is typically double-stranded, whereas the catalytic intermediate for self-depurination is a single-stranded stem-loop. Hence, the self-depurination mechanism requires that the inverted repeat sequence harboring the self-depurinating loop first extrude as a cruciform (half of which contains the single-stranded self-depurinating stem-loop), and that the cruciform have a sufficient lifetime for the reaction to occur. That such cruciform extrusion can take place has been demonstrated previously (Alvarez et al. 2002; Inagaki et al. 2009; Kim et al. 1998; Shlyakhtenko et al. 1998). We have recently performed experiments under physiological conditions in vitro with stem-loop-forming sequences for G-residue self-depurination embedded in supercoiled plasmids, in which such extrusion has been directly shown to be crucial for the self-catalytic depurination (Amosova et al. 2011b).

As Fig. 1.1 indicates, the essential features of the stem-loops are very similar, i.e., both have tetra-loops with different highly specific sequences: 5′G-A/T-G-G for depurination of the extreme 5′ G-residue and 5′G-A-G-A for depurination of the second residue in from the 5′ end, which is an A. Interestingly, both can form a homopurine base pair within the loop, G^{+}•G and A^{+}•A, in which the residue to be depurinated is protonated at N3 of the base (Lavelle and Fresco, in preparation). It is this base pair formation that likely explains why acid-catalyzed depurination can actually occur in the neutral pH environment of most cells.

The G-residue self-depurination activity exhibits very limited tolerance for loop sequence variation (Amosova et al. 2011a). For example, the three G-residues in the loop are replaceable only by hypoxanthine, a closely related purine analogue, with only modest reduction in the activity; and the A-residue in the G-self-depurinating sequence is replaceable only by a T-residue, in this case with activity enhancement. In contrast, there is total tolerance for variation in the complementary base pairs of the helical stem, except for the first one at the base of the loop. Even a single base pair mismatch or a single extrahelical base elsewhere in the stem can be tolerated. Apparently, the role of the stem is to stabilize and maintain the loop in some strained configuration favorable to glycosyl bond cleavage. Thus, the more stable the stem, as affected by length (Blake and Fresco 1973; Brahms et al. 1967), G•C content (Marmur and Doty 1959), base pair sequence (Ornstein and Fresco 1983), base pair mismatches (Lomant and Fresco 1973), and the presence of extrahelical bases (Lomant and Fresco 1973), the faster the rate of self-depurination. The nature of the first base pair is somewhat restricted, possibly because it orients the water molecule involved in the hydrolysis of the glycosyl bond of the residue to be depurinated.

1.3 Biological Relevance of the Self-Depurination Mechanism

The self-depurination reaction was discovered in the course of working with 29-residue long complementary deoxyoligonucleotide strands of the human β-globin gene that contained the sickle cell anemia mutation site. Whereas the duplex formed from those strands, as well as the noncoding strand by itself, were characteristically stable, the coding strand was not. Rather, under solvent, pH, and temperature that mimic physiological conditions, it rapidly self-fragmented not randomly, but in a unique way. This fragmentation was found to arise from backbone cleavage that was ultimately traced to spontaneous β-elimination at the apurinic site caused by self-catalyzed depurination mediated by stem-loop formation (Amosova et al. 2006). The occurrence of such a reaction in a strand segment of a significant human gene, immediately upstream of the sickle cell mutation site, provided the impetus for trying to understand the significance of what was obviously a DNA self-catalyzed reaction.

With the finding that the stem base-paired sequence is tolerant of great variation, it was decided to determine the number of potential stem-loops for self-depurination of G-residues in the human genome, and their distribution across the phyla. The numbers proved to be surprisingly large, and indicative of substantial overrepresentation relative to random expectation, which was not the case for stem-loops of similar size with non-self-depurinating loop sequences. The overrepresentation was therefore taken to be indicative of some important bio-functionality of the stem-loops for self-depurination. The whole-genome search for self-depurinating stem-loops was complemented first by identifying all human genes containing those sequences, including the loci of their occurrence, i.e., exons, introns, control elements, untranslated regions, intergenic regions, etc. This was followed by analysis of G-residue consensus sequence occurrence within more than 100 individual genes and their degree of overrepresentation relative to random expectation (see Table 1.3 as an example). These searches, together with the uncovering of the stem-loop consensus sequence for self-depurination of A-residues by way of its overrepresentation in the human genome, led us to the indications of the role of self-depurinating sequences in evolution; and it is from this vantage point that we present the findings that follow.

1.4 Self-Depurination in the Human β-Globin Gene

This gene, which has 148 codons (in its message), contains no A-residue self-depurination site, and but one site for G-residue self-depurination, of which the first three residues of the loop correspond to codon 6. It is the second residue of the loop that is the site of the sickle cell anemia mutation. Figure 1.2 shows a plot of the number of independent variations (i.e., those in different haplotypes) per codon of this gene. The plot is based upon data obtained from two databases, HB Var (http://globin.cse.psu.edu), which includes all human hemoglobin variants and β-thalassemia mutations reported in the literature over more than half a century, and the recently assembled Human Gene Mutation Database (HGMD) (http://www.hgmd.cf.ac.uk). As such, the plot represents a compendium of all unique β-globin alleles revealed in most of the human populations in the world. It is striking that the most prominent mutation site in the plot in Fig. 1.2, that at codon 6, corresponds to three of the four loop residues of that single self-depurinating consensus sequence in this gene. In this glutamic acid codon, residue #1, the self-depurinating G-residue, and residue #2, the site of the sickle cell mutation, are the sites of readily detectable

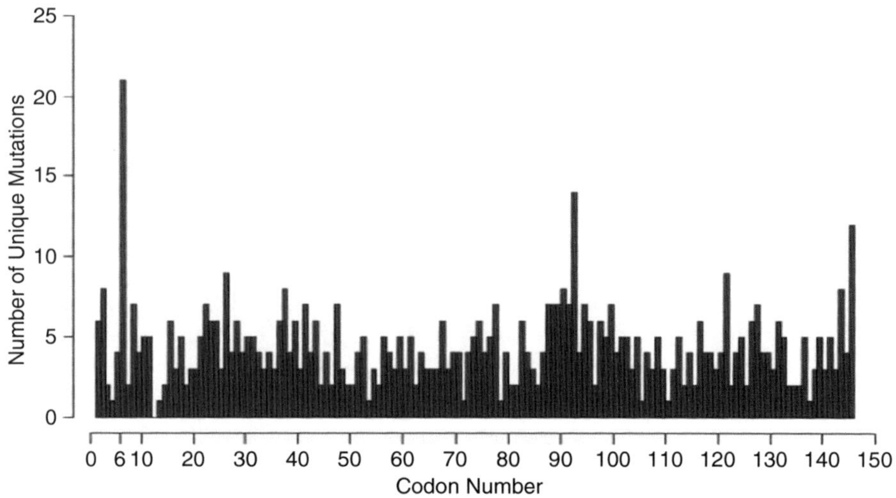

Fig. 1.2 Distribution of mutations appearing in independent haplotypes, i.e., unique mutations, among the codons of the β-globin gene. Codon 6, with the highest mutation frequency, is the only site in this gene capable of forming a stem-loop for self-depurination. Codon 6 and an additional 3′G-residue constitute the loop, which is preceded and followed by residues that form the stem shown in Fig. 1.1 (*left*). As can be deduced from Table 1.1, error-prone repair of the apurinic site resulting from self-depurination of the 5′G-residue of the loop must give rise to the observed substitution and deletion mutations, and thereby to the various anemias and β-thalassemias listed

Table 1.1 Coincidence of mutations reported in codon 6 of the β-globin gene with the first three loop residues of its only stem-loop-forming G-residue self-depurination consensus sequence[a]

		Loop residues		
Type of human hemoglobin	Reference	#1	#2	#3
Hb A (wild-type)		G	A	G
Hb C	Siriboon et al. (1993)	A	A	G
Hb Machida	Harano et al. (1982)	C	A	G
Hb Grignoli	Grignoli et al. (2000)	T	A	G
Hb S (Sickle cell)	Engelke et al. (1988)	G	T	G
Hb Lavagna	Tanca et al. (2008)	G	G	G
Hb G Makassar	Sangkitporn et al. (2002)	G	C	G
β-Thalassemia	Kazazian et al. (1983)	G	D	G
β-Thalassemia	Dejong et al. (1968); Juricic et al. (1983)	D	D	D

[a]A, G, C, T are standard base symbols; D indicates a deleted residue

substitutions (Table 1.1). Due to the partial degeneracy in the glutamic acid codon, a transition mutation in residue #3 would be silent, i.e., result in no amino acid change; but a transversion would encode aspartic acid, with little effect on hemoglobin function. Such a transversion should be detectable by the protein sequencing used to identify most of these mutations, and both types should be by DNA sequence analysis. So far, SNPs at that codon residue have not been reported. Codon 6 is also the site of a single base deletion at residue #2 and a deletion of the entire codon. As noted earlier, this range of substitutions and short deletions at an apurinic site is just what is expected as a consequence of error-prone repair. All these are known to be inherited germline mutations that must have occurred over evolutionary time, and are responsible for different anemias and β-thalassemias. As such, the coincidence of these inherited mutations with the self-depurination consensus sequence at codon 6 in the β-globin gene provides strong support for the occurrence of the self-depurination mechanism in vivo, at least in germline cells. In this connection, it is worthy of mention that several haplotypes of the sickle cell mutation have been identified (Chebloune et al. 1988; Wailoo 1991), each likely representing an independent occurrence traceable to at least four different places in the Indian/Saudi Arabian subcontinent and in Africa (Kulozik et al. 1986; Lapouméroulie et al. 1992; Pagnier et al. 1984; Schroeder et al. 1989). These occurrences speak to the mutagenicity associated with the error-prone repair of the apurinic sites created by the self-depurinating mechanism.

1.5 Coincidence of Substitution and Short Deletion Mutations with the Consensus Sequence for Self-Depurination of G-Residues in Some Human Genes

Having discovered the self-depurination mechanism at the site of codon 6 in the β-globin gene, and then found that nearly all detectable mutations anticipated for this mechanism have occurred over time at this site, it was of interest to see whether such mutations have occurred as well in other genes by this mechanism. With ~1.25×10^6 potential G-residue self-depurinating sites in the human genome, and given their fairly regular distribution among exons, introns, control elements, etc., numbers of those sites should coincide with mutations detected in exons of other genes. A number of other such coincidences (see below) have indeed been found. At this early stage of human gene sequencing and the analysis of SNPs and short deletions, the finding of such coincidences in some cases can be viewed as more significant than their absence in others.

Table 1.2 provides details of nine such examples in five different genes. In each example, the entire consensus sequence is capable of forming a stem-loop catalytic intermediate. This sampling includes examples with the two loop sequences, 5′G-A-G-G and 5′G-T-G-G, each with T•A or G•C as the complementary pair at the base of the loop. Substitutions are seen to occur in the depurinated first loop residue, in the second loop residue, in the third, and in the fourth; there are three examples of deletion, one of the depurinated G-residue, another of the two consecutive G residues in the loop, and one of two stem residues.

As was apparent from the very limited tolerance for stem-loop sequence variation, it is reasonable to assume that any site in a genome which meets the sequence criteria for the self-depurinating mechanism is a potential site for mutation. That this is in fact the case is made even more convincing based upon the findings that follow.

Table 1.2 Coincidence of self-depurination consensus sequence and mutation sites in some human genes

Gene	Loop + first base-pair sequence[a]	Loop codons	Codon at mutation site	Mutation	Reference
TTR, transthyretin	T-**G**AG-GAG	61–62	61	**G**AG → AAG	Rosenzweig et al. (2007)
TTR, transthyretin	T-**G**AG-GAG	61–62	61	GAG → GGG	Shiomi et al. (1993)
SERPINC1, antithrombin	G-**G**AGG-C	42–43	42	G-**G**AG → TAG	Jochmans et al. (1998)
SERPINC1, antithrombin	T-**G**TGG-A	224–225	225	G-TGG → CGG	Lane et al. (1997)
BRCA1, breast cancer 1	T-**G**AGG-A	1,004–1,005	1,003	2 nt deletion of stem residues	Wagner et al. (1999)
BRCA1, breast cancer 1	T-**G**AGG-A	1,004–1,005	1,004	Loop G deleted	Goelen et al. (1999)
BRCA1, breast cancer 1	T-**G**AGG-A	1,219–1,220	1,219	**G**AG → **G**AC	Durocher et al. (1996)
Tp53, tumor protein 53	T-**G**AGG-A	190–191	190	deletion of 2 loop Gs	Varley et al. (1997)
MAPT, tau protein	G-**G**AGG-C	271–272	272	GGC → GTC	Hutton et al. (1998)

[a]The G-residue in bold is the self-depurination site. In each case, the entire consensus sequence is capable of forming a stable stem-loop catalytic intermediate.

1.6 Stem-Loop-Forming Consensus Sequences for Site-Specific Self-Catalyzed Depurination of G-Residues Are Highly Overrepresented in Some Human Genes

Although genomes have not evolved as strictly random sequences relative to some defined base pair composition, it is not unreasonable to assume that sequences of the size of the consensus sequences for self-depurination (14 nt minimal length) should be present at frequencies that do not deviate too far from random unless they have some biofunctional role to play that led to their selection. Besides, the frequency of random sequence stem-loops can be used as a control or basis for comparison.

Table 1.3 gives “random probabilities” of the consensus sequence for each gene listed, and for comparison, the actual or “observed” number of consensus sequences in each. It will be noticed that the random probabilities are quite close to the “observed” number for non-depurinating stem-loops for each gene (the ratios of the “observed” to “calculated” are in the range of 0.7–2.1), in contrast to the wide variation of this ratio for the G-residue self-depurinating stem-loops (from 4 to 48). If the random probabilities are generally slightly less than the observed numbers, it is probably because these stem-loop-forming sequences represent a subclass of inverted repeats. These are known to be overrepresented in the genome (Cox and Mirkin 1997), which, as noted, is not a true random sequence. This overrepresentation of inverted repeats could also contribute to the number of “observed”

Table 1.3 Overrepresentation of the stem-loop-forming consensus sequence for site-specific, self-catalyzed depurination of G-residues in some human genes[a]

		Self-depurinating consensus sequences			Non-self-depurinating stem-loop sequences		
Gene symbol	Total gene size (#bp)[b]	Observed[b]	Random expectation	Ratio of observed to random expectation	Observed[b]	Random expectation	Ratio of observed to random expectation
HRAS	3,309	0	0.08	0.0	81	38.05	2.1
CDKN1A	8,623	4	0.22	18.4	119	99.16	1.2
IGF2	9,653	1	0.24	4.1	132	111.01	1.2
CYP7A1	9,984	3	0.25	11.9	199	114.82	1.7
THBS1	16,389	3	0.41	7.3	238	188.47	1.3
TP53	19,179	22	0.48	45.5	156	220.56	0.7
CDKN2A	26,740	5	0.67	7.4	554	307.51	1.8
THRA	31,058	32	0.78	40.9	410	357.17	1.1
MDM2	32,235	31	0.81	38.2	600	370.70	1.6
HPRT1	40,440	28	1.02	27.5	705	465.06	1.5
Erbb2	40,523	5	1.02	4.9	699	466.01	1.5
KRAS	45,675	19	1.15	16.5	896	525.26	1.7
BARD1	81,090	32	2.04	15.7	1,530	932.54	1.6
BRCA1	81,155	99	2.05	48.4	879	933.28	0.9
BRCA2	84,193	50	2.12	23.6	965	968.22	1.0
APC	108,353	40	2.73	14.6	1,872	1246.06	1.5
ATM	146,268	68	3.69	18.4	2,639	1682.08	1.6

[a]Consensus and non-self-depurinating stem-loops were identified based upon the presence and absence, respectively, of the consensus loop and first base pair (T•A or G•C for the self-depurinating stem-loops and any base pair for the non-self-depurinating one) plus a minimum of four base pairs within the adjacent five base pair stretch.
[b]Exons + introns + control elements

depurinating consensus sequences, and its effect should not be vastly different from what is observed for the non-depurinating stem-loops, i.e., it can contribute at most a factor of 2, not of 48. The fact that G-residue self-depurination sites throughout the genome occur far in excess of random expectation, by a factor of more than 5, suggests that they probably have some significant or essential biological role(s) that might be connected to the potential they create for mutation. Some critical insights in this regard followed upon the discovery of the consensus sequence for self-depurination of A-residues.

1.7 Discovery of the Consensus Sequence for Self-Depurination of A-Residues

It is the recognition of the overrepresentation of the consensus sequence for G-residue self-depurination, and the assumption that the consensus stem-loop sequence for similar self-depurination of A-residues might also be overrepresented that led us to proceed with a search for it in the human genome. An important clue in that search was obtained from the work on the mechanism of the toxic protein ricin, which had been found to depurinate an adenine residue at a unique site, position 4,324, in 23S ribosomal RNA (Endo and Tsurugi 1987). The sequence at that site had been shown to form a stem-loop with the same number of loop and stem residues as the stem-loop for DNA self-depurination of G-residues, but with a different loop sequence, 5′G-A-G-A instead of 5′G-T/A-G-G (Amukele et al. 2005). Why then the requirement for a protein to depurinate the RNA target but not a DNA target? Our explanation was that the deoxyriboglycosyl bond, in being some three orders of magnitude more susceptible to acid-catalyzed hydrolysis than the riboglycosyl bond (Shabarova and Bogdanov 1994), did not require an enzyme catalyst. Based on this reasoning, we initiated a search in the human genome for stem-loop sequences with a 5′G-A-G-A loop and all four possible first base pairs. Once they were found, calculations were made to determine which if any were overrepresented. Interestingly, three of the four base pairs at the base of the loop were found to be similarly overrepresented significantly, which was taken to mean that they were likely self-catalytic for self-depurination of A-residues. This was confirmed in preliminary experiments.

More informative searches of the entire human genome followed to determine whether there are any genes with common functional annotation among the top 100 genes in which the A-residue self-depurinating consensus sequence is most highly overrepresented. This information was sought in order to learn whether there are any groups of genes that might have exploited the A-residue consensus sequence for self-depurination. It was indeed gratifying to find that there are, in fact, two groups of such genes, each with a common functional annotation, for which a built-in mutagenic mechanism to create sequence diversity is consistent with their function. The group with the greatest overrepresentation is the one encoding the hypervariable regions of immunoglobulins, while that with the second highest overrepresentation consists of the olfactory receptor genes.

1.8 Self-Depurination of A-Residues and Antibody Diversity

Why should the genes for the hypervariable regions of immunoglobulins encode sequences particularly disposed to undergoing mutation? In order for antibodies to perform their function, they must recognize a wide variety of antigens, including many that have not been encountered previously. Rather than the immune response mechanism having encoded what, in effect, would have to be so large a number of antibody genes in the germline as would be wasteful, biologically unproductive, and require a genome of excessive size, antibody diversity is generated by a capacity for targeted mutagenesis within a relatively small number of genes in the genome of certain somatic cells (Tonegawa 1983). From the diversity of somatic antibody genes so created by mutagenic mechanisms, clonal selection can enable production of a group of antibodies appropriate to any new antigenic challenge.

Two mechanisms were uncovered previously in (somatic) B cells to generate the requisite diversity of the different polypeptide components of the Y-shaped antibody structure (see Fig. 1.3). One such mechanism involves recombination of certain immunoglobulin gene sequence elements (Roth et al. 1992). A second mechanism, occurring during B cell division, involves somatic hypermutation that targets a different segment of antibody structure by way of enzymatic deamination of cytosine to uracil, thereby creating transition mutations (Neuberger et al. 2003). With our finding that immunoglobulin genes are the most significantly overrepresented with the A-residue self-depurination consensus sequence, and that they contain G self-depurination sites as well, it would appear that nature has selected this additional mutagenic mechanism to create antibody diversity.

In this connection, it is interesting that in the constant regions of immunoglobulin genes, the self-depurinating consensus sequence for A-residues is not highly overrepresented. This is a further indication of its selection for the purpose of creating the hypervariable regions. It is apparent then that these genes evolved so as to exploit the self-depurinating mechanism in their contemporary function.

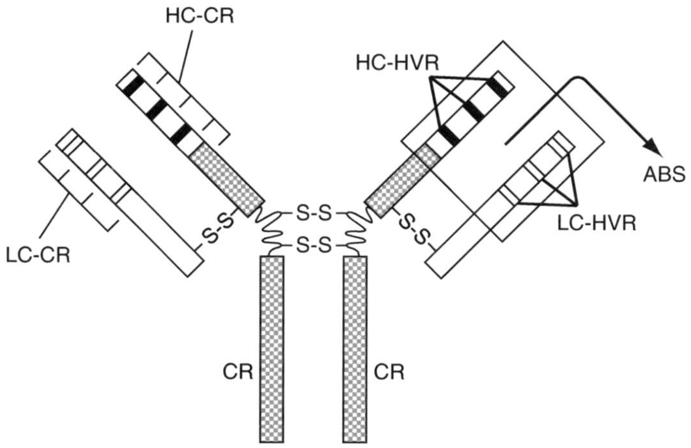

Fig. 1.3 A schematic diagram of the Y-shaped structure of a typical human antibody (protein) molecule showing the location of the constant regions (CR) that form the skeletal framework from which are extended various combinations of hypervariable heavy chain (HC-HVR) and light chain (LC-HVR) segments. Those hypervariable sequence segments, three in each heavy and each light chain, are interspersed by constant segments (HC-CR and LC-CR), and together form antigen-binding sites (ABS). Previously, those hypervariable segments had been found to attain great sequence diversity by a combination of genetic recombination and C-residue deamination. Now we have discovered that over the course of evolution, nature selected the self-depurination mechanism for this purpose as well, since the hypervariable regions are very heavily endowed with A-residue self-depurinating sites. Hence, the gene sequences for those sites provide another major mechanism to enable them to undergo mutation in order to create the extraordinary sequence diversity required to meet the challenges to the immune response

1.9 Self-Depurination of A-Residues in the Evolution of Olfactory Receptor Genes

As with the immunoglobulins, the function of the structurally similar olfactory receptors requires them to be able to recognize a large number of different odorific molecules. However, in contrast to the genes encoding the hypervariable regions of the immunoglobulins, the olfactory receptor genes are all already encoded in the genome, and there is no indication that their somatic DNA sequences undergo hyper mutation in their coding regions to any significant extent (Sharon et al. 1998, 1999). In effect, they are currently utilized with the diversity with which they evolved over time. Of the ~850 such human genes, of which some 55–60% are pseudogenes (Olender et al. 2004), a small fraction are singlets, but the majority are in clusters distributed among all chromosomes but #20 and Y. The gene clusters are likely to have arisen by repeated duplication of individual genes and clusters (Glusman et al. 1996), as many of the genes in tandem arrays are closely related (Niimura and Nei 2003). At the same time, during their evolution, they apparently exploited mutagenesis, at least by the self-depurination of A-residues to differentiate in function.

With such high overrepresentation of the self-depurination consensus sequence in olfactory receptor genes, it is reasonable to expect that over evolutionary time, the same mutagenic mechanism may have also resulted in the loss of the self-depurinating capacity in some of those genes. In that case, corresponding nonfunctional or pseudogenes would have accumulated that arose, e.g., as a consequence of substitution mutations in the loop residues of the catalytic intermediate. Evidence for this possibility was sought by searching these gene clusters for highly overrepresented pseudogenes whose consensus loop sequences had been mutated from 5′ G-A-G-A to 5′G-T-G-A and 5′ G-G-G-A, some of the gene products expected at those self-depurination sites. Such a mechanism could account, then, for the very high fraction of these clustered genes that were found to be pseudogenes (Rouquier et al. 1998a, b). In fact, pseudogenes with these very A → T and A → G mutations at the self-depurinating sites were found to be overrepresented. These observations do indicate, then, that over the course of evolution the same mutagenic mechanism has served to create mutations leading both to olfactory receptor diversity and to erasure of such genes.

1.10 Discussion

Our goal in this chapter has been to summarize that aspect of our knowledge of self-catalyzed DNA depurination that particularly relates to ways in which the mechanism played a role in the evolution of certain biological processes and some inherited diseases.

Thus, we have presented several types of observations that support the notion that the self-depurination mechanism is likely to have played a role in molecular evolution. One relates to the distribution of the consensus sequence for self-depurination of G-residues. Not only is that consensus sequence found in every double-stranded genome we have examined from various Archaea species to Homo sapiens, but its frequency has been found to be very high, in the neighborhood of once every 3,000–5,000 base pairs. In mitochondrial genomes, which are also double-stranded, they are present in most species at a frequency similar to those in the genomes of lower forms. While they appear to be lacking in the mitochondrial genomes of some species, it must be kept in mind that we have not yet searched them for the consensus sequence for self-depurination of A-residues. Nor, in fact, have we yet searched any genomes other than that of Homo sapiens for the A-residue consensus sequence.

It is for this same reason that we are not yet able to interpret the absence of G-residue self-depurination sequences in half the single-stranded viral genomes recorded in the viral genome database. If the viral genomes with no G-residue self-depurination sites are in fact lacking in those for A-residues as well, that would provide important evidence that evolution selected against the self-depurination mechanism for these species. Such negative selection would arise because once an apurinic site resulted in backbone fragmentation of the single-stranded genome, the fragments might have no template complementary strand to enable their repair.

Another type of evidence supporting a role for self-depurination in molecular evolution relates to the highly error-prone repair of apurinic sites in all species in which it has been examined, from viruses (whose hosts carry the repair mechanism) and bacteria (Shearman and Loeb 1979) to lower and higher eukaryotes (Chakravarti et al. 2000). Moreover, in a number of cases, such sites have been specifically identified as mutational hot spots (Kunkel 1984).

We have additionally presented three examples of genes for which the evidence would seem to be compelling that self-depurination must have played a role in their evolution, in their loss of function or conversion to pseudogenes, and/or in their contemporary functioning.

In the case of the human olfactory receptor genes, the self-depurination consensus sequence, at least for A-residues, is clearly associated with the evolutionary development of their diversity; and just as these sites are significantly overrepresented in this group of genes, so are their mutation products overrepresented in their nonfunctional pseudogenes. Yet, in the human species, these olfactory receptor genes do not appear to be evolving now at any detectable rate. They are already all encoded in the genome, with no evidence of somatic cell DNA rearrangements or somatic mutations in the coding regions of those genes. So it is reasonable to conclude that the mechanism played a role in the course of the evolution of those genes to their current state of diversity, but that it does not play a significant role contemporaneously.

The case of the genes encoding the hypervariable regions of the immunoglobulins would appear to be somewhat different. Here we have genes for which mechanisms for variation are essential elements for their contemporary function. The consensus sequences for both G- and A-residue self-depurination are again highly overrepresented in these genes. The consequent mutagenic mechanism may have played a role in their evolution, something which our analysis cannot say with certainty at the present time. What is much more apparent, however, is that it is one of the several mechanisms by which antibody diversity is contemporaneously achieved in response to confrontation with an antigen.

Finally, we have presented the case of the unique occurrence of six different anemias due to substitutions and two β-thalassemias due to short deletions, all within codon 6 of the human β-globin gene. That codon is the only self-depurination consensus sequence site for G-residues in the entire gene, and there are none for A-residues. While these mutations are all explicable as a consequence of the self-depurination mechanism, there are no indications that they readily occur as somatic mutations (and even if they do, they are unlikely to be detected, since a majority of hemoglobin-producing cells would still have the wild-type DNA sequence). Consequently, they appear as germline mutations, evolutionarily retained, no doubt, as a consequence of the resistance to malaria that at least the sickle cell anemia mutation confers (Friedman 1978), and possibly some of the others as well.

Taken together, the foregoing examples serve as a strong indication that the self-depurination mechanism, coupled to the process of error-prone repair, has played an evolutionary role. But why has a mechanism capable of causing DNA damage been selected for and even concentrated in genomes in the course of evolution? Perhaps the advantages accruing from the availability of a mechanism inherent in the sequence and structural dynamics of DNA to help create gene sequence diversity far outweighs the negative consequences of genetic instability, at least where the products of gene sequence diversification have positive value. It is interesting in this connection that we have found that some genes appear to be favored by the presence of either G-residue or of A-residue self-depurinating consensus sequences; this might possibly indicate different evolutionary origins for such genes. Moreover, genes that are highly conserved, e.g., those encoding HOX, histone core, ribosomal proteins, contain very few if any such consensus sequences. It appears then, that these two consensus sequences have sometimes been selected against, but other times selected for, which would indicate that self-depurination is a significant mutagenic mechanism that has been taken note of and sometimes harnessed by evolution.

Acknowledgments This research was supported in part by a grant from the U.S. Army Research Office (W911NF-07-1-0152) and by fellowships to D. Glumcher from Dickinson College and Princeton University, to R. Torres from the Howard Hughes Medical Institute, to J. Alvarez-Dominguez and P. Wei from NIH grant P50 GM071508.

References

Alvarez D, Novac O, Callejo M, Ruiz M, Price G, Zannis-Hadjopoulos M (2002) 14-3-3 sigma is a cruciform DNA binding protein and associates in vivo with origins of DNA replication. J Cell Biochem 87:194–207

Amosova O, Coulter R, Fresco JR (2006) Self-catalyzed site-specific depurination of guanine residues within gene sequences. Proc Natl Acad Sci USA 103:4392–4397

Amosova O, Smith A, Fresco J (2011a) The consensus sequence for self-catalyzed, site-specific G-residue depurination in DNA. J Biol Chem, in press

Amosova O, Kumar V, Deutsch A, Fresco J (2011b) Self-catalyzed, site-specific depurination of G-residues mediated by cruciform extrusion in closed circular DNA plasmids. J Biol Chem, in press

Amukele TK, Roday S, Schramm VL (2005) Ricin A-chain activity on stem-loop and unstructured DNA substrates. Biochemistry 44:4416–4425

Blake RD, Fresco JR (1973) Polynucleotides. XI. Thermodynamics of $(A)_N{\bullet}2(U)_\infty$ from the dependence of Tm_N on oligomer length. Biopolymers 12:775–786

Boiteux S, Guillet M (2004) Abasic sites in DNA: repair and biological consequences in Saccharomyces cerevisiae. DNA Repair (Amst) 3:1–12

Brahms J, Maurizot JC, Michelson AM (1967) Conformation and thermodynamic properties of oligocytidylic acids. J Mol Biol 25:465–480

Chakravarti D, Mailander PC, Cavalieri EL, Rogan EC (2000) Evidence that error-prone DNA repair converts dibenzo[a, l]pyrene-induced depurinating lesions into mutations: formation, clonal proliferation and regression of initiated cells carrying H-ras oncogene mutations in early preneoplasia. Mutat Res 456:17–32

Chebloune Y, Pagnier J, Trabuchet G, Faure C, Verdier G, Labie D, Nigon V (1988) Structural analysis of the 5′ flanking region of the beta-globin gene in African sickle cell anemia patients: further evidence for three origins of the sickle cell mutation in Africa. Proc Natl Acad Sci USA 85:4431–4435

Cox R, Mirkin SM (1997) Characteristic enrichment of DNA repeats in different genomes. Proc Natl Acad Sci USA 94:5237–5242

Dejong WWW, Went LN, Bernini LF (1968) Haemoglobin leiden - deletion of beta6 or 7 glutamic acid. Nature 220:788–790

Durocher F, ShattuckEidens D, McClure M, Labrie F, Skolnick MH, Goldgar DE, Simard J (1996) Comparison of BRCA1 polymorphisms, rare sequence variants and/or missense mutations in unaffected and breast/ovarian cancer populations. Hum Mol Genet 5:835–842

Endo Y, Tsurugi K (1987) RNA N-glycosidase activity of ricin A-chain. Mechanism of action of the toxic lectin ricin on eukaryotic ribosomes. J Biol Chem 262:8128–8130

Engelke DR, Hoener PA, Collins FS (1988) Direct sequencing of enzymatically amplified human genomic DNA. Proc Natl Acad Sci USA 85:544–548

Fresco JR, Alberts BM (1960) The accommodation of noncomplementary bases in helical polyribonucleotides and deoxyribonucleic acids. Proc Natl Acad Sci USA 46:311–321

Friedman M (1978) Erythrocytic mechanism of sickle cell resistance to malaria. Proc Natl Acad Sci USA 75:1994–1997

Glusman G, Clifton S, Roe B, Lancet D (1996) Sequence analysis in the olfactory receptor gene cluster on human chromosome 17: recombinatorial events affecting receptor diversity. Genomics 37(2):147–160

Goelen G, Teugels E, Bonduelle M, Neyns B, De Greve J (1999) High frequency of BRCA1/2 germline mutations in 42 Belgian families with a small number of symptomatic subjects. J Med Genet 36:304–308

Grignoli CR, Carvalho MH, Kimura EM, Sonati MF, Arruda VR, Saad ST, Costa FF (2000) Beta0-thalassemia resulting from a novel mutation: beta66/u– > stop codon. Eur J Haematol 64:137–138

Harano T, Harano K, Ueda S, Shibata S, Imai K, Seki M (1982) Hemoglobin Machida [beta-6 (A3) Glu-]Gln], a new abnormal hemoglobin discovered in a Japanese family - structure, function and biosynthesis. Hemoglobin 6:531–535

Hutton M, Lendon CL, Rizzu P, Baker M, Froelich S, Houlden H, Pickering-Brown S, Chakraverty S, Isaacs A, Grover A, Hackett J, Adamson J, Lincoln S, Dickson D, Davies P, Petersen RC, Stevens M, de Graaff E, Wauters E, van Baren J, Hillebrand M, Joosse M, Kwon JM, Nowotny P, Che LK, Norton J, Morris JC, Reed LA, Trojanowski J, Basun H, Lannfelt L, Neystat M, Fahn S, Dark F, Tannenberg T, Dodd PR, Hayward N, Kwok JBJ, Schofield PR, Andreadis A, Snowden J, Craufurd D, Neary D, Owen F, Oostra BA, Hardy J,

Goate A, van Swieten J, Mann D, Lynch T, Heutink P (1998) Association of missense and 5′-splice-site mutations in tau with the inherited dementia FTDP-17. Nature 393:702–705

Inagaki H, Ohye T, Kogo H, Kato T, Bolor H, Taniguchi M, Shaikh TH, Emanuel BS, Kurahashi H (2009) Chromosomal instability mediated by non-B DNA: cruciform conformation and not DNA sequence is responsible for recurrent translocation in humans. Genome Res 19:191–198

Jochmans K, Lissens W, Seneca S, Capel P, Chatelain B, Meeus P, Osselaer JC, Peerlinck K, Seghers J, Slacmeulder M, Stibbe J, van de Loo J, Vermylen J, Liebaers I, De Waele M (1998) The molecular basis of antithrombin deficiency in Belgian and Dutch families. Thromb Haemost 80:376–381

Juricic D, Ruzdic I, Beer Z, Efremov GD, Casey R, Lehmann H (1983) Hemoglobin Leiden [beta 6 or 7 (A3 or A4) Glu leads to O] in a Yugoslavian woman arisen by a new mutation. Hemoglobin 7:271–277

Kazazian HH, Orkin SH, Boehm CD, Sexton JP, Antonarakis SE (1983) Beta-Thalassemia due to a deletion of the nucleotide which is substituted in the beta-S-globin gene. Am J Hum Genet 35:1028–1033

Kim E, Peng H, Esparza F, Maltchenko S, Stachowiak M (1998) Cruciform-extruding regulatory element controls cell-specific activity of the tyrosine hydroxylase gene promoter. Nucleic Acids Res 26:1793–1800

Korolev VG (2005) Base excision repair: AP endonucleases and DNA polymerases. Russ J Genet 41:1063–1070

Kulozik A, Wainscoat J, Serjeant G, Kar B, Al-Awamy B, Essan G, Falusi A, Haque S, Hilali A, Kate S et al (1986) Geographical survey of beta S-globin gene haplotypes: evidence for an independent Asian origin of the sickle-cell mutation. Am J Hum Genet 39:239–244

Kunkel TA (1984) Mutational specificity of depurination. Proc Natl Acad Sci USA 81:1494–1498

Lane DA, Bayston T, Olds RJ, Fitches AC, Cooper DN, Millar DS, Jochmans K, Perry DJ, Okajima K, Thein SL, Emmerich J (1997) Antithrombin mutation database: 2nd (1997) update. For the Plasma Coagulation Inhibitors Subcommittee of the Scientific and Standardization Committee of the International Society on Thrombosis and Haemostasis. Thromb Haemost 77:197–211

Lapouméroulie C, Dunda O, Ducrocq R, Trabuchet G, Mony-Lobé M, Bodo J, Carnevale P, Labie D, Elion J, Krishnamoorthy R (1992) A novel sickle cell mutation of yet another origin in Africa: the Cameroon type. Hum Genet 89:333–337

Lhomme J, Constant JF, Demeunynck M (1999) Abasic DNA structure, reactivity, and recognition. Biopolymers 52:65–83

Lindahl T, Nyberg B (1972) Rate of depurination of native deoxyribonucleic acid. Biochemistry 11:3610–3618

Lomant AJ, Fresco JR (1973) Polynucleotides. 13. Stoichiometric and thermodynamic studies of polynucleotide helices with non-complementary residues. Biopolymers 12:1889–1903

Marmur J, Doty P (1959) Heterogeneity in deoxyribonucleic acids. I. Dependence on composition of the configurational stability of deoxyribonucleic acids. Nature 183:1427–1429

Neuberger M, Harris R, Di Noia J, Petersen-Mahrt S (2003) Immunity through DNA deamination. Trends Biochem Sci 28(6):305–312

Niimura Y, Nei M (2003) Evolution of olfactory receptor genes in the human genome. Proc Natl Acad Sci USA 100(21):12235–12240

Olender T, Feldmesser E, Atarot T, Eisenstein M, Lancet D (2004) The olfactory receptor universe–from whole genome analysis to structure and evolution. Genet Mol Res 3:545–553

Ornstein RL, Fresco JR (1983) Correlation of Tm and sequence of DNA duplexes with ΔH computed by an improved empirical potential method. Biopolymers 22:1979–2000

Pagnier J, Mears JG, Dunda-Belkhodja O, Schaefer-Rego KE, Beldjord C, Nagel RL, Labie D (1984) Evidence for the multicentric origin of the sickle cell hemoglobin gene in Africa. Proc Natl Acad Sci 81:1771–1773

Rosenzweig M, Skinner M, Prokaeva T, Théberge R, Costello C, Drachman BM, Connors LH (2007) A new transthyretin variant (Glu61Gly) associated with cardiomyopathy. Amyloid 14:65–71

Roth DB, Nakajima PB, Menetski JP, Bosma MJ, Gellert M (1992) V(D)J recombination in mouse thymocytes: double-strand breaks near T cell receptor delta rearrangement signals. Cell 69:41–53

Rouquier S, Friedman C, Delettre C, van den Engh G, Blancher A, Crouau-Roy B, Trask B, Giorgi D (1998a) A gene recently inactivated in human defines a new olfactory receptor family in mammals. Hum Mol Genet 7(9):1337–1345

Rouquier S, Taviaux S, Trask B, Brand-Arpon V, van den Engh G, Demaille J, Giorgi D (1998b) Distribution of olfactory receptor genes in the human genome. Nat Genet 18(3):243–250

Sangkitporn S, Rerkamnuaychoke B, Mitrakul C, Sutivigit Y (2002) Hb G Makassar (beta 6:Glu-Ala) in a Thai family. J Med Assoc Thai 85:577–582

Schroeder W, Powars D, Kay L, Chan L, Huynh V, Shelton J, JR S (1989) Beta-cluster haplotypes, alpha-gene status, and hematological data from SS, SC, and S-beta-thalassemia patients in southern California. Hemoglobin 13:325–353

Shabarova Z, Bogdanov A (1994) Advanced organic chemistry of nucleic acids. VCH (UK) Ltd, Cambridge, England, p 60

Sharon D, Glusman G, Pilpel Y, Horn-Saban S, Lancet D (1998) Genome dynamics, evolution, and protein modeling in the olfactory receptor gene superfamily. Olfaction Taste Xii: 182–193

Sharon D, Glusman G, Pilpel Y, Khen M, Gruetzner F, Haaf T, Lancet D (1999) Primate evolution of an olfactory receptor cluster: diversification by gene conversion and recent emergence of pseudogenes. Genomics 61(1):24–36

Shearman CW, Loeb LA (1979) Effects of depurination on the fidelity of DNA synthesis. J Mol Biol 128:197–218

Shiomi K, Nakazato M, Matsukura S, Ohnishi A, Hatanaka H, Tsuji S, Murai Y, Kojima M, Kangawa K, Matsuo H (1993) A basic transthyretin variant (Glu(61)-]Lys) causes familial amyloidotic polyneuropathy - protein and DNA-sequencing and PCR-induced mutation restriction analysis. Biochem Biophys Res Commun 194:1090–1096

Shlyakhtenko LS, Potaman VN, Sinden RR, Lyubchenko YL (1998) Structure and dynamics of supercoil-stabilized DNA cruciforms. J Mol Biol 280:61–72

Simonelli V, Narciso L, Dogliotti E, Fortini P (2005) Base excision repair intermediates are mutagenic in mammalian cells. Nucleic Acids Res 33:4404–4411

Siriboon W, Srisomsap C, Winichagoon P, Fucharoen S, Svasti J (1993) Identification of Hb C [beta 6(A3)Glu- > Lys] in a Thai male. Hemoglobin 17:419–425

Sugiyama H, Fujiwara T, Ura A, Tashiro T, Yamamoto K, Kawanishi S, Saito I (1994) Chemistry of thermal degradation of abasic sites in DNA. Mechanistic investigation on thermal DNA strand cleavage of alkylated DNA. Chem Res Toxicol 7:673–683

Tanca D, Devoto G, Deiana F, Luciano B, Lisi A, Ivald G (2008) Characterization of a new hemoglobin variant. Ital J Lab Med 4:142

Tonegawa S (1983) Somatic generation of antibody diversity. Nature 302:575–581

Topal MD, Fresco JR (1976) Complementary base pairing and the origin of substitution mutations. Nature 263:285–289

Varley JM, Evans DGR, Birch JM (1997) Li-Fraumeni syndrome - a molecular and clinical review. Br J Cancer 76:1–14

Wagner T, Stoppa-Lyonnet D, Fleischmann E, Muhr D, Pagès S, Sandberg T, Caux V, Moeslinger R, Langbauer G, Borg A, Oefner P (1999) Denaturing high-performance liquid chromatography detects reliably BRCA1 and BRCA2 mutations. Genomics 62:369–376

Wailoo K (1991) "A disease sui generis": the origins of sickle cell anemia and the emergence of modern clinical research, 1904–1924. Bull Hist Med 65:185–208

Chapter 2
Stochastic Processes Driving Directional Evolution

Sean H. Rice, Anthony Papadopoulos, and John Harting

Abstract Evolution is a stochastic process, resulting from a combination of deterministic and random factors. We present results from a general theory of directional evolution that reveals how random variation in fitness, heritability, and migration influence directional evolution. First, we show how random variation in fitness produces a directional trend toward phenotypes with minimal variation in fitness. Furthermore, we demonstrate that stochastic variation in population growth rate amplifies the expected change due to directional selection in small populations. Second, we show that the evolutionary impacts of migration depend on the entire distribution of migration rates such that increasing the variance in migration rates reduces the impact of migration relative to selection. This means that changing the variance in migration rates, holding the mean constant, can substantially change the potential for local adaptation. Finally, we show that covariation between stochastic selection and stochastic heritability can drive directional evolutionary change, and that this can substantially alter the outcome of evolution in variable environments.

2.1 Introduction

Evolutionary biologists have long recognized the importance of stochastic processes in the mechanics of evolution. The best studied stochastic evolutionary process is genetic drift – change in allele frequency resulting from random variation in fitness and segregation – which plays a critical role in the modern theory of molecular evolution. Drift is nondirectional, meaning that the expected change in allele frequency due to drift alone is zero, and it is often assumed that this will be true of any stochastic evolutionary process.

S.H. Rice • A. Papadopoulos • J. Harting
Department of Biological Sciences, Texas Tech University, Lubbock, TX 79409, USA
e-mail: sean.h.rice@ttu.edu; anthony.papadopoulos@ttu.edu; john.harting@ttu.edu

P. Pontarotti (ed.), *Evolutionary Biology – Concepts, Biodiversity, Macroevolution and Genome Evolution*, DOI 10.1007/978-3-642-20763-1_2,

In fact, the potential of stochastic variation in fitness to contribute to directional evolution was noted by a number of authors in the 1970s (Hartl and Cook 1973; Karlin and Liberman 1974; Gillespie 1974). These authors recognized that differences in the variances of individual fitness distributions could contribute to directional change, just as differences in the mean values can.

Differential fitness is not the only factor influencing evolution. Both migration between populations and the process of inheritance itself can drive directional evolution – and, like selection, both of these are inherently stochastic processes (this is most obvious in the case of inheritance, since both mutation and recombination are chemical processes subject to quantum uncertainty). In this chapter, we will discuss some of the ways in which random variation in fitness, migration, or inheritance can lead to directional evolutionary change.

2.2 Modeling Stochastic Evolution

Introducing stochasticity into our models of evolution requires that we treat values like fitness and migration rate as random variables. For our purposes, a random variable differs from an ordinary variable in a mathematical equation in that a random variable has a distribution of possible values, rather than a single value.

It is important to note that saying that fitness, migration, or anything else, is stochastic is not the same as just saying that it varies over time. If we specify that fitness values will alternate, across generations, between specific values, we are defining a deterministic (not stochastic) process in which the value varies in a predictable manner over time. By contrast, making fitness stochastic means that we cannot say what value it will have at any particular time, only that it has a distribution of possible values at that time.

This distinction is important; treating a variable as deterministic but temporally variable can yield very different results than does treating it as a stochastic random variable. This is illustrated, for the case of fitness, in Fig. 2.1. In Fig. 2.1a, fitness is treated as an ordinary variable that fluctuates over time – alternating between 1 and 2. In Fig. 2.1b, fitness is a random variable that, in any particular generation, has a 50% chance of being 1 and a 50% chance of being 2.

Though we might be tempted to treat the deterministic case (Fig. 2.1a) as an "average" instance of the stochastic case (Fig. 2.1b), that would be misleading. After four generations, a population following the deterministic case will have increased in size by a factor of 4, corresponding to a per generation change of $\sqrt{2} \approx 1.414$, which is the geometric mean of 1 and 2. By contrast, the expected size of a population following the stochastic case for four generations is just over 5 (specifically, 5.0625), corresponding to a per generation change of 1.5 – the arithmetic mean of 1 and 2. The different outcomes illustrated in Fig. 2.1 are not results of the short time interval considered; the same effective fitness values arise if we consider an arbitrary number of generations. This should be kept in mind when evaluating arguments about the utility of geometric mean fitness.

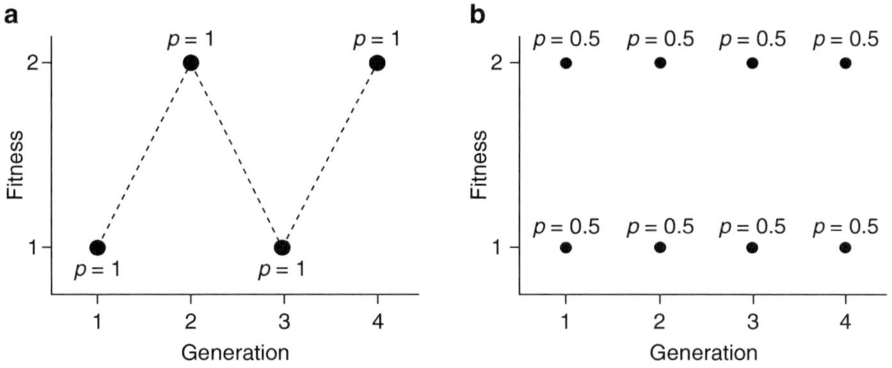

Fig. 2.1 Illustration of the difference between treating fitness as a deterministic variable that changes over time (**a**), and as a random variable (**b**)

Treating fitness, migration, and other values as random variables requires that we consider the variances and covariances of their distributions. This can lead to notational confusion, since we are also concerned with means, variances, and covariances of the same values within a population. For example, we will be concerned with both the mean fitness of an individual (the mean of its fitness distribution) and the mean fitness in the entire population.

We thus will distinguish between two different sets of statistical operators: frequency and probability. Frequency operators, denoted by straight symbols ($\bar{a}$ for mean, $[[^2a]]$ for variance, and $[[a, b]]$ for covariance), describe operations over some collection of things. For instance, $\overline{w}$ is the average fitness across individuals in a population, and $[[\phi, w]]$ is the covariance, across all individuals in the population, between phenotype and fitness.

Probability operators, denoted by angled symbols ($\hat{a}$ for mean, $\langle\langle^2a\rangle\rangle$ for variance, and $\langle\langle a, b\rangle\rangle$ for covariance), describe operations over distributions of random variables. For example, $\widehat{w}$ is the expected fitness of an individual – the mean of its fitness distribution – while $\langle\langle^2w\rangle\rangle$ is the variance of the same distribution (the variance in fitness values that the individual might have). A detailed discussion of these two kinds of operators, and the rules for manipulating them, is given in Rice and Papadopoulos (2009). Table 2.1 lists the main symbols that we will use in this chapter.

2.3 Stochastic Fitness

An individual's fitness is the number of descendants that it has after some chosen time interval. We often choose the time interval to be a single generation and think of fitness as simply the number of offspring, but in the general case, we need to consider all descendants, including grand offspring and the individual itself at the future time.

Table 2.1 Symbols and notation

Symbol	Meaning
N	Population size
ϕ	Phenotype of an individual
δ	Difference between the mean phenotype of an individual's offspring and that individual's phenotype
$\widehat{\overline{\delta}}$	Expected mean value of δ in the population
w	Fitness of an individual
Ω	$\frac{w}{\overline{w}}$conditional on $\overline{w} \neq 0$
ξ	Number of immigrants divided by deme size
ε	Number of emigrants divided by deme size
Ξ	Number of immigrants divided by per capita deme growth rate
R	Per capta deme growth rate
$\mathrm{H}(\overline{w})$	Harmonic mean of $\overline{w}$
$\overline{X}$ or Ave(X)	Average value of X across some set of individuals
$\widehat{A}$ or E(A)	Expected value of random variable A
$[[^2X]]$	Variance in the value of X across some set of individuals
$\langle\langle ^2A\rangle\rangle$	Variance in random variable A
$[[X,Y]]$	Covariance, over a set of objects, between the values of X and Y
$\langle\langle A,B\rangle\rangle$	Covariance, across all possible outcomes, between random variables A and B

Because we cannot know with certainty how many descendants each individual in a population will have, we need to treat fitness as a random variable – having a distribution of possible values. The vast majority of evolutionary models consider only the mean of this distribution – the expected number of descendants – and in fact fitness is often defined as this expected value. We will see, though, that accurately describing evolution requires consideration of the entire distribution.

Using the notation in Table 2.1, the general equation for evolution in a closed population (no migration in or out) with stochastic fitness and inheritance is (Rice 2008):

$$E(\Delta\bar{\phi}) = [[\phi,\hat{\Omega}]] + [[\hat{\delta},\hat{\Omega}]] + \overline{\langle\langle \delta,\Omega\rangle\rangle} + \hat{\bar{\delta}} \tag{2.1}$$

Fitness enters into Eq. (2.1) through the term Ω, which is the ratio of an individual's fitness to mean population fitness. We will refer to this as "relative fitness" (note, though, that the term "relative fitness" is sometimes used in other ways). Strictly, Ω is defined under the condition that $\overline{w} \neq 0,$ for both mathematical and biological reasons. Mathematically, the ratio is undefined if $\overline{w} = 0$. Biologically, mean population fitness being zero corresponds to extinction, and change in mean phenotype is undefinable when the population ceases to exist.

It is common to treat mean population fitness ($\overline{w}$) as a constant, even when individual fitness (w) is a random variable. This is done, for example, in both the Wright-Fisher and Moran models of genetic drift. This assumption, though, is made purely for the sake of simplifying the mathematics – nobody expects $\overline{w}$ to be constant in most real populations. We will thus treat mean population fitness as another random variable. As we show below, relaxing this seemingly inoffensive assumption exposes an entire class of evolutionary processes that are otherwise invisible.

Acknowledging that both w and $\overline{w}$ are random variables complicates our interpretation of $\widehat{\Omega}$. The expected value of a ratio of random variables can behave in surprising ways. For instance, given random variables a and b, it can be the case that the expected value of $\frac{a}{b}$ and that of $\frac{b}{a}$ are both greater than 1. Rice (2008) showed that expected relative fitness can be written as an infinite series, the terms of which contain moments of the individual fitness distributions. When the fitness values of different individuals are independent, then this series can be written as:

$$\widehat{\Omega} = \frac{\widehat{w}}{\mathrm{H}(\overline{w})} - \frac{\langle\langle {}^2w \rangle\rangle}{N\widehat{\overline{w}}^2} + \frac{\langle\langle {}^3w \rangle\rangle}{N^2\widehat{\overline{w}}^3} - \frac{\langle\langle {}^4w \rangle\rangle}{N^3\widehat{\overline{w}}^4} + \cdots \tag{2.2}$$

Substituting this series into the first term on the right-hand side of Eq. (2.1) yields:

$$\widehat{\Delta\bar{\phi}} = \frac{[[\phi, \hat{w}]]}{\boldsymbol{H}(\bar{w})} - \frac{[[\phi, \langle\langle {}^2w \rangle\rangle]]}{N\widehat{\overline{w}}^2} + \frac{[[\phi, \langle\langle {}^3w \rangle\rangle]]}{N^2\widehat{\overline{w}}^3} - \frac{[[\phi, \langle\langle {}^4w \rangle\rangle]]}{N^3\widehat{\overline{w}}^4} + \cdots \tag{2.3}$$

The $\langle\langle {}^iw \rangle\rangle$ terms in Eqs. (2.2) and (2.3) are the central moments of an individual's fitness distribution ($\langle\langle {}^2w \rangle\rangle$ being the variance, and $\langle\langle {}^3w \rangle\rangle$ being the third central moment, etc.). Eq. (2.3), thus, shows how different aspects of the shapes of individual fitness distributions contribute to directional evolution.

The standard interpretation of selection is captured by the first term on the right-hand side of Eq. (2.3), which contains the covariance between phenotype and expected fitness ($[[\phi, \hat{w}]]$). All of the subsequent terms, involving the various moments of the fitness distribution, represent directional stochastic effects in evolution. Note that these disappear if we treat fitness as a fixed value, since, in that case, all of the $\langle\langle {}^iw \rangle\rangle$ terms are zero.

To illustrate how these directional stochastic effects influence evolution, we consider the second term, containing the covariance between phenotype and variance in fitness ($\left[\left[\phi, \left\langle\left\langle {}^2w \right\rangle\right\rangle\right]\right]$). This term is negative (as are all terms involving even-valued moments) – telling us that there is a force pulling the population toward phenotypes with minimum variance in fitness.

Figure 2.2 illustrates schematically how a population can be pulled toward phenotypes that minimize variance in fitness. The figure shows a case in which two different phenotypic values have the same expected fitness (i.e., the same $\widehat{w}$), but different variances in their fitness distributions (i.e., different values of $\langle\langle {}^2w \rangle\rangle$). The key is to note that the magnitude of change in mean phenotype is inversely proportional to mean population fitness ($\bar{w}$). Thus, in case (B), when individuals with $\phi = 0$ have low fitness (and thus they decrease in frequency), the change in mean phenotype is relatively large, because $\bar{w}$ is low in that case. By contrast, when

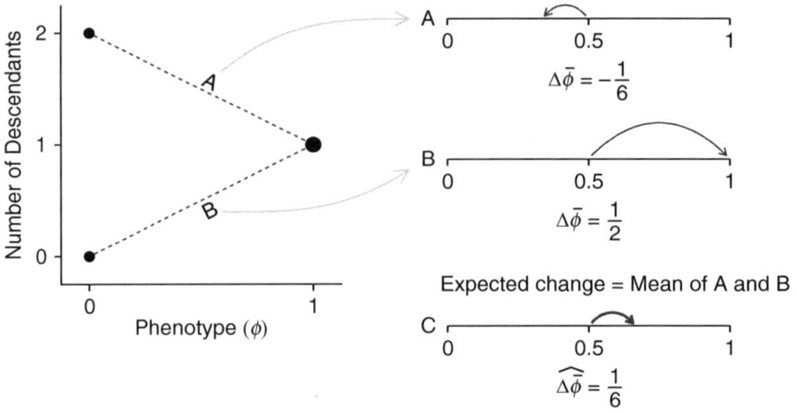

Fig. 2.2 A simple case of directional stochastic evolution. Individuals with phenotype (ϕ) of 0 leave two descendants or none, each with probability 0.5. Individuals with phenotypic value 1 always leave one descendant. Initial mean phenotype ($\bar{\phi}$) is 0.5. If individuals with phenotype 0 leave 2 offspring each, then the mean phenotype changes to $\bar{\phi} = \frac{1}{3}$, so $\Delta\bar{\phi} = -\frac{1}{6}$. If these individuals leave no offspring, then $\Delta\bar{\phi} = \frac{1}{2}$. Since these two outcomes occur with equal probability, the expected change in mean phenotype is $\widehat{\Delta\bar{\phi}} = \frac{1}{2}(-\frac{1}{6}) + \frac{1}{2}\frac{1}{2} = \frac{1}{6}$

$\phi = 0$, individuals are doing well (and thus increasing in frequency), their increase is relatively small because $\bar{w}$ is larger in this case (A). The result is that, even though mean phenotype increases half of the time and decreases half of the time, the step sizes are different – large when $\bar{\phi}$ increases and small when it decreases – leading to a net positive expected change.

This example illustrates the basic principle underlying directional stochastic evolution. When population growth rate (here captured by $\bar{w}$) is large, the step size in evolutionary change tends to be smaller than when growth rate is low. Strategies that contribute disproportionately to variation in population growth rate (such as the strategy $\phi = 0$ in Fig. 2.2) thus tend to take smaller steps when they increase than when they decrease in frequency. In Fig. 2.2, the expected fitness values ($\widehat{w}$) are the same for the two strategies, so there is no directional selection acting. If this is not the case, then the expected change is a function of both selection and the directional stochastic effects.

Note that, in Eq. (2.3), the terms on the right-hand side are each divided by increasing powers of population size (N). This is because Eqs. (2.2) and (2.3) assume that each individual's realized fitness is independent of that of other individuals. This is what we expect when variation in fitness is due to pure demographic stochasticity. In such cases, the strength of directional stochastic evolutionary effects declines with increasing population size. By contrast, when variation in fitness is due to stochastic environmental variation, such that all individuals with a particular phenotype either do well or poorly together, then directional stochastic effects remain strong even in large populations (Rice 2008).

Though we have discussed only the effects of variance, it is clear from Eq. (2.1) that all of the moments of an individual's fitness distribution can influence evolution. To see the general pattern, note that the terms containing even moments (2nd, 4th, etc.) are all negative, while those containing odd moments are positive. Since even moments measure symmetrical spread about the mean, and odd moments measure asymmetry, we can say that directional stochastic evolution tends to shift populations toward phenotypes with minimum symmetrical variation in fitness and maximum positive skewness in fitness.

Finally, we note that even the selection term in Eq. (2.1) is influenced by stochasticity. The $H(\overline{w})$ in the denominator of the first term on the right-hand side represents the harmonic mean of $\bar{w}$. Because the harmonic mean is strongly influenced by small values, this term will get smaller as the variance in $\bar{w}$ increases – thus amplifying the selection differential (Rice 2008). Since $\bar{w}$ is the mean of a finite set of individuals, its variance is expected to go up as population size declines (corresponding to taking the mean of a smaller sample). Thus, the expected change due to selection will tend to increase as population size gets very small. Note, though, that the variance in $\Delta\overline{\phi}$ will also increase in small populations, so it will be necessary to examine a large number of cases to see the amplifying effect on the mean.

2.4 Stochastic Migration

Migration, like fitness, influences population growth. We thus might expect that stochastic variation in migration rates will generate the same kinds of directional evolutionary effects that we saw with stochastic fitness. The general equation for change in mean phenotype in an open population (one subject to immigration and emigration) is (Rice and Papadopoulos 2009):

$$\widehat{\Delta\overline{\phi}} = \left[\left[\phi, {}^{d}\widehat{\Omega}\right]\right] + \left[\left[\widehat{{}^{d}\delta, {}^{d}\Omega}\right]\right] + \overline{\langle\langle {}^{d}\delta, {}^{d}\Omega \rangle\rangle} + \widehat{\overline{\delta}} + \langle\langle \gamma, \Xi \rangle\rangle + \widehat{\Xi}\left(\widehat{\gamma} - \widehat{\overline{\delta}}\right). \tag{2.4}$$

Here, the left superscript ^{d}s indicate that the values are measured within a deme – a subpopulation subject to migration.

The various terms in Eq. (2.4) capture all of the ways that selection, transmission, and migration can influence directional change. We will focus here only on the last term on the right-hand side, $\widehat{\Xi}\left(\widehat{\gamma} - \widehat{\overline{\delta}}\right)$ (Rice and Papadopoulos (2009) present the full derivation, and discuss the meaning of each of the terms). $\left(\widehat{\gamma} - \widehat{\overline{\delta}}\right)$ is simply the difference between the expected phenotype of immigrants and that of native offspring who stay in the deme. The expected relative immigration rate is captured by $\widehat{\Xi}$, which is the expected value of the number of immigrants divided by the total deme growth rate.

Note that the number of immigrants directly influences the deme growth rate; so $\widehat{\Xi}$, like $\widehat{\Omega}$, is the expectation of the ratio of correlated random variables. Defining ξ

and ε as the numbers of immigrants and emigrants divided by deme size, we can expand $\widehat{\Xi}$ to yield:

$$\widehat{\Xi} \approx \frac{\widehat{\xi}}{\boldsymbol{H}(R)} - \frac{\langle\langle^2\xi\rangle\rangle}{\widehat{R}^2} - \frac{\langle\langle\bar{w},\xi\rangle\rangle}{\widehat{R}^2} + \frac{\langle\langle\varepsilon,\xi\rangle\rangle}{\widehat{R}^2}. \tag{2.5}$$

Equation 2.5 shows only the first- and second-order terms in the expansion, but already we can see that directional evolution will be influenced not only by the expected immigration rate ($\widehat{\xi}$), but also by the variance in immigration ($\langle\langle^2\xi\rangle\rangle$), the covariance between immigration and fitness within the deme ($\langle\langle\bar{w},\xi\rangle\rangle$), and the covariance between immigration and emigration ($\langle\langle\varepsilon,\xi\rangle\rangle$).

Focusing on the variance in immigration rate ($\langle\langle^2\xi\rangle\rangle$), the fact that the second term on the right-hand side of Eq. (2.5) is negative shows that increasing the variance in migration reduces the impact of migration on directional change. This effect is illustrated in Fig. 2.3 (the numbers are from Fig. 2.3a in Rice and Papadopoulos (2009)).

Figure 2.3 shows the consequences of changing the variance in immigration rate in a continent–island model, where evolution on the island is a consequence of both local selection and migration from the continent. Here, selection on the continent favors phenotypic value zero (white) while selection on the island favors phenotypic value 1 (black). In all cases, the expected number of migrants from the continent to the island is two per generation. The distribution of migration rates varies though, from a case with relatively low variance (case 1, in which either one or three immigrants arrive, each with probability 0.5) to a case of very high variance (case 5, in which 200 immigrants may arrive at once, but none arrive in most generations).

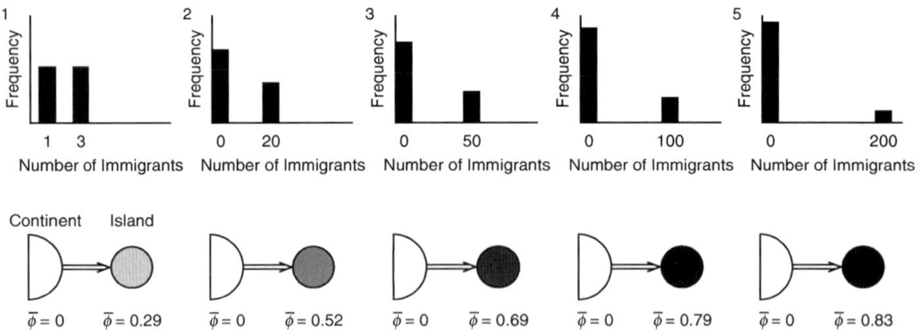

Fig. 2.3 The effect of changing the variance in immigration rates in a continent–island model. The *bar graphs* show the distribution of immigration rates to the island, ranging from low variance (case 1) to very high variance (case 6). In each case, the mean immigration rate is two individuals per generation. Selection on the continent favors a phenotypic value of 0, while selection on the island favors a phenotypic value of 1. In the *lower figures*, mean phenotype is indicated by *shading*, with 0 being *white* and 1 being *black*

In the example shown, selection on the island favors a phenotypic value of 1 (black), but with a mean immigration rate of two individuals per generation and low variance, the equilibrium mean phenotype on the island is only 0.29. Increasing the variance in migration rate (while holding the mean rate constant) significantly increases the degree to which the island population can diverge from the continental population.

We thus see that the potential for local adaptation is a function not only of expected migration rates, but of the entire distribution of rates at which individuals arrive or depart. The reason that high variance in immigration rate reduces the impact of migration relative to selection is that when many immigrants arrive together, deme growth rate (here defined as R, which combines reproduction within the deme with immigration and emigration) is large. Large R (just like large $\bar{w}$ in the example in Fig. 2.2) reduces the magnitude of change in mean phenotype.

This result also has consequences for speciation. A number of authors (Schluter 2001; Rundle and Nosil 2005; Fitzpatrick et al. 2009) have noted that pure allopatric or pure sympatric speciation are extreme cases, and that many actual cases of speciation will involve alternation of allopatry and sympatry. The result presented above illustrates that even if the average migration rate is held constant, lengthening the time between immigration pulses (even when those pulses involve more individuals) greatly increases the opportunity for the fixation of traits that facilitate reproductive isolation.

Finally, these results have consequences for our interpretation of traditional migration models. In nearly all natural populations, there will be variation in migration rates. The fact that such variation reduces the impact of migration relative to selection means that models that treat migration as a single parameter (and thus assume no variance in migration rates) will always tend to overestimate the relative importance of migration as an evolutionary force.

2.5 Stochastic Inheritance

Unlike reproduction and migration, the degree to which offspring resemble their parents does not directly influence population size. Stochastic variation in inheritance, thus, does not lead to the kind of directional stochastic evolution that we see in the cases of fitness and migration. In fact, so long as inheritance is independent of fitness or migration rates, we need not know anything more than the expected phenotype of offspring in order to calculate $\widehat{\Delta\bar{\phi}}$. The situation changes, though, if inheritance covaries with selection.

Though it is generally treated as a fixed parameter in quantitative genetic models, heritability often changes as a function of the environment in which organisms develop (Merila and Sheldon 2001; Charmantier and Garant 2005; Wilson et al. 2006). We thus expect that, to the extent that the environment is stochastic, so will be heritability. More significantly, if any of the environmental factors that influence heritability also influence fitness, then we expect heritability and fitness to covary. That this happens in many natural populations is suggested by the observation that environments that confer low fitness also tend to confer low heritability (Charmantier and Garant 2005; Wilson et al. 2006).

To illustrate how covariation between heritability and selection can influence evolution, we consider a simple case. Following the standard assumption of quantitative genetics, we assume that the mean phenotype of offspring is a linear function of parental phenotype (or, properly, midparent phenotype). This assumption allows us to capture inheritance with a single term, heritability (h^2), defined as the slope of the linear regression of offspring phenotype on midparent phenotype. We could just as well use the covariance between offspring and midparent phenotype – which is the "additive genetic variance". (Note that, for sexually reproducing organisms, an "individual" parent with respect to Eq. (2.1) is really a mated pair, with ϕ being the mean phenotype of the pair. We were thus tacitly already using "midparent" phenotype.)

Inheritance enters into Eq. (2.1) through the term δ, the difference between the mean phenotype of an individual's offspring and that individual's own phenotype. Under the standard quantitative genetics assumptions, δ can be derived from the parent's phenotype and population wide heritability, as shown in Fig. 2.4.

The key results, derivable from Fig. 2.4, are that, under the quantitative genetics assumptions, $\delta = (h^2 - 1)(\phi - \overline{\phi})$ and $\widehat{\overline{\delta}} = 0$. Substituting these results into Eq. (2.1) and simplifying yields:

$$\widehat{\Delta\overline{\phi}} = \widehat{h^2}\left[\left[\phi, \widehat{\Omega}\right]\right] + \left[\left[\phi, \langle\langle h^2, \Omega\rangle\rangle\right]\right] \tag{2.6}$$

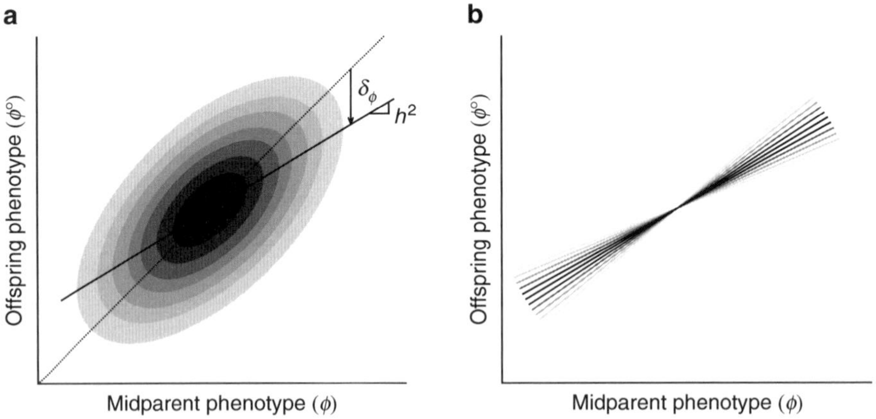

Fig. 2.4 (**a**) The relationship between δ and heritability (h^2) under the assumptions of quantitative genetics. Heritability is the slope of the regression of offspring phenotype on midparent phenotype. For a given midparent phenotype (ϕ), δ_ϕ is the vertical distance between the 45° line (defined by $\phi^\circ = \phi$) and the regression line. (**b**) When heritability is stochastic, the regression of offspring on parents becomes a random variable – having a distribution of possible slopes, rather than a single slope

The first term on the right-hand side of Eq. (2.6), $\widehat{h^2}[[\phi, \widehat{\Omega}]]$, is just the expected heritability multiplied by the selection differential. This is written "h^2S" in quantitative genetics (S is the selection differential), so the first term is just the standard "breeder's equation".

The second term on the right of Eq. (2.6), $[[\phi, \langle\langle h^2, \Omega\rangle\rangle]]$, captures the evolutionary consequences of covariation between heritability and selection. This term is read as: the covariance across the population (frequency covariance) between an individual's phenotype (ϕ) and the covariance (probability covariance) between that individual's relative fitness and heritability of ϕ. (Note that the distinction between frequency and probability covariance is critical here).

Figure 2.5 shows one way that covariation between heritability and selection can influence the outcome of evolution. In this example, the environment that a population experiences varies unpredictably across generations such that, in any one generation, there is a 50% chance of experiencing Environment 1 and a 50% chance of experiencing Environment 2. (e.g., these might correspond to wet and dry years experienced by an annual plant). The solid lines show the fitness in each environment as a function of phenotype, and the dashed line shows the expected fitness across both environments.

If heritability is equal in both environments, then the evolutionary equilibrium is (in this case) the strategy that maximizes mean fitness. This outcome changes when heritability covaries with the selective regime. If heritability is higher in

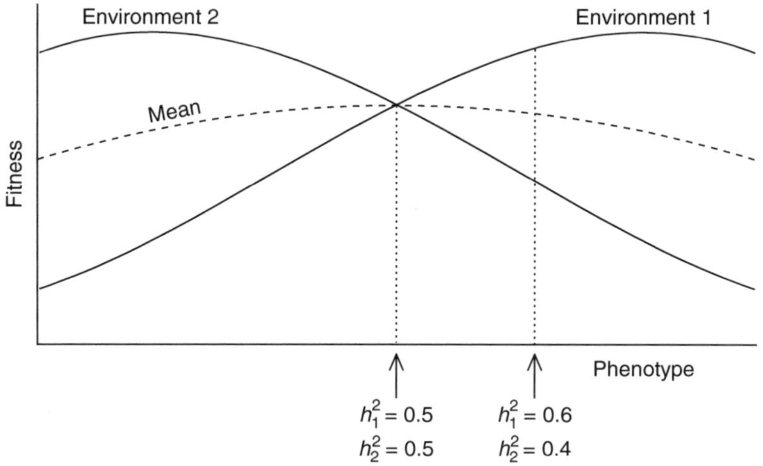

Fig. 2.5 Illustration of one consequence of covariation between selection and heritability. Heritabilities in Environments 1 and 2 are represented by h_1^2 and h_2^2, respectively. The *arrows* and *vertical dotted lines* indicate the equilibrium phenotype for the case of constant heritability ($h_1^2 = h_2^2 = 0.5$) and the case in which heritability covaries with the environment ($h_1^2 = 0.6, \quad h_2^2 = 0.4$). The equilibria were calculated from (2.6) under the assumption that the population variance is low enough that the regressions of fitness on phenotype are approximated by the slopes of the fitness curves

Environment 1 than in Environment 2, then the equilibrium shifts to a point where the population is much better adapted to Environment 1.

Biologically, this is because selection is more efficient at driving evolution in the environment in which heritability is higher. Mathematically, we can see how this result follows from Eq. (2.6) by noting how $\langle\langle h^2, \widehat{\Omega}\rangle\rangle$ varies with phenotype. For individuals with large values of the trait, near the optimum for Environment 1, high fitness (Environment 1) co-occurs with high heritability (also Environment 1), so $\langle\langle h^2, \widehat{\Omega}\rangle\rangle > 0$. By contrast, for individuals with lower values of the trait, near the optimum for Environment 2, high fitness co-occurs with low heritability, so $\langle\langle h^2, \widehat{\Omega}\rangle\rangle < 0$. Thus $[[\phi, \langle\langle h^2, \Omega\rangle\rangle]]$ is positive, shifting the population toward higher phenotypic values.

Two points should be noted from this example: First, covariance between heritability and selection shifts the equilibrium substantially toward the optimal phenotype for Environment 1, and substantially away from the optimum for environment 2. Thus, even though the population encounters Environment 2 roughly half of the time, and has significant heritability in that environment, it ends up rather poorly adapted to Environment 2. Second, the equilibrium does not maximize $\bar{w}$. This example thus illustrates that, even with frequency-independent selection, mean population fitness is not necessarily maximized when fitness and heritability are stochastic.

2.6 Conclusions

Many of the factors that influence evolution, including individual fitness, migration, and genetic transmission, are inherently stochastic. For the sake of mathematical simplicity, many evolutionary models treat some or all of these factors as deterministic, on the assumption that any stochasticity in real systems will simply add noise to the outcome, without changing the expected value.

We have demonstrated that, contrary to the common assumption, stochastic variation in fitness or migration, even when that variation is completely symmetrical, imposes a directionality on evolution that is not apparent in deterministic models. Stochastic heritability, while not directional itself, can significantly influence adaptation when heritability covaries with selection.

In this chapter, we have chosen only a few examples for the sake of illustration. However, the number of terms in Eqs. (2.1) and (2.4), and the fact that each of these terms can be expanded as in Eqs. (2.2) and (2.5), suggest that we have only scratched the surface in the study of directional stochastic effects in evolution.

References

Charmantier A, Garant D (2005) Environmental quality and evolutionary potential: lessons from wild populations. Proc Biol Sci 272(1571):1415–1425

Fitzpatrick BM, Fordyce JA, Gavrilets S (2009) Pattern, process and geographic modes of speciation. J Evol Biol 22:2342–2347

Gillespie JH (1974) Natural selection for within-generation variance in offspring number. Genetics 76(3):601–606

Hartl DL, Cook RD (1973) Balanced polymorphisms of quasineutral alleles. Theor Popul Biol 4 (2):163–172

Karlin S, Liberman U (1974) Random temporal variation in selection intensities: case of large population size. Theor Popul Biol 6(3):355–382

Merila J, Sheldon BC (2001) Avian quantitative genetics. In: Nolan V Jr (ed) Current ornithology, Chapter 4, vol 16. Kluwer/Plenum, New York, p 179

Rice SH (2008) A stochastic version of the Price equation reveals the interplay of deterministic and stochastic processes in evolution. BMC Evol Biol 8:262

Rice SH, Papadopoulos A (2009) Evolution with stochastic fitness and stochastic migration. PLoS ONE 4(10):e7130

Rundle HD, Nosil P (2005) Ecological speciation. Ecol Lett 8(17):336–352

Schluter D (2001) Ecology and the origin of species. Trends Ecol Evol 16(7):372–380

Wilson AJ, Pemberton JM, Pilkington JG, Coltman DW, Mifsud DV, Clutton-Brock TH, Kruuk LEB (2006) Environmental coupling of selection and heritability limits evolution. PLoS Biol 4 (7):e216

Chapter 3
Evolution of Self-Fertile Hermaphrodites

Ronald E. Ellis and Yiqing Guo

Abstract Self-fertile hermaphrodites have evolved in several independent lineages of nematodes. Surprisingly, both *C. elegans* and *C. briggsae* have recruited members of the large family of F-box genes to promote hermaphrodite development. However, *C. elegans* FOG-2 and *C. briggsae* SHE-1 have different biochemical functions, and each was created by a unique series of gene duplications. Despite these differences, they share a common target – the transmembrane receptor TRA-2, which plays a central role in the sex-determination pathway. When *tra-2* activity is knocked down in the male/female species *C. remanei*, some of the animals develop as hermaphrodites, but are unable to self-fertilize. This defect is due to the inability of their sperm to auto-activate, since knocking down a second gene that blocks sperm activation leads to self-fertility. Based on these results, we propose that hermaphroditic reproduction is a complex trait, because it requires the independent coordination of different regulatory pathways, one controlling sexual development and the other controlling sperm activation. Further analysis of the evolution of these hermaphrodites should reveal how novel traits first arise during evolution.

3.1 Animal Species with Self-Fertile Hermaphrodites Are Rare

3.1.1 The Androdioecious Lifestyle Is Adapted for Colonization

In androdioecious species, some individuals are males and others are hermaphrodites. Early models for the origin of androdioecy focused on flowering plants, where evidence suggests that ancestral populations consisted entirely of hermaphrodites

R.E. Ellis • Y. Guo
Department of Molecular Biology, The UMDNJ School of Osteopathic Medicine, B303 Science Center, 2 Medical Center Drive, Stratford, NJ 08084, USA
e-mail: ron.ellis@umdnj.edu

P. Pontarotti (ed.), *Evolutionary Biology – Concepts, Biodiversity, Macroevolution and Genome Evolution*, DOI 10.1007/978-3-642-20763-1_3,

(reviewed by Charlesworth and Charlesworth 1978). Since most hermaphroditic plants can either cross-pollinate or self-pollinate, it seemed likely that males subsequently arose by acquiring mutations that eliminated female reproductive structures and their associated costs. This idea fit in well with other analyses of the cost of sexual reproduction.

Work with animals has reopened this problem (reviewed by Pannell 2002; Weeks et al. 2006a). First, hermaphrodites from androdioecious species appear to be modified females, which lack the male reproductive structures needed to inseminate other individuals. Hence, these androdioecious species must have arisen from male/female ancestors through modification of the female sex. Second, these hermaphrodites cannot cross-fertilize, but instead only self or mate with males. As a result, a population of purely hermaphroditic animals should become completely inbred.

Given this scenario, what advantage could self-fertility confer on animals that would overcome the cost of inbreeding depression? Darwin (1876) suggested that selfing might aid in colonization, and this idea was elaborated on by Baker (1955, 1967). A beautiful example involves the European tadpole shrimp, which has male/female, male/hermaphrodite and purely hermaphroditic populations (Zierold et al. 2007). Phylogenetic studies indicate that much of the continent was repopulated by hermaphrodites following the retreat of glaciers at the end of the last ice age. By contrast, male/female populations remain more common in ancient refuges in Iberia. These data support the commonsense notion that hermaphrodites should excel at colonization, since a single individual can found a new population.

3.1.2 In Some Taxa, Androdioecy Has Evolved Repeatedly

Although the advantages of self-fertility might seem beneficial to all species that frequently colonize new environments, androdioecy is extremely rare in both plants and animals (Charlesworth and Charlesworth 1978; Pannell 2002). However, despite this general trend, certain taxa have exhibited repeated, parallel evolution of this mating system. For example, nematodes have undergone independent evolution of androdioecious mating systems on numerous occasions (Kiontke and Fitch 2005), and even in the genus *Caenorhabditis*, androdioecy has evolved in at least three different species (Fig. 3.1, Cho et al. 2004; Kiontke et al. 2004, K. Kiontke, M.-A. Félix, M. Ailion, and D. H. A. Fitch, pers. comm.). Branchiopod crustaceans have also produced many androdioecious species, like the clam shrimp *Eulimnadia texana* (Sassaman and Weeks 1993) and other members of its genus (Weeks et al. 2006b), or the tadpole shrimp *Triops cancriformis* (Zaffagnini and Trentini 1980). Phylogenetic studies imply that androdioecy also evolved independently in several of these cases (Weeks et al. 2009).

By contrast, only one self-fertilizing hermaphrodite is known among the vertebrates – the fish *Kryptolebias* (formerly *Rivulus*) *marmoratus* (Turner et al. 1992). Thus, certain taxa are predisposed to the origin of self-fertilizing hermaphrodites, whereas other taxa are not.

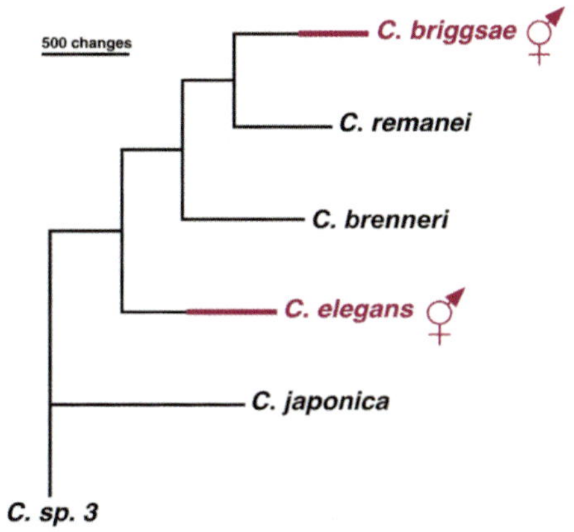

Fig. 3.1 Hermaphrodites evolved independently in *C. elegans* and *C. briggsae*. Maximum Likelihood tree prepared by Cho et al. (2004). Androdiocious lineages are maroon. All branches are significantly positive ($P < 0.01$)

The parallel evolution of androdioecy makes it an ideal model for the origin of complex traits, since theories about how change occurs can be tested by comparative analyses of related species. This task is facilitated in nematodes by the recent origin of selfing in several *Caenorhabditis* lineages. Finally, the lack of androdioecy in other taxa makes it an excellent model for probing the role that developmental biases play in evolutionary change.

3.2 *C. elegans* Is a Model for Self-Fertile Hermaphrodites

The nematode *C. elegans* was originally selected as a model for studying development and neurobiology (Brenner 1974). One of the key advantages Sydney Brenner considered was the fact that these animals are androdioecious (Nigon 1949), since the ability of hermaphrodites to self-fertilize dramatically simplifies mutant screens. Eventually, *C. elegans* also became a model for the genetic control of sexual identity (reviewed by Goodwin and Ellis 2002), and for the population dynamics of hermaphrodites (reviewed by Barriere and Felix 2005).

3.2.1 The Hermaphrodite Soma Is Essentially Female in Structure

In *C. elegans,* about a third of all cells are sexually dimorphic (Fig. 3.2, reviewed by Zarkower 2006). For example, some hypodermal cells produce the vulva in hermaphrodites, but do not divide in males (Fig. 3.2, purple structure). Many muscles are also sexually dimorphic, such as the hermaphrodite sex muscles, which control the opening of the vulva (Fig. 3.2, orange cells). Moreover, some neurons are specific to one sex or the other; in particular, the HSN neurons innervate the sex

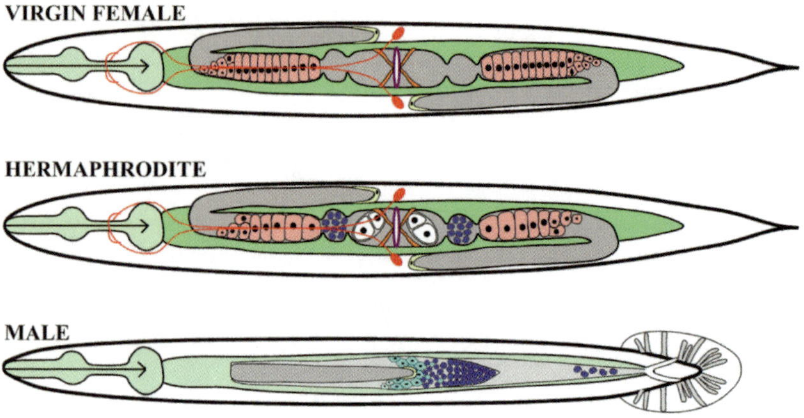

Fig. 3.2 Nematode hermaphrodites are modified females. Ventral views of each adult sex, with anterior to the left. The digestive system is colored *light green*, except for the female and hermaphrodite intestines, which are *dark green* to denote the production of yolk. The HSN neurons are *red*, the sex muscles *orange*, and the vulva *purple*. In each sex, the somatic gonad is *grey*, with oocytes colored *pink* and male germ cells *light blue* for spermatocytes and *dark blue* for sperm. For clarity, the mitotic and early meiotic germ cells are not included in the diagram, since they appear similar in each sex

muscles and control egg laying in hermaphrodites, but die in males (Fig. 3.2, red cells). As one might expect, the gonads of the two sexes also differ dramatically; the hermaphrodite ovotestes is a bilobed structure devoted to nurturing oocytes, whereas the male testes has a single lobe that connects to the cloaca and is specialized for nurturing spermatocytes (Fig. 3.2, gray structures). Even the intestine differs between the sexes. In hermaphrodites, it produces yolk (Fig. 3.2, dark green) but in males it does not (Fig. 3.2, light green).

Examination of females from related species of *Caenorhabditis* reveals that they are almost identical to *C. elegans* hermaphrodites, with one exception – the hermaphrodites produce sperm when they are young, but females do not.

3.2.2 The Hermaphrodite Germ Line Produces Sperm as well as Oocytes

In hermaphroditic nematodes, the first germ cells begin to differentiate during the fourth and final larval stage, and become spermatocytes. Later, the animals switch to oogenesis after molting into adults. Since this switch is irrevocable, each hermaphrodite must produce all of the self sperm it will need during larval development. Furthermore, the timing of the switch is of critical importance. Mutants that undergo extended spermatogenesis cannot compete with the wild type, despite the fact that they produce more self-progeny, because the delay in beginning oogenesis is too costly (Hodgkin and Barnes 1991).

3.2.3 Genes That Regulate Sex-Determination Influence Hermaphrodite Development

The genetic control of sex determination in *C. elegans* is now understood in great detail (Fig. 3.3, reviewed by Goodwin and Ellis 2002; Zarkower 2006). Briefly, it involves three steps.

First, a group of genes responds to the difference in *X* chromosome dose to regulate XOL-1. In males, XOL-1 is active and represses the SDC genes, thus blocking dosage compensation and allowing the expression of the secreted protein HER-1. In hermaphrodites, XOL-1 is inactive, which allows dosage compensation to proceed and blocks the production of HER-1. Second, sexual identity is coordinated by HER-1. In males, HER-1 diffuses throughout the body and inactivates the TRA-2 receptor. In hermaphrodites, the absence of HER-1 allows TRA-2 to direct female development in each cell. Third, a signal transduction cascade controls the activity of the master transcription factor TRA-1 in each cell. In males, the FEM proteins promote the ubiquitinylation and degradation of TRA-1 (Starostina et al. 2007), whereas in hermaphrodites, TRA-2 downregulates the FEM proteins, so that

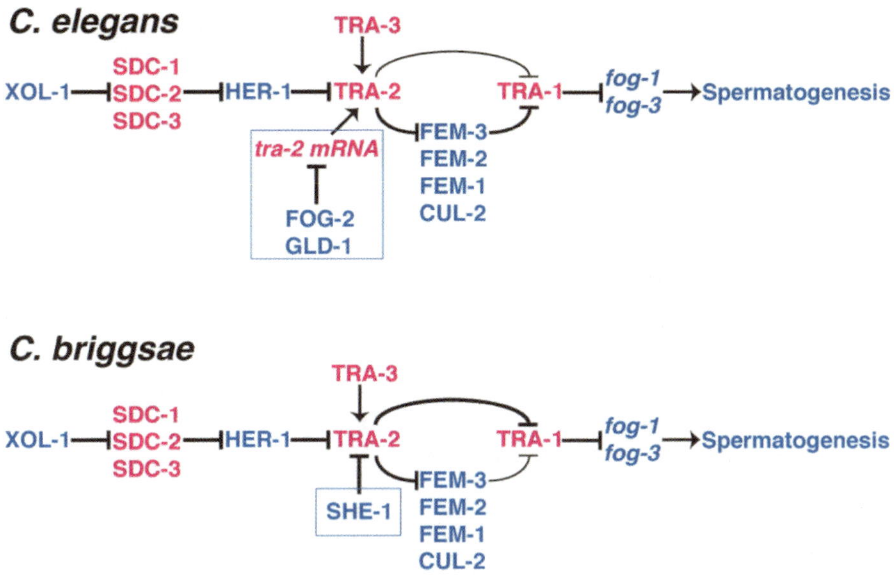

Fig. 3.3 Nematode sex determination pathways. Genes promoting spermatogenesis are shown in *blue*, and those promoting oogenesis in *pink*. Positive interactions are indicated with an *arrow*, negative interactions with a *barred line*, and relatively weak interactions with a *thin line*. Regulatory circuits that specifically promote hermaphrodite spermatogenesis are *boxed*. For simplicity, some genes that regulate sexual development in germ cells are not pictured (reviewed by Ellis 2008)

TRA-1 can be processed into a repressor that blocks male cell fates (Schvarzstein and Spence 2006).

These sex-determination genes act throughout the body, so additional regulatory interactions are needed in the germ line, to allow hermaphrodites to make sperm. These germline regulators control the expression of *fog-3* during larval development, which must occur in *XX* hermaphrodites, but is not observed in *XX* females (Chen et al. 2001). FOG-3 itself cooperates with FOG-1 to promote spermatogenesis (Barton and Kimble 1990; Ellis and Kimble 1995).

3.2.4 The F-Box Protein FOG-2 Specifies Hermaphrodite Development in C. elegans

The *fog-2* gene plays a critical role in hermaphrodite development, since the mutants form male/female strains (Schedl and Kimble 1988). FOG-2 interacts with the translational regulator GLD-1 (Clifford et al. 2000), which in turn binds *tra-2* messenger RNAs (Fig. 3.3, Jan et al. 1999). Because FOG-2 and GLD-1 repress the translation of *tra-2,* they lower the overall activity of the gene enough to allow spermatogenesis to proceed during larval development.

Several questions about these genes remain unanswered. First, since GLD-1 is a STAR protein that normally represses its RNA targets (Jones and Schedl 1995), why does it require FOG-2 as a cofactor? Along the same lines, since FOG-2 is an F-box protein that can interact with SKR-1, why doesn't it cause the degradation of GLD-1? Finally, is either GLD-1 or FOG-2 activity modulated in adult hermaphrodites to allow the beginning of oogenesis?

3.3 In *C. briggsae,* a Novel F-Box Protein Specifies Hermaphrodite Development

Although *fog-2* plays a central role in hermaphrodite development in *C. elegans,* it has no ortholog in *C. briggsae* (Nayak et al. 2005). To learn how *C. briggsae XX* animals become hermaphrodites, we screened for mutations that create male/female strains (Guo et al. 2009). All of the mutations we identified were recessive and failed to complement each other. Since they mapped to a new location, we named the gene *she-1,* for *s*permless *he*rmaphrodites.

By constructing double mutants with alleles of other genes that control sex determination, we showed that *she-1* acts upstream of *tra-2* to repress its activity, much as *fog-2* does in *C. elegans* (Fig. 3.3, Guo et al. 2009). However, *she-1* seemed unlikely to work directly with *gld-1,* because *gld-1* promotes oogenesis in *C. briggsae* (Nayak et al. 2005), whereas it promotes spermatogenesis in *C. elegans.*

3.3.1 SHE-1 Is a Novel F-Box Protein

Thus, to learn how *she-1* controls sexual development, we used SNP mapping to clone the gene (Guo et al. 2009). After narrowing the region that could contain *she-1* to about 100 kb, we used RNA interference to test candidate genes, and confirmed our identification by sequencing DNA from mutant strains, all of which turned out to have lesions in *she-1*. This work showed that the sequenced genome (Stein et al. 2003) and existing SNP database (Hillier et al. 2007) could be used to identify *C. briggsae* genes that had been known only through mutations.

We found that *she-1* encodes a novel F-box protein. Genetic tests had indicated that SHE-1 was unlikely to interact with GLD-1, and yeast two-hybrid data confirmed this prediction. Thus, SHE-1 and FOG-2 have distinct biochemical activities. Two results suggest that SHE-1 functions like most other F-box proteins. First, the *v35* missense mutation alters a conserved residue in the F-box, which implies that this domain is essential for function. Second, yeast two-hybrid assays show that SHE-1 can bind to SKR-1, a component of E3 ubiquitin-ligase complexes. Moreover, this interaction is abolished by the *v35* missense mutation. Thus, SHE-1 is likely to regulate development by controlling the ubiquitinylation and degradation of a target protein. So far, the direct target remains unknown, although it must regulate TRA-2 activity.

3.3.2 *In* she-1 *Null Mutants, the Environment Determines Sexual Development*

Some of the *she-1* alleles we identified are molecular null alleles; in particular, *v49* is an early stop mutation, and *vDf2* deletes most of the gene. Like other *she-1* mutations, these alleles are temperature sensitive – at 25° all *XX* animals develop as females, but at 15° about half of them become hermaphrodites. Thus, prior to the origin of *she-1*, the development of *XX* animals in *C. briggsae* might have responded to environmental conditions, so that they became females in some circumstances and hermaphrodites in others. If so, this ability might have helped them navigate the period of inbreeding depression that should occur during the transition from a male/female species to a purely androdioecious one.

3.3.3 *The* she-1 *Gene Was Created by a Recent Gene Duplication Event*

One of the most common paradigms for evolutionary change involves the alteration of the *cis* regulatory sequences that control the expression of critical genes (Weatherbee et al. 1999; Prud'homme et al. 2006; Williams et al. 2008; Chan et al. 2010).

Such changes can create new regions of expression, while at the same time preserving all of the ancestral functions of the gene. Although gene duplications are known to play a significant role in evolution, the most striking examples of recent duplications involve structural genes (e.g., Zhang et al. 2002).

However, our studies show that hermaphrodite development in *C. briggsae* depends on *she-1*, a gene created by a recent duplication event (Fig. 3.4). Furthermore, hermaphrodite development in *C. elegans* also relies on a gene created by a

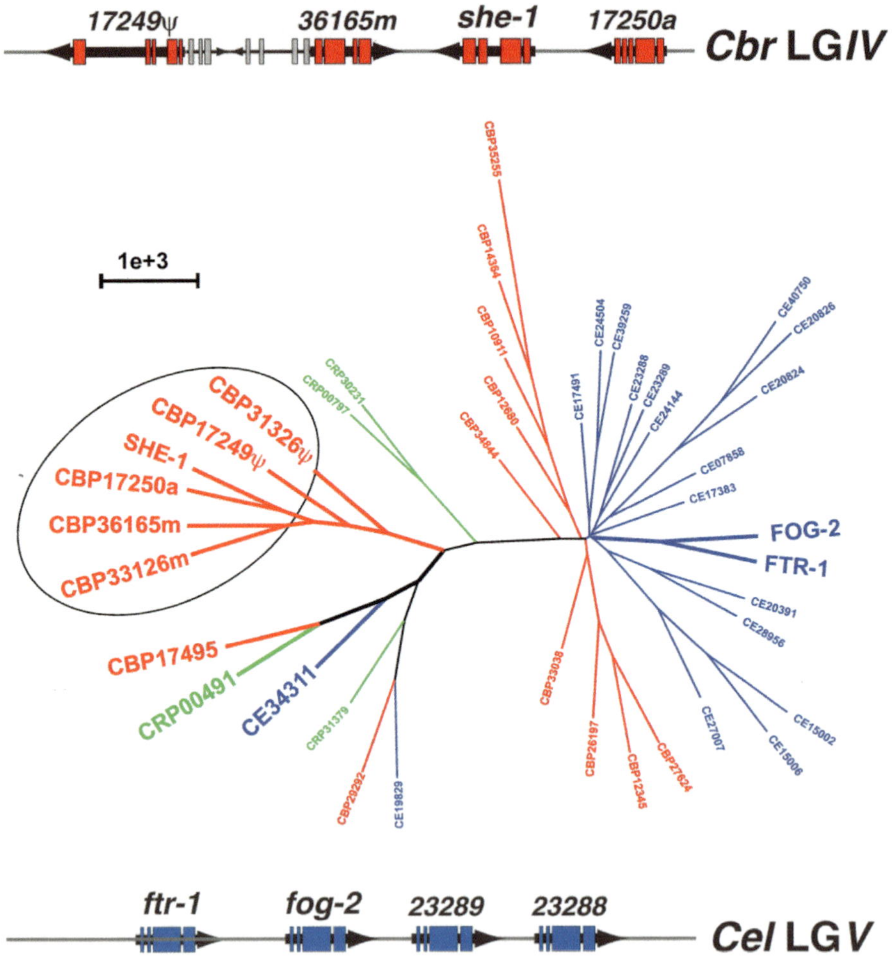

Fig. 3.4 Independent gene duplications regulate hermaphrodite development. Maximum likelihood tree of select F-box genes (Guo et al. 2009), flanked by diagrams of the *she-1* and *fog-2* genomic regions (above and below, respectively). In the phylogeny, *C. briggsae* proteins are shown in *red*, *C. elegans* proteins in *blue*, and *C. remanei* proteins in *green*. In the diagrams, the chromosomes are *gray*, *C. briggsae* F-box genes are *red*, *C. elegans* F-box genes are *blue*, and non-F-box genes are *gray*. *Arrows* connect exons and point toward the 3′ end of each gene

recent duplication (Fig. 3.4, Clifford et al. 2000). Surprisingly, these genes were both recruited from the F-box family, even though they appear to have distinct biochemical functions. Why should this one family have produced evolutionary novelties in two independent lineages? In nematodes, the F-box genes form one of the largest and most rapidly diversifying families (Thomas 2006). If a large number of F-box genes are constantly being created by duplication and altered by mutation and drift, these high numbers should increase the odds that some family members will adopt novel functions.

In addition, the structure and function of regulatory pathways might also influence the types of change that occur during evolution. Many of the models now being studied involve changes in spatial patterning (Weatherbee et al. 1999; Prud'homme et al. 2006; Williams et al. 2008; Chan et al. 2010). In one respect, the recruitment of tissue-specific regulatory genes like *fog-2* or *she-1* is analogous to the tissue-specific changes in enhancers seen in other systems. However, hermaphrodite development requires more than altering the sex-determination pathway in the germ line but not the soma; it also involves precise temporal regulation.

To allow for self-fertilization, the hermaphrodite germ line first produces male cells, and later makes female cells. Thus, the relative activity of genes in the sex-determination pathway must be able to flip around the time the animals molt into adults. Perhaps the recruitment of genes like *she-1* and *fog-2* allowed this flip by lowering the activity of *tra-2* without eliminating it. Thus, investigating additional evolutionary changes that involve developmental timing or heterochronic phenotypes might broaden our understanding of how regulatory pathways evolve.

3.3.4 The tra-2 *Gene Might Be a Hot Spot for Changes to the Sex-Determination Pathway*

In fruit flies, almost all of the changes that have altered trichome patterns during evolution affect the transcription factor Shavenbaby (McGregor et al. 2007). This gene probably plays a privileged role because it acts at a nexus in the pathway that controls trichome development (reviewed by Stern 2007; Stern and Orgogozo 2009). Upstream genes that regulate the expression of *shavenbaby* are highly pleiotropic, so mutations that affect them are likely to be deleterious. Downstream mutations affect the structure of trichomes themselves, and might damage them. However, mutations in the *shavenbaby* promoter affect where the gene is expressed, and thus where the developmental subprogram that produces trichomes is active. These favorable conditions appear to make it a hotspot for evolutionary change.

Since independent genes that regulate hermaphrodite development impinge on the sex-determination pathway at *tra-2* (Fig. 3.3), it also appears to be a hotspot. However, TRA-2 is not a transcription factor, so the reason it is favored must be different. We note that the genes that act upstream of TRA-2 influence tissues throughout the body through the secreted protein HER-1, whereas TRA-2 is a cell-autonomous receptor. Hence, only *tra-2* or downstream genes are likely to be altered to create the hermaphrodite germ line. Perhaps *tra-2* is the most likely target because it controls two separate branches of the sex-determination pathway (Fig. 3.3).

3.4 The Origin of Self-Fertility Requires Two Separate Adaptations

To test our model that small decreases in *tra-2* activity can create self-fertile hermaphrodites, we studied the male/female species *C. remanei* (Baldi et al. 2009).

3.4.1 Lowering the Activity of tra-2 *Allows Sperm Production in* XX *Females*

Although the complete inactivation of the sex-determination gene *tra-2* transforms *XX* animals into imperfect males (Hodgkin and Brenner 1977), we used a low dose for RNA interference that only partially knocked down its activity. In *C. remanei,* this treatment produced a range of weaker phenotypes among the *XX* progeny. Most importantly, we found individuals with a normal female soma that produced sperm while young, and oocytes when older (Baldi et al. 2009). Although these animals strongly resembled *C. elegans* hermaphrodites, they were not self-fertile (Fig. 3.5). Thus, we refer to them as pseudohermaphrodites.

3.4.2 Altering tra-2 *Activity Is Not Sufficient to Activate Sperm in* XX *Animals*

Closer analysis of the pseudohermaphrodites showed that their sperm neither activated nor moved into the spermatheca, but were lost following the first ovulation (Fig. 3.5). In *C. elegans,* hermaphrodite sperm must be activated to fertilize oocytes and avoid being lost when pushed into the uterus during ovulation (reviewed by L'Hernault 2006). Thus, we dissected individual pseudohermaphrodites, and found that their sperm were indeed inactive, but could be activated by treatment with pronase. Taken together, these results implied that pseudohermaphrodites were not self-fertile because their spermatids could not self activate.

In *C. elegans,* male seminal fluid can activate sperm (reviewed by L'Hernault 2006). Thus, we tested our hypothesis by crossing pseudohermaphrodites with sterile males or *C. elegans* males. In both crosses, male seminal fluid activated sperm in the pseudohermaphrodites, leading to fertilization and the production of self progeny. These results imply that altering the sex-determination pathway is necessary to create hermaphrodites, but not sufficient for self-fertility.

3.4.3 *The* swm-1 *Genes Plays a Conserved Role in the Activation of* Caenorhabditis *Sperm*

In *C. elegans,* the *swm-1* gene prevents the premature activation of sperm in males (Stanfield and Villeneuve 2006). It encodes a protease inhibitor with two TIL domains, which suggests that the ability of proteases to activate sperm in vitro reflects normal regulation in vivo. In addition, genetic tests indicated that *swm-1* also plays a weak role in hermaphrodites. Thus, we analyzed the function of *swm-1* in the male/female species *C. remanei.*

Each of the nematode species in Fig. 3.1 has a single *swm-1* gene with a highly conserved sequence (Baldi et al. 2009). In addition, the two hermaphroditic species have independent and highly divergent duplications of *swm-1* that have no known function. We found that using RNA interference to knock down *swm-1* in *C. remanei* promotes the activation of male sperm, which implies that its function has been conserved in nematodes.

Moreover, when both *tra-2* and *swm-1* were knocked down in *C. remanei,* some of the *XX* animals developed as self-fertile hermaphrodites (Fig. 3.5, Baldi et al. 2009). Thus, *tra-2(RNAi) XX* pseudohermaphrodites are sterile because their sperm

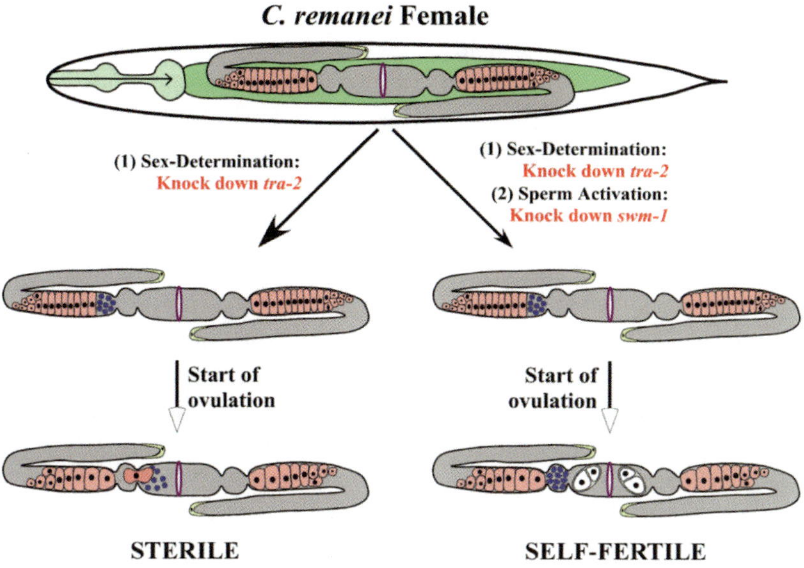

Fig. 3.5 Two independent changes are required to produce hermaphrodites. Summary of experiments that use RNA interference to alter the development of *C. remanei* females. The diagrams use the same conventions as Fig. 3.2, but focus on the germ line, since it plays a central role in self-fertilization. If only *tra-2* is knocked down, the animals produce inactive sperm, which are pushed into the uterus during ovulation and lost (*left*). If *swm-1* is also knocked down, the sperm activate; they can fertilize oocytes, and crawl back into the spermatheca to avoid being lost (*right*)

are unable to activate. Moreover, the *swm-1* gene is expressed in these *XX* animals and is capable of blocking sperm activation.

3.4.4 Two Independent Pathways Must Be Altered to Produce Self-Fertile Hermaphrodites

These results imply that two independent pathways were altered during evolution to create self-fertile hermaphrodites. One set of changes affected the sex-determination pathway, and led to the production of sperm during larval development in otherwise female animals. The *fog-2* gene plays a critical role in this process in *C. elegans*, and the *she-1* gene does so in *C. briggsae*. Additional genes that have not yet been identified might assist them.

The second set of changes allowed the sperm produced by *XX* animals to activate and fertilize oocytes. Although these changes could have been caused by mutations in *swm-1*, the process of sperm activation involves a complex signal transduction pathway (reviewed by L'Hernault 2006), and many other candidates exist. This topic remains a wide-open area for research.

3.5 A Model for the Origin of Self-Fertility

Because the origin of hermaphrodites required the coordination of changes in at least two independent pathways, self-fertility is a complex trait. The analysis of intraspecies hybrids between *C. briggsae* and *C. sp. 9* supports our conclusion that multiple genes were involved in the origin of hermaphroditism (Woodruff et al. 2010). Thus, it provides a model for how other complex traits originated. Two types of explanations seem possible.

In the first, the initial genetic change was neutral, and set the stage for mutations affecting the second trait to sweep through the population. For example, mutations that caused the expression of proteases in the female spermatheca might have conferred the ability to active sperm. In a male/female population, these mutations would probably be selectively neutral. They might have accumulated to low frequencies, or become fixed in small, isolated populations. If so, mutations that altered the sex-determination pathway could then have immediately led to self-fertility.

In the second, these changes proceeded through a selectively favorable intermediate. For example, if nematodes of different species copulate as frequently in the wild as they do in the laboratory, it is possible that "pseudohermaphrodites" would have been able to reproduce using male seminal fluid to activate their own sperm. If so, then an initial change that altered the sex-determination pathway might have been advantageous on its own, even before the acquisition of a second mutation that allowed incipient hermaphrodites to activate self-sperm on their own.

In either scenario, the origin of self-fertility would push nematodes into a niche that should favor additional changes. For example, the population would need to eliminate recessive lethal mutations to minimize inbreeding depression (Dolgin et al. 2007), the number of hermaphrodite sperm would have to be optimized (Hodgkin and Barnes 1991), and the transition from spermatogenesis to oogenesis would need to be sharpened to prevent the production of sexually ambiguous cells. In addition, genomic databases imply that a large-scale reduction in genome size occurs in androdioecious species, and comparative studies show the size of spermatocytes declines in both androdioecious sexes (LaMunyon and Ward 1999). Finally, the structure of the sex-determination pathway itself might drift in hermaphrodites, as shown by the differing importance of the *fem* genes in *C. elegans* and *C. briggsae* (Hill et al. 2006).

To date, we know very little about which characteristics set the stage for the independent, parallel evolution of hermaphrodites in many species of nematodes and branchiopod crustaceans. One factor that might be important in nematodes is their *XO* mating system. Because of it, *XX* females contain all of the genetic information needed to make male tissues. By contrast, many animal species use an *XY* mating system, so the *XX* females lack crucial genes on the *Y* chromosome that might be needed for spermatogenesis. We know less about the sex-determination system in branchiopod crustaceans, but it seems likely that in the ancestral state males were homozygous *ZZ* and females were heterozygous *ZW* (Weeks et al. 2010), which is consistent with our hypothesis. Other traits, such as a gonad that would facilitate the ability of sperm in newly evolving hermaphrodites to find and fertilize oocytes, might be critical as well.

Three factors should make this decade a golden age for evolutionary studies in nematodes. We can now analyze the origin of a complex trait, and recreate it in the laboratory. Moreover, a large suite of changes occurred in response to the origin of hermaphrodites, and can be probed with genetic and genomic tools. Finally, these changes happened recently, in several parallel lineages.

Acknowledgments We thank the National Institutes of Health for support (Grant GM085282), and Eric Haag, Eric Moss, Steve Weeks, and Pierre Pontarotti for comments on this manuscript.

References

Baker HG (1955) Self-compatibility and establishment after "long-distance" dispersal. Evolution 9:347–348

Baker HG (1967) Support for Baker's law – as a rule. Evolution 21:853–856

Baldi C, Cho S, Ellis RE (2009) Mutations in two independent pathways are sufficient to create hermaphroditic nematodes. Science 326(5955):1002–1005

Barriere A, Felix MA (2005) Natural variation and population genetics of Caenorhabditis elegans. WormBook 1–19

Barton MK, Kimble J (1990) *fog-1*, a regulatory gene required for specification of spermatogenesis in the germ line of *Caenorhabditis elegans*. Genetics 125(1):29–39

Brenner S (1974) The genetics of *Caenorhabditis elegans*. Genetics 77(1):71–94

Chan YF, Marks ME, Jones FC, Jr Villarreal G, Shapiro MD, Brady SD, Southwick AM, Absher DM, Grimwood J, Schmutz J, Myers RM, Petrov D, Jonsson B, Schluter D, Bell MA, Kingsley DM (2010) Adaptive evolution of pelvic reduction in sticklebacks by recurrent deletion of a Pitx1 enhancer. Science 327(5963):302–305

Charlesworth B, Charlesworth D (1978) A model for the evolution of dioecy and gynodioecy. Am Nat 112:975–997

Chen PJ, Cho S, Jin SW, Ellis RE (2001) Specification of germ cell fates by FOG-3 has been conserved during nematode evolution. Genetics 158(4):1513–1525

Cho S, Jin SW, Cohen A, Ellis RE (2004) A phylogeny of *Caenorhabditis* reveals frequent loss of introns during nematode evolution. Genome Res 14(7):1207–1220

Clifford R, Lee MH, Nayak S, Ohmachi M, Giorgini F, Schedl T (2000) FOG-2, a novel F-box containing protein, associates with the GLD-1 RNA binding protein and directs male sex determination in the *C. elegans* hermaphrodite germline. Development 127(24):5265–5276

Darwin C (1876) The effects of cross- and self-fertilization in the vegetable kingdom. Murray, London

Dolgin ES, Charlesworth B, Baird SE, Cutter AD (2007) Inbreeding and outbreeding depression in Caenorhabditis nematodes. Evolution 61(6):1339–1352

Ellis RE (2008) Sex determination in the *Caenorhabditis elegans* germ line. Curr Top Dev Biol 83:41–64

Ellis RE, Kimble J (1995) The *fog-3* gene and regulation of cell fate in the germ line of *Caenorhabditis elegans*. Genetics 139(2):561–577

Goodwin EB, Ellis RE (2002) Turning clustering loops: sex determination in *Caenorhabditis elegans*. Curr Biol 12(3):R111–R120

Guo Y, Lang S, Ellis RE (2009) Independent recruitment of F box genes to regulate hermaphrodite development during nematode evolution. Curr Biol 19(21):1853–1860

Hill RC, de Carvalho CE, Salogiannis J, Schlager B, Pilgrim D, Haag ES (2006) Genetic flexibility in the convergent evolution of hermaphroditism in *Caenorhabditis* nematodes. Dev Cell 10(4): 531–538

Hillier LW, Miller RD, Baird SE, Chinwalla A, Fulton LA, Koboldt DC, Waterston RH (2007) Comparison of C. elegans and C. briggsae genome sequences reveals extensive conservation of chromosome organization and synteny. PLoS Biol 5(7):e167

Hodgkin J, Barnes TM (1991) More is not better: brood size and population growth in a self-fertilizing nematode. Proc Biol Sci 246(1315):19–24

Hodgkin JA, Brenner S (1977) Mutations causing transformation of sexual phenotype in the nematode *Caenorhabditis elegans*. Genetics 86(2 Pt. 1):275–287

Jan E, Motzny CK, Graves LE, Goodwin EB (1999) The STAR protein, GLD-1, is a translational regulator of sexual identity in *Caenorhabditis elegans*. EMBO J 18(1):258–269

Jones AR, Schedl T (1995) Mutations in *gld-1*, a female germ cell-specific tumor suppressor gene in *Caenorhabditis elegans*, affect a conserved domain also found in Src-associated protein Sam68. Genes Dev 9(12):1491–1504

Kiontke K, Fitch DH (2005) The phylogenetic relationships of *Caenorhabditis* and other rhabditids. In: WormBook (ed) The *C. elegans* Research Community

Kiontke K, Gavin NP, Raynes Y, Roehrig C, Piano F, Fitch DH (2004) *Caenorhabditis* phylogeny predicts convergence of hermaphroditism and extensive intron loss. Proc Natl Acad Sci USA 101(24):9003–9008

L'Hernault SW (2006) Spermatogenesis. WormBook 1–14

LaMunyon CW, Ward S (1999) Evolution of sperm size in nematodes: sperm competition favours larger sperm. Proc Biol Sci 266(1416):263–267

McGregor AP, Orgogozo V, Delon I, Zanet J, Srinivasan DG, Payre F, Stern DL (2007) Morphological evolution through multiple cis-regulatory mutations at a single gene. Nature 448(7153): 587–590

Nayak S, Goree J, Schedl T (2005) *fog-2* and the evolution of self-fertile hermaphroditism in *Caenorhabditis*. PLoS Biol 3(1):e6

Nigon V (1949) Les modalités de la réproduction et le déterminisme de sexe chez quelques Nématodes libres. Ann Sci Nat Zool 11:1–132

Pannell JR (2002) The evolution and maintenance of androdioecy. Annu Rev Ecol Syst 33: 397–425

Prud'homme B, Gompel N, Rokas A, Kassner VA, Williams TM, Yeh SD, True JR, Carroll SB (2006) Repeated morphological evolution through cis-regulatory changes in a pleiotropic gene. Nature 440(7087):1050–1053

Sassaman C, Weeks SC (1993) The genetic mechanism of sex determination in the conchostracan shrimp Eulimnadia texana. Am Nat 141(2):314–328

Schedl T, Kimble J (1988) *fog-2,* a germ-line-specific sex determination gene required for hermaphrodite spermatogenesis in *Caenorhabditis elegans*. Genetics 119(1):43–61

Schvarzstein M, Spence AM (2006) The *C. elegans* sex-determining GLI protein TRA-1A is regulated by sex-specific proteolysis. Dev Cell 11(5):733–740

Stanfield GM, Villeneuve AM (2006) Regulation of sperm activation by SWM-1 is required for reproductive success of *C. elegans* males. Curr Biol 16(3):252–263

Starostina NG, Lim JM, Schvarzstein M, Wells L, Spence AM, Kipreos ET (2007) A CUL-2 ubiquitin ligase containing three FEM proteins degrades TRA-1 to regulate *C. elegans* sex determination. Dev Cell 13(1):127–139

Stein LD, Bao Z, Blasiar D, Blumenthal T, Brent MR, Chen N, Chinwalla A, Clarke L, Clee C, Coghlan A, Coulson A, D'Eustachio P, Fitch DH, Fulton LA, Fulton RE, Griffiths-Jones S, Harris TW, Hillier LW, Kamath R, Kuwabara PE, Mardis ER, Marra MA, Miner TL, Minx P, Mullikin JC, Plumb RW, Rogers J, Schein JE, Sohrmann M, Spieth J, Stajich JE, Wei C, Willey D, Wilson RK, Durbin R, Waterston RH (2003) The genome sequence of *Caenorhabditis briggsae:* a platform for comparative genomics. PLoS Biol 1(2):E45

Stern DL (2007) The developmental genetics of microevolution. Novartis Found Symp 284: 191–200, discussion 200–206

Stern DL, Orgogozo V (2009) Is genetic evolution predictable? Science 323(5915):746–751

Thomas JH (2006) Adaptive evolution in two large families of ubiquitin-ligase adapters in nematodes and plants. Genome Res 16(8):1017–1030

Turner BJ, Jr Elder JF, Laughlin TF, Davis WP, Taylor DS (1992) Extreme clonal diversity and divergence in populations of a selfing hermaphroditic fish. Proc Natl Acad Sci USA 89 (22):10643–10647

Weatherbee SD, Nijhout HF, Grunert LW, Halder G, Galant R, Selegue J, Carroll S (1999) Ultrabithorax function in butterfly wings and the evolution of insect wing patterns. Curr Biol 9(3):109–115

Weeks SC, Benvenuto C, Reed SK (2006a) When males and hermaphrodites coexist: a reiew of androdioecy in animals. Integr Comp Biol 46(4):449–464

Weeks SC, Sanderson TF, Reed SK, Zofkova M, Knott B, Balaraman U, Pereira G, Senyo DM, Hoeh WR (2006b) Ancient androdioecy in the freshwater crustacean Eulimnadia. Proc Biol Sci 273(1587):725–734

Weeks SC, Chapman EG, Rogers DC, Senyo DM, Hoeh WR (2009) Evolutionary transitions among dioecy, androdioecy and hermaphroditism in limnadiid clam shrimp (Branchiopoda: Spinicaudata). J Evol Biol 22(9):1781–1799

Weeks SC, Benvenuto C, Sanderson TF, Duff RJ (2010) Sex chromosome evolution in the clam shrimp, Eulimnadia texana. J Evol Biol 23(5):1100–1106

Williams TM, Selegue JE, Werner T, Gompel N, Kopp A, Carroll SB (2008) The regulation and evolution of a genetic switch controlling sexually dimorphic traits in Drosophila. Cell 134(4): 610–623

Woodruff GC, Eke O, Baird SE, Felix MA, Haag ES (2010) Insights into species divergence and the evolution of hermaphroditism from fertile interspecies hybrids of Caenorhabditis nematodes. Genetics 186(3):997–1012

Zaffagnini F, Trentini M (1980) The distribution and reproduction of *Triops cancriformis* (Bosc) in Europe (Crustacea Notostraca). Monitor Zool Ital (NS) 14:1–8

Zarkower D (2006) Somatic sex determination. In: Wormbook (ed) The *C. elegans* Research Community
Zhang J, Zhang YP, Rosenberg HF (2002) Adaptive evolution of a duplicated pancreatic ribonuclease gene in a leaf-eating monkey. Nat Genet 30(4):411–415
Zierold T, Hanfling B, Gomez A (2007) Recent evolution of alternative reproductive modes in the "living fossil" Triops cancriformis. BMC Evol Biol 7:161

Chapter 4
Insights into Eukaryotic Interacting Protein Evolution

Sandip Chakraborty, Soumita Podder, Bratati Kahali, Tina Begum, Kamalika Sen, and Tapash Chandra Ghosh

Abstract The overall molecular architecture of all organisms is mainly mediated through the sophisticated coordination of protein–protein interactions. It has immensely prospered the arena of systems biology providing an inclusive perspective of the interrelationships between proteins. The evolutionary mechanisms of protein–protein interaction networks are now being appreciated as a major factor in shaping their present-day structures and properties. This chapter provides a systematic computational framework for identifying important evolutionary forces within protein–protein interaction network by considering *Saccharomyces cerevisiae* and *Homo sapiens* as model organisms. In summary, our works enlighten that in yeast protein interaction network, evolutionary rate discrimination of date and party hub proteins are ascertained by protein disorderness and presence of buried residue. For protein complex in interaction network, evolutionary rate is mostly governed by complex forming ability though the role of connectivity and expression level is also established. However, in case of human, no such role of connectivity has been detected, rather multifunctionality, protein domain coverage, as well as expression level are regarded as a major determining evolutionary forces in protein interaction network. Intriguingly, human complex forming units offer an exclusive facet of non-hub proteins, which with a high disorderness play a hub-like nature and involve in large complex formation.

4.1 Introduction

Networks of interacting proteins offer a global understanding of cellular functions and biological processes. Over the last few years, proteome-wide studies for model organisms such as *Helicobacter pylori* (Rain et al. 2001), *Saccharomyces*

All authors contributed equally to this chapter.

S. Chakraborty • S. Podder • B. Kahali • T. Begum • K. Sen • T.C. Ghosh
Bioinformatics Centre, Bose Institute, P 1/12, C.I.T. Scheme VII M, Kolkata 700 054, India
e-mail: tapash@boseinst.ernet.in

P. Pontarotti (ed.), *Evolutionary Biology – Concepts, Biodiversity, Macroevolution and Genome Evolution*, DOI 10.1007/978-3-642-20763-1_4,

cerevisiae (Ito et al. 2001; Uetz et al. 2000), *Caenorhabditis elegans* (Li et al. 2004), and *Drosophila melanogaster* (Giot et al. 2003) have been performed to have a profound vista of their respective biological systems. A systematic human protein interactions was also screened by automated yeast two-hybrid (Y2H) interaction mating (Stelzl et al. 2005) and are seen in orchestrating the metabolic, signaling, and regulatory pathways in a cell (Raman 2010). With the development of high-throughput interaction detection methods, rapid accumulation of large amounts of protein interaction data has been now possible (Bader et al. 2003; Salwinski et al. 2004; Prasad et al. 2009; Ceol et al. 2010), providing a comprehensive insight into the assessment of cellular world via protein interactions. The interaction networks exhibit "a small world property" (Watts and Strogatz 1998) and affirm the presence of hubs, the elements connected to many other elements in the network (Jeong et al. 2001; Kim et al. 2006). The spatial versus temporal complexities in networks were factored out after designating the hubs as single-interface hubs and multiple-interface hubs (Yeates and Beeby 2006) along with a parallel subdivision of date hubs and party hubs correlating cellular expression with their interacting partners (Han et al. 2004).

The topological features of interaction networks endow with a framework to study protein deregulation in complex diseases (Rhodes et al. 2005; Goh et al. 2007; Sam et al. 2007; Kar et al. 2009) and disease, non-disease gene inheritance (Gandhi et al. 2006; Xu and Li 2006). Evolutionary considerations of the interacting proteins unveil a negative correlation between connectivity and the rate of evolution of well-conserved proteins (Fraser et al. 2002), and the latter was seen to be influenced by the characteristic features of protein–protein interactions (Makino and Gojobori 2007).

Here, we described the evolutionary features of protein–protein interaction network under the perspective of disorderness, hydrophobicity, multifunctionality, and tissue specificity. We also shed light on the dependence of evolutionary rates of monogenic and polygenic disease genes on their cellular coexpression pattern in interaction network. Moreover, in the context of protein–protein interaction network, we here, also analyzed the evolutionary pattern and network properties of the proteins residing in protein complexes.

4.2 An Integrated View of Network Protein Evolution

4.2.1 Heterogeneous Rate of Evolution Between Party and Date Hub Proteins in Yeast Protein–Protein Interaction (PPI) Network

Proteins are the key components of cellular machinery, and most biological functions are executed by groups of proteins acting in concert. The yeast PPI network exhibits organized modularity where most proteins interact with few partners and a small proportion of proteins the "hubs" – interact with many partners. Hubs can be classified into two varieties: "party" hubs, which interact with most of their partners simultaneously, and "date" hubs, which bind different partners at different locations and times (Han et al. 2004). Earlier, dissimilar evolutionary mechanisms have been proposed for the party hub and date hub by relating the three-dimensional structures to protein networks (Kim et al. 2006). Estimating the number of synonymous nucleotide substitutions per synonymous site (dS) and the number of non-synonymous nucleotide substitutions per non-synonymous site (dN) for the hub proteins of *Saccharomyces cerevisiae* with its orthologs *Saccharomyces pombe*, it was also evident that the average value of dN/dS is significantly lower in party hub proteins than in date hub proteins (dN/dS of date hubs $= 0.2038$; party hubs $= 0.1581$; Mann–Whitney U test, $P = 0.015$) (Kahali et al. 2009). According to Drummond et al. (2006), gene expression level is a major factor in determining the protein evolutionary rate and protein–protein interactions showed little or no influence on the rate of protein evolution. It has been observed that the party hubs have significantly higher average expression level (3.11) than date hub proteins (1.76) (Mann–Whitney U test, $P = 0.001$) (Kahali et al. 2009). In order to confirm whether the evolutionary rate of a protein depends on its type of interaction in the PPI network and is independent of its expression level, we normalized the dN and dS values by dividing them by their corresponding gene expression level. The corresponding dN/dS values were used to perform one way ANOVA with unequal number of observations for the date and party hub proteins. The average values of dN/dS for the date hub proteins were still significantly higher from that of the party hub proteins ($F_{1,332} = 11.61489$, $P < 0.001$) (Kahali et al. 2009). A recent report also suggests that contact density (measured by the fraction of buried sites in the protein) correlates positively with evolutionary rate and these correlations do not seem to be confounded by gene expression level (Zhou et al. 2008).

4.2.2 *Influence of Amino Acid Substitutions of Buried and Exposed Residues in Determining Party and Date Hubs Evolutionary Rate*

The selective constraints imposed on amino acid residues in a protein due to the various three-dimensional structural contexts include the solvent accessibility of the residues as one of the important factors (Choi et al. 2006). An early hypothesis states that residues that interact with other proteins tend to be highly conserved (Dickerson 1971). Evolutionary studies on network component suggest that residues at the interface of obligate complexes tend to evolve at a relatively slower rate than those involved in transient interaction (Mintseris and Weng 2005). Of late, it has been shown that the substitution rates are lower for the buried residues than the residues on the solvent-exposed surfaces (Tseng and Liang 2006). Nevertheless, another contradictory report suggested that proteins with a higher fraction of buried residues evolve their sequences more rapidly (Bloom et al. 2006). Calculating the dN/dS of buried and exposed residue of the corresponding party and date hub proteins shows that the average dN/dS values of buried residues are significantly lower for the party hub proteins than the date hub proteins (dN/dS of date hub = 0.3631; party hub = 0.2627; Mann–Whitney U test, $P = 0.017$) (Kahali et al. 2009). However, the differences in average values of dN/dS are not significant between the date and party hub proteins for the solvent-exposed residues (dN/dS of date hub = 0.2651; party hub = 0.2022; Mann–Whitney U test, $P = 0.060$). From this observation, it could be iterated that the buried residues contribute significantly in lowering the evolutionary rate of the party hub proteins than the date hub proteins (Kahali et al. 2009).

The substantial decrease in the evolutionary rate of buried amino acids of party hub proteins than the date hub proteins led us to investigate the degree to which changes in amino acid composition of solvent inaccessible buried regions could influence the evolution of the encoded proteins. The amino acid substitution matrices constructed for the regions of buried residues were analyzed for the various substitutions of hydrophobic, hydrophilic, and amphipathic amino acid residues between *Saccharomyces pombe* and *Saccharomyces cerevisiae* in the party hub and date hub proteins (Figs. 4.1 and 4.2). Increase in cumulative frequencies of hydrophobic amino acids was observed in the *Saccharomyces cerevisiae* sequences for the buried regions of party hub proteins, whereas no

	Leu	Ala	Val	Cys	Phe	Ile	Met	Arg	Lys	Asn	Gln	His	Glu	Asp	Ser	Thr	Pro	Gly	Tyr	Trp
Leu	5739	254	653	88	403	865	396	125	148	98	159	83	146	77	209	211	104	50	163	29
Ala	322	4431	530	234	91	228	137	153	224	299	218	55	374	211	1177	471	296	468	59	23
Val	989	482	4496	163	185	1566	163	153	188	86	169	57	153	118	316	421	121	86	86	3
Cys	630	1037	556	4944	278	315	74	130	19	167	74	74	74	167	741	333	148	148	93	0
Phe	1105	169	342	89	5314	398	164	108	108	103	80	117	140	61	178	150	80	66	1081	150
Ile	1544	269	1603	88	260	4280	311	116	170	119	107	79	141	68	240	317	102	51	116	20
Met	1842	387	437	93	294	631	4096	118	261	219	252	34	160	109	370	353	101	101	126	17
Arg	244	105	140	0	52	105	87	5846	1344	175	471	140	279	192	314	175	87	70	87	87
Lys	641	128	0	0	0	128	128	1282	4872	641	513	0	385	513	128	256	256	0	128	0
Asn	368	301	201	67	201	134	33	201	368	3913	301	268	334	736	1271	669	134	301	201	0
Gln	291	631	194	49	97	97	243	583	388	291	4320	146	631	388	728	291	243	146	194	49
His	221	240	166	37	203	129	92	406	424	535	590	4963	332	240	369	277	92	111	535	37
Glu	274	274	137	137	0	548	137	274	274	411	274	0	5616	685	411	274	137	0	137	0
Asp	79	317	79	0	0	159	79	159	79	476	317	79	1429	5238	714	397	159	238	0	0
Ser	260	1103	322	186	93	155	74	235	229	452	322	124	465	347	3897	936	192	489	99	19
Thr	362	615	600	146	92	500	162	223	377	469	338	85	369	223	1131	3862	169	185	62	31
Pro	250	406	234	16	47	156	47	125	203	172	281	47	203	281	531	281	6594	94	31	0
Gly	93	646	86	65	36	65	29	129	151	258	129	86	237	244	567	143	151	6801	86	0
Tyr	493	190	255	125	1247	190	48	196	125	202	119	297	148	77	255	190	83	59	5588	113
Trp	194	172	108	65	517	86	129	86	65	86	0	86	22	0	151	86	86	43	431	7586

Fig. 4.1 Amino acid substitution matrix showing the pattern of sequence divergence between the aligned sequences of *Saccharomyces pombe* and *Saccharomyces cerevisiae* orthologs for buried residues of party hub proteins. In this matrix, the columns and rows, respectively, represent the *Saccharomyces pombe* and *Saccharomyces cerevisiae* sequence data; and the diagonal represents the invariant sites. Values in the matrix were scaled to represent the number of amino acid substitutions per 10,000 sites. The regions highlighted in *green* represent the hydrophilic to hydrophobic and amphipathic to hydrophobic amino acid substitutions. The regions highlighted in *red* represent the hydrophobic to hydrophilic and hydrophobic to amphipathic substitutions

	Leu	Ala	Val	Cys	Phe	Ile	Met	Arg	Lys	Asn	Gln	His	Glu	Asp	Ser	Thr	Pro	Gly	Tyr	Trp
Leu	4481	378	660	97	563	1022	430	169	197	137	209	129	233	109	382	314	145	72	213	60
Ala	539	3359	422	117	203	336	148	227	367	289	234	125	500	305	1180	586	383	516	148	16
Val	1239	556	3237	176	232	1471	204	134	225	176	148	155	317	162	500	612	218	91	120	28
Cys	569	736	702	4482	268	535	0	201	201	134	67	134	301	167	535	468	100	268	134	0
Phe	1144	248	381	80	4504	488	204	115	142	222	89	160	151	168	275	230	177	71	1002	151
Ile	1678	354	1398	114	400	3231	314	211	263	240	143	80	205	103	360	382	217	103	160	46
Met	1593	496	549	71	265	743	3150	230	283	230	301	159	248	195	372	336	283	142	248	106
Arg	302	129	129	86	43	0	86	4612	1638	517	431	259	345	216	517	172	259	172	86	0
Lys	667	444	0	0	0	0	0	444	3778	667	222	0	1111	444	889	1111	222	0	0	0
Asn	227	303	76	0	227	379	76	303	303	3788	833	379	606	530	758	606	227	76	227	76
Gln	392	392	392	0	98	196	294	392	882	294	3529	98	588	588	588	882	196	0	196	0
His	315	405	270	0	225	225	90	541	315	450	450	4054	360	586	541	315	180	180	450	45
Glu	0	303	303	0	0	606	0	303	0	606	303	606	3636	1212	606	909	0	606	0	0
Asp	0	0	179	0	0	0	0	1071	357	893	357	179	536	5357	536	179	0	179	179	0
Ser	519	897	366	165	213	236	71	224	413	484	295	189	354	390	3471	791	401	390	83	47
Thr	443	611	656	214	153	427	122	168	397	595	382	183	550	321	1374	2748	305	244	76	31
Pro	219	533	313	63	282	251	0	157	219	219	125	63	282	408	658	313	5737	63	63	31
Gly	207	602	120	34	172	103	69	120	379	482	241	34	258	327	637	310	224	5577	69	34
Tyr	654	160	481	111	1136	469	74	185	235	210	173	358	259	173	358	272	222	99	4210	160
Trp	313	78	313	156	664	313	0	78	39	195	156	117	39	39	234	117	39	117	469	6523

Fig. 4.2 Amino acid substitution matrix showing the pattern of sequence divergence between the aligned sequences of *Saccharomyces pombe* and *Saccharomyces cerevisiae* orthologs for buried residues of date hub proteins. In this matrix, the columns and rows, respectively, represent the *Saccharomyces pombe* and *Saccharomyces cerevisiae* sequence data; and the diagonal represents the invariant sites. Values in the matrix were scaled to represent the number of amino acid substitutions per 10,000 sites. The regions highlighted in *green* represent the hydrophilic to hydrophobic and amphipathic to hydrophobic amino acid substitutions. The regions highlighted in *red* represent the hydrophobic to hydrophilic and hydrophobic to amphipathic substitutions

such enrichment in hydrophobic residues is observed for the buried regions in date hub proteins. For the buried regions of party hub proteins, the hydrophilic to hydrophobic (3.98%) or amphipathic to hydrophobic (5.33%) transition is always greater than the values for hydrophobic to hydrophilic (3.33%) or hydrophobic to amphipathic (4.68%) transition, but this is not followed for the buried regions of date hub proteins. Thus, the overall increased hydrophobicity in the buried regions of party hub proteins influence the overall lowering of evolutionary rates of the corresponding proteins (Kahali et al. 2009).

4.2.3 *Evolutionary Rate of Disordered and Ordered Regions in Party Hub and Date Hub Proteins*

Intrinsically disordered proteins or protein regions that lack rigid three-dimensional structures under physiological conditions in vitro are known as a key manifestation among other sequence features in hub proteins and are resulted in their specialization in network evolution (Dunker et al. 2002; Dosztanyi et al. 2006). Indeed, several hub proteins have been shown to be completely or almost completely disordered in solution including alpha-synuclein, HMGA, and synaptobrevin as appraised by Dunker et al. (2005). Extending these observations to protein–protein interaction networks, these authors suggested that intrinsically disordered hub proteins and regions could serve for multiple and distinct signaling networks (Dunker et al. 2005). In some protein families, it has been demonstrated that the disordered regions evolve at a significantly higher rate than the ordered regions (Brown et al. 2002). These studies incited us to reinvestigate the contribution of ordered and/or disordered regions in the evolution of hub proteins. Estimating the evolutionary rates of the disordered and ordered regions for the date and party hub proteins separately, it was observed that the disordered regions have higher evolutionary rate than the ordered regions both for the date hub (Mann–Whitney U test, $P = 0.001$) as well as party hub proteins (Mann–Whitney U test, $P = 1.0 \times 10^{-4}$). However, no significant differences were observed in case of the evolutionary rate (dN/dS) for the disordered regions between the date and party hub proteins (dN/dS of date hub = 0.4220; party hub = 0.3898; Mann–Whitney U test, $P = 0.867$). Alternatively, evolutionary rate (dN/dS) for ordered residues yields significant lower value for the corresponding party hub proteins than the date hub proteins (dN/dS of date hub = 0.2261; party hub = 0.1582; Mann–Whitney U test, $P = 0.020$). Thus, the evolutionary rate difference between date hub and party hub proteins might be attributed due to the ordered regions of proteins (Kahali et al. 2009).

4.2.4 Factors Defining Differential Evolutionary Rate Between Housekeeping and Tissue-Specific Interacting Proteins in Human PPI Networks

The study of networks formed by protein–protein interactions facilitated to uncover how the complex functionality of cells emerges from simple biochemistry. Several preceding studies endeavored to correlate protein evolutionary rates with different parameters such as gene essentiality (Hurst and Smith 1999; Hirsh and Fraser 2001; Jordan et al. 2002; Wall et al. 2005; Zhang and He 2005), gene expression level (Pal et al. 2001; Akashi 2003; Rocha and Danchin 2004; Subramanian and Kumar 2004; Drummond et al. 2006), tissue specificity (Hastings 1996; Duret and Mouchiroud 2000; Winter et al. 2004; Zhang and Li 2004), gene duplicability (Nembaware et al. 2002; Castillo-Davis and Hartl 2003; Yang et al. 2003), properties in the protein–protein interaction network (Fraser et al. 2002, 2003, 2004, 2005; Hahn and Kern 2005; Makino and Gojobori 2006; Kahali et al. 2009), and pleiotropy (He and Zhang 2006). However, some of the above factors are confounding to some extent as a correlate of evolutionary rate (Pal et al. 2006). Focusing on the human PPI network, it was observed that a significant distinction exhibits between the rate of housekeeping (Hk) and tissue-specific (Ts) interacting proteins evolution, measured using the mouse orthologs (dN of Hk interactome = 0.09 and dN of Ts interactome = 0.11, Mann–Whitney U test, $P < 1 \times 10^{-3}$) (Podder et al. 2009).

Previous studies (Winter et al. 2004; Zhang and Li 2004) also have demonstrated that Hk genes are evolutionary slower than Ts genes due to their broad expression pattern in all tissues. It will be interesting to figure out the features of the Hk and Ts interacting proteins and more precisely to study how these features individually can explain the evolutionary rate of proteins in PPI network. On the contrary to the previous reports (Fraser et al. 2002, 2003) that entailed proteins with more interactors evolve more slowly than the proteins with fewer interactors, it was observed that no significant differences of average connectivity exhibit between Hk and Ts interactome (average connectivity of Hk interactome = 8.92 and average connectivity of Ts interactome = 8.59, Mann–Whitney U test, $P < 7.2 \times 10^{-2}$) (Podder et al. 2009).

Previously it has been ascertained that rate of mutation of a hub protein is constrained by the amount of the protein surface involved in the interactions with other proteins and not simply by the number of proteins with which it interacts (Kim et al. 2006). Examining the interacting domain coverage of proteins (percentages of the length of protein interfaces that are involved in protein–protein interactions to the whole protein length), a significant (Mann–Whitney U test, $P < 4.1 \times 10^{-2}$) difference was found between the length of interfaces taking part in the interaction of Hk interactome (average percentage of interacting length = 46.54) and Ts interactome (average percentage of interacting length = 40.82). Thereafter, protein expression level, which has always been a keen correlate of evolutionary rate (Drummond et al. 2005), also have varied within Hk and Ts interacting proteins (average expression rank in Hk = 3.92; Ts = 3.01; Mann–Whitney U test, $P < 1 \times 10^{-3}$). Intriguingly, a significant difference in protein multifunctionality (the number of biological processes in which a protein is involved) between the two separate interactome indicates that multifunctionality may guide protein evolutionary rate (average multifunctionality in Hk = 11.50; Ts = 9.63; Mann–Whitney U test, $P < 1 \times 10^{-3}$) (Podder et al. 2009), which has been rejected to influence protein evolution in yeast (Salathe et al. 2006). Here, we carry out a combined analysis of four biological predictors [connectivity, domain coverage, expression level (using both microarray data and EST data) and multifunctionality] with evolutionary rate of interacting proteins present in PPI network and found except protein connectivity, the other three factors have a significant contributions in guiding protein evolutionary fate (Fig. 4.3). It is imperative to examine whether all these factors independently influence evolutionary rate; we computed partial correlation analysis. In the partial correlation analysis, we focused on the correlation between the evolutionary rate and 1 of the 4 factors (i.e., connectivity, protein interacting length, protein expression level, protein multifunctionality), by controlling the other three factors. All the factors except connectivity have significant partial correlation with the protein evolutionary rate (Table 4.1). Therefore, the relative importance of the factors in determining the evolutionary rate of proteins residing in housekeeping as well as tissue-specific PPI Network is multifunctionality > protein expression level > interacting protein length (Podder et al. 2009).

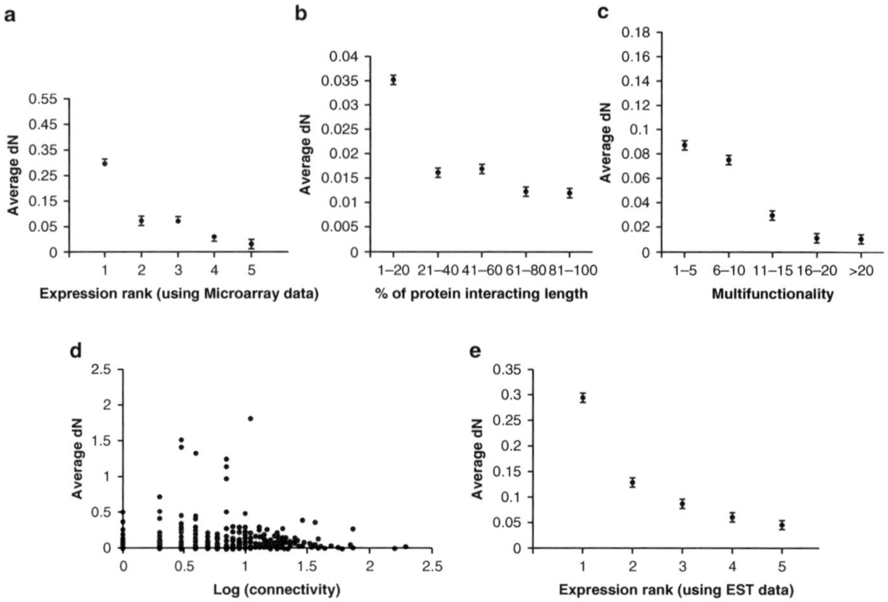

Fig. 4.3 Scatter plot showing correlation between evolutionary rate and (**a**) expression rank using microarray data, (**b**) % of interacting length, (**c**) multifunctionality, (**d**) connectivity, (**e**) expression rank using EST data

Table 4.1 Spearman's rank correlation coefficient (ρ) and partial correlation coefficient (ρ′) between various factors and dN

	dN		dN	
	Spearman's rank correlation		Partial rank correlation	
Factors	ρ	*P* value	ρ′	*P* value
	−0.024	5.69×10^{-1}	−0.007	8.74×10^{-1}
Connectivity	−0.007#	7.79×10^{-1}#	−0.010#	6.79×10^{-1} #
	−0.775	1.00×10^{-3}	−0.098	1.90×10^{-2}
% of protein's interacting length	−0.276#	1.00×10^{-3}#	−0.069#	0.8×10^{-3}#
	−0.785	1.00×10^{-3}	−0.228	1.00×10^{-3}
Expression level	−0.316#	1.00×10^{-3}#	−0.159#	1.00×10^{-3}#
	−0.977	1.00×10^{-3}	−0.275	1.00×10^{-3}
Multifunctionality	−0.964#	1.00×10^{-3}#	−0.492#	1.00×10^{-3}#

Correlations (#) by using EST expression data rests are by using microarray data

4.2.5 Evolutionary Forces in Determining Multi and Singlish-Interface Protein Evolution in Human

Parallel to the concept of party and date hubs, Kim et al. (2006) introduced two different types of interacting proteins, namely, singlish-interface and multi-interface

Table 4.2 Multiple regression analysis between various factors and dN of multi-interface proteins and singlish-interface proteins in housekeeping and tissue-specific PPI network

Factors	Multi-interface proteins *P* value	Singlish-interface proteins *P* value
% of protein's interacting length	6.75×10^{-1}	1.00×10^{-2}
	2.16×10^{-1} #	1.00×10^{-3} #
	2.09×10^{-1}	1.00×10^{-3}
Expression level	1.47×10^{-1}#	1.00×10^{-3} #
	2.60×10^{-2}	1.00×10^{-3}
Multifunctionality	1.00×10^{-3}#	1.00×10^{-3}#

Correlations (#) by using EST expression data rests are by using microarray data

hubs according to their number of binding interfaces. In our dataset, we generalized the term as singlish and multi-interface proteins since we observed proteins with low to high connectivity have different number of binding interfaces (Podder et al. 2009). Alike party hubs, multi-interface hubs are evolutionary conserved than singlish-interface hubs or date hubs (Kim et al. 2008). We intended to describe whether the above biological parameters (protein expression level, multifunctionality, domain coverage) can explain such evolutionary differences between singlish and multi-interface proteins across Hk and Ts interactome. In this regard, multivariate regression analysis was performed to look at the influence of all potential predictor variables and at the same time can eliminate step by step those predictors that contribute least to the regression model (Drummond et al. 2006). It was noticed that among three determinants only multifunctionality independently influences the evolutionary rate of multi-interface proteins across Hk and Ts interacting proteins (Table 4.2). On the other hand, for singlish-interface proteins, all the three determinants independently influence evolutionary rate across these two sets of interacting proteins (Table 4.2). From the results, it could be inferred that multifunctionality is the main driving force for the evolutionary rate differences in two types of interacting proteins (multi and singlish-interface) (Podder et al. 2009).

4.2.6 No Influence of Connectivity in the Regulation of Protein Evolutionary Rate: Evidence from Human Disease PPI Network

Genetic basis of disease causing phenomenon asserts to classify disease genes into two groups: (1) monogenic disease, caused merely by the modification of a single gene; and (2) polygenic disease, caused by several substitutions in a number of genes. Like regular protein interaction network, influence of connectivity has also not been observed in disease protein interaction network since it was observed that the polygenic disease genes, which are evolved at a faster rate than monogenic disease genes (dN of monogenic disease = 0.091, polygenic disease = 0.193; Mann–Whitney *U* test, $P = 4.5 \times 10^{-2}$), are associated with higher number of interacting partners [average connectivity of monogenic disease = 11.28 (Large Network); 7.22 (High Confidence Network), and polygenic disease = 15.79 (Large Network); 9.90 (High Confidence Network)] (Podder and Ghosh 2010). Our findings are also echoed in the study of Feldman et al. (2008), which ensured connectivity of polygenic disease genes is significantly higher than monogenic disease genes. According to Jeong et al. (2001), hub proteins play more crucial role in the architecture of the protein network. To examine whether hub proteins in the disease and non-disease protein-protein interaction network are responsible for causing evolutionary rate differentiation, we excluded all non-hub proteins from three datasets and again measured evolutionary rate. We found that evolutionary rate differences between three groups remain significant at least at 0.05 levels in Mann–Whitney *U* test. By calculating protein-interacting interface using Pfam, we reasoned that the overabundance of singlish-interface hub proteins in polygenic disease protein interaction network (32/40, i.e., 86%) compared to monogenic disease protein interaction network (70%, i.e., 298/426) facilitate the higher evolutionary rate in polygenic disease genes (Podder and Ghosh 2010). It was also verified that interacting partners of monogenic disease proteins are mostly coexpressed with each other whereas a small number of interacting partners are coexpressed in case of polygenic disease proteins (Fig. 4.4) (Podder and Ghosh 2010).This fact evoked that a stable interaction pattern exhibits in the monogenic disease PPI network, whereas in polygenic disease PPI network, most of them are involved in transient interactions. From this observation, it could be confirmed again that the number of binding interfaces through which they establish stable/ transient interaction fashion but not the simple number of interacting partners are

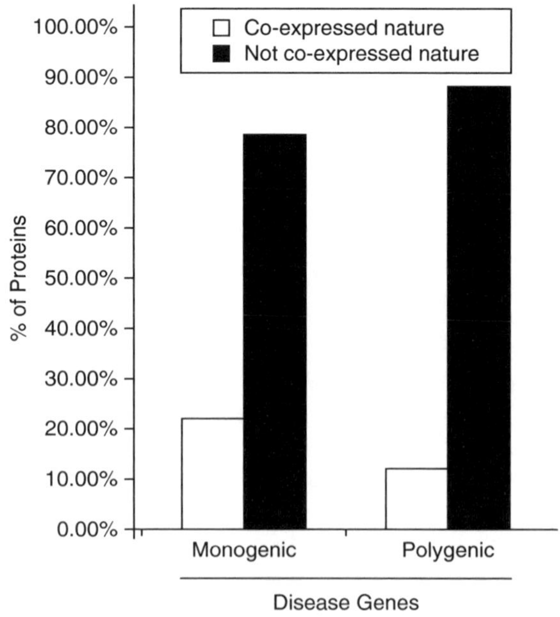

Fig. 4.4 Expression profile similarity/dissimilarity between hub and its corresponding partners for the two classes of disease genes

important for determining heterogeneous evolutionary rate between non-disease as well as disease PPI network.

4.3 Evolutionary Attributes of Complex Forming Proteins in PPI Networks

4.3.1 Protein Complex Forming Ability Influences the Evolutionary Rate

Studies on the evolution of protein complex have a great importance in understanding cellular life. Current developments in the analysis of protein complexes suggest that the internal subunit arrangement in complexes is crucial for depicting their functional details (Dziembowski and Seraphin 2004). Recently, structural characterization of mammalian protein complex organization provided evidence that the protein complexity is negatively associated with protein evolution (Wong et al. 2008). Till date, various aspects of the protein complexes in PPI network have been studied. Analysis on interaction networks so far has revealed the global centrality (*viz.* closeness, betweenness) as one of the most influencing factors in protein evolution (Hahn and Kern 2005). Over and above, protein expression level is the major determinant of protein evolution (Drummond et al. 2005, 2006; Warringer and Blomberg 2006; Chakraborty et al. 2010). Recently it has been suggested that protein evolutionary rate is related to the features of interacting partners in a protein–protein interaction network, *viz.*, same (SF) or different functional (DF) proteins – based on the coefficient of functionality (the SF and DF proteins are distinguished by the higher and lower coefficient of functionality, respectively) (Makino and Gojobori 2006). Likewise, another constraint on protein evolution is its complex forming nature (Mintseris and Weng 2005; Manna et al. 2009), proteins involved in formation of stable complexes have much more sequence identity with their orthologs than those involved in the transient interactions (Teichmann 2002). We have investigated the evolutionary distances in yeast (*Saccharomyces cerevisiae* with *Saccharomyces paradoxus* by Kimura's method 1983) proteins by taking into account the various evolutionary forces including the complex number (i.e., the total number of the protein-complexes in which the particular protein takes part). Multivariate regression analysis revealed that the expression level, protein–protein interaction network as well as the complex number independently control the evolutionary distances (Table 4.3) (Chakraborty et al. 2010). Principal Component Analysis (PCA) was used to disentangle the contributions of various factors. The first principal component accounts for 43% of the total variance. Its main

Table 4.3 Multiple regression analysis between various factors and evolutionary rate in yeast PPI network

Factors	*P* value
Expression level	9.0×10^{-26}
Connectivity	1.4×10^{-9}
Complex number	1.0×10^{-4}

contribution comes from the complex number (≈0.77), expression level (≈0.75), and connectivity (≈0.40). Moreover, the first principal component generated by PCA is significantly negatively correlated with the evolutionary distances (Spearman's $\rho = -0.439$, $P = 1.00 \times 10^{-6}$) (Chakraborty et al. 2010). Thus, the complex forming ability of proteins emerged as a significant contributor of evolutionary rate variation followed by expression level and protein connectivity.

4.3.2 Complex Forming Proteins Are Conserved in Yeast PPI Network

Previously, Makino and Gojobori (2006) showed DF proteins evolve slower than the SF proteins in yeast PPIs network irrespective of connectivity. We also observed that the DF proteins evolve slower than SF proteins (Fig. 4.5). In our dataset, there exists a higher ratio (0.82) of complex forming proteins. Thus, the dataset was grouped as complex forming proteins and non-complex forming proteins and investigated their evolutionary distances. It was found that complex forming proteins are strongly conserved than the non-complex forming proteins (Mann–Whitney U test, $P = 1.50 \times 10^{-28}$). In non-complex forming proteins no significant difference was attained in SF and DF proteins, whereas in the complex forming group, SF proteins evolve faster than the complex forming DF proteins (Fig. 4.5) (Chakraborty et al. 2010). Thus it would be interesting to explore the relationship between the complex forming ability of the DF and SF proteins with their evolutionary rates. For this, we have counted the number of complexes for each DF/SF protein in which it can participate as a subunit and labeled this number as the complex number for this protein. Spearman's rank correlation analysis

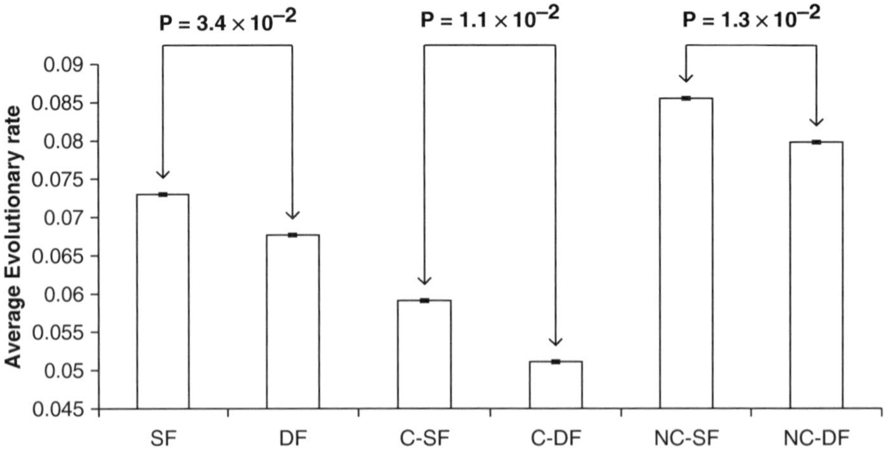

Fig. 4.5 Average values of evolutionary rates of SF and DF proteins; C and NC denote the complex and the non-complex proteins, respectively

revealed that the complex number correlates negatively with the protein distance (Spearman's $\rho = -0.156$, $P = 1.10 \times 10^{-5}$) as well as with the coefficient of functionality (Spearman's $\rho = -0.083$, $P = 2.00 \times 10^{-2}$). Thus, we infer that the DF proteins are more likely to be part of protein complexes, which might be a decisive factor in lowering their evolutionary rates.

We also analyzed the contribution of expression level since it is one of the most important determinants in determining the evolutionary rate (Drummond et al. 2005, 2006). According to our expectation, the complex forming SF proteins have significantly lower average expression level (Mann–Whitney U test, $P = 3.0 \times 10^{-4}$) than their DF counterparts, which is not observed for the non-complex forming SF and DF proteins (Chakraborty et al. 2010). Taking together the complex and non-complex forming proteins, it was established that the SF proteins showed lower expression level than the DF proteins (Mann–Whitney U test, $P = 4.0 \times 10^{-3}$). The classification of SF and DF proteins was done by considering the functional class assignment of the proteins and their partners in the PPIs. Interestingly, we found a negative correlation between functional coefficient and protein connectivity (Spearman's $\rho = -0.145$, $P = 1.00 \times 10^{-6}$). This correlation suggests that coefficient of functionality decreases with increasing connectivity, i.e., the DF proteins should have higher connections than SF proteins. Accordingly, we observed that DF proteins have higher connections than SF proteins in the datasets (Mann–Whitney U test, $P = 5.4 \times 10^{-9}$). Thus in yeast the coefficient of functionality is related to the protein connectivity in the overall PPI network. Being a unicellular simple organism, yeast interactome possess a contradicting feature with human interactome since a significant contribution of protein connectivity on evolutionary rate was found in yeast (Spearman's $\rho = -0.166$, $P = 1.00 \times 10^{-6}$). The significant positive correlation (Spearman's $\rho = 0.267$, $P = 1.00 \times 10^{-6}$) between the complex number and the expression level for the DF and SF proteins signifies that the evolutionary rate of the DF proteins is more constrained. This is perhaps due to their greater ability to be a part of protein complexes. Subsequently, the increase in the expression levels for the DF proteins is possibly due to their participation in larger number of complexes. Thus, the difference in functional features of interacting partners is not the sole reason of evolutionary rate variation rather the interrelationship between the features, viz., the expression level, complex forming ability, which guide the difference in evolutionary rates of DF and SF proteins.

4.3.3 Evolutionary Impact of Protein Complex Forming Property on Human Hub and Non-hub Proteins

Studies relating protein complex formation with the properties of PPI network attained a much focus in the recent research era. Protein complex is a group of two or more associated proteins formed by protein–protein interactions that is stable over times. Using protein interactions in interologous networks, it has been proved that protein complexes are preferentially conserved in their sequence

Table 4.4 Average values of dN for hub and non-hub proteins present in protein complex, not present in protein complex, and in overall (in complex/not in complex)

	Hub	Non-hub	Significance level (P value)
In complex	0.052026 ($n = 825$)	0.052012 ($n = 306$)	0.516 (NS)
Not in complex	0.074126 ($n = 1{,}396$)	0.090906 ($n = 2{,}577$)	1×10^{-4}
Overall (present/absent in complex)	0.065428 ($n = 2{,}221$)	0.086722 ($n = 2{,}882$)	1×10^{-4}

NS difference not significant, *dN* non-synonymous substitution per site, *n* number of proteins used for the analysis

(Brown and Jurisica 2007). Propositions exist that complex forming proteins are more evolutionary conserved since they are involved in a greater number of interactions (Teichmann 2002; Makino and Gojobori 2006). In order to investigate whether the complex forming proteins have any role in dictating the evolutionary rate of hub and non-hub proteins, we assessed the evolutionary rate of complex forming hub and non-hub proteins as well as proteins which are not present in protein complexes. No significant difference has been observed between evolutionary rate of complex-forming hub and non-hub proteins, while a significant difference was noticed in the evolutionary rate between hub and non-hub proteins, which are not present in protein complexes (Table 4.4) (Manna et al. 2009). From the result, it appears that complex forming ability of the hub and non-hub proteins might be an important factor in shaping their evolutionary rates.

4.3.4 Interrelation Between Structural Disorderness, Connectivity, Complex Forming Property, and Evolutionary Rate of Hub and Non-hub Proteins

It was known that the binding ability of hub proteins to their partners is facilitated by the global flexibility provided by their disordered domains (Patil and Nakamura 2006). Their disordered structure serves as flexible linker and plays a great role in protein–protein interaction and also in protein assembly (Dyson and Wright 2002). Structural disorder also promotes transient interactions in hub proteins (Singh et al. 2007). However, the assembly of protein complexes is also closely linked with the protein disorderness in *Escherichia coli* and in *Saccharomyces cerevisiae* (Hegyi et al. 2007). This prompted us to estimate the disorderness of the complex forming hub and non-hub proteins and surprisingly, we found that there is no significant difference (Mann–Whitney U test, $P = 0.812$) in disorderness between these two groups of proteins. The disorderness of hub proteins does not vary even when they are classified as complex forming and non-complex forming ones. This is different for the non-hub proteins where the complex forming non-hub proteins are significantly more disordered than their non-complex forming counterparts.

Hence, we ask what drives this increase in disorderness in non-hub complex forming proteins alike the disorderness of hub proteins? It has already been proposed that in *Saccharomyces cerevisiae* and *Escherichia coli*, the protein complexes having a higher amount of disorderness, promote the assembly of the subunits of the protein complexes (Hegyi et al. 2007). Thus, disorderness facilitates large complex assembly with the non-hub proteins though they have few network connections. Such "hub-like nature" of the non-hub proteins present in protein complexes suggests that this behavior of non-hub proteins may be a genuine property of the protein complexes to which they belong and non-reflective of the network connectivity of that particular protein. From this, it can be speculated that the non-hub proteins, which have very high disorderness (say, $\geq 90\%$), should belong to large protein complexes (less than or equal to ten subunits). Interestingly, the precise fact reflected in Fig. 4.6 incited that the non-hub proteins belonging to protein complexes, which are greater than or equal to 90% disordered have a higher number of average unique subunits (average unique subunits = 35) than those complex forming non-hub proteins, which are less than or equal to 10% disordered (average unique subunits of less disordered non-hubs = 27). That is why the non-hub proteins which are not components of protein complexes are the least disordered with an average of 24.71% disorderness. Again, if the disorderness is the result of higher connectivity and complex assembly of the complex forming hub proteins, then these hub proteins should have the highest disorderness. But, in the case of hub proteins, the disorderness is similar for the complex forming and non-complex forming proteins (Mann–Whitney U test, $P = 0.066$). A negative correlation between the network connectivity and number of subunits (Spearman's $\rho = -0.177$, $P < 0.01$) for the hub proteins further suggests that proteins with higher number of interactions mostly belong to small complexes proteins, and

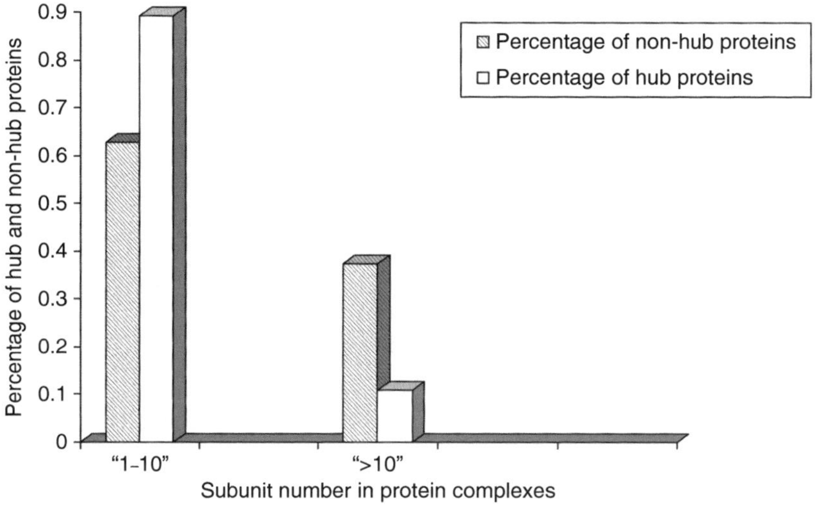

Fig. 4.6 Complex forming ability of hub and non-hub proteins

disorderness might be promoting network connectivity as well as subunit assembly in hub proteins in a *collectively exhaustive* manner. Since, the evolutionary rate of protein having more protein–protein interactions is much slower than that of proteins having fewer interacting partners (Makino et al. 2006), non-hub proteins, which are not members of protein complexes, are more relaxed in constraining their evolutionary rate. Hence, connectivity (intra or inter) facilitated by disordered regions in proteins governs the evolutionary rates of the hub and non-hub proteins present/absent in protein complexes.

4.4 Conclusion

To comprehend life as biological system, evolutionary understanding is indispensable. Taken together, we here, analyzed the evolutionary aspects of the eukaryotic interactome and delved out some interesting outlines associating network features and protein evolution. We showed that the evolutionary rate differences of the party hubs and date hubs in *Saccharomyces cerevisiae* protein interaction network are preferentially guided by the percentage of buried residues and the ordered regions of proteins. Concentrating on human proteome, we did not find out any dependence of connectivity on the evolutionary pattern of the housekeeping and tissue specific proteins in human protein-protein interaction network, rather reliance with interacting fashion was explored out. Again, the monogenic and polygenic disease protein interaction network exhibit evolutionary rate variation due to the differential population of party hubs and date hubs showing no differences in connectivity. Expanding our search on interacting protein evolution, we also pondered on the complex forming proteins in *Saccharomyces cerevisiae* and found an association of evolutionary rate with complex forming ability, gene expression, and connectivity. Moreover, in our analysis on human proteome, we observed that, within complex, the non-hubs execute hub-like properties and consequently show no variation in evolutionary rate. Providing a number of interesting inferences, our analyses thus illuminate the field of network protein evolution and open up a new window enabling novel insights on eukaryotic complex forming interacting proteins to spring up.

References

Akashi H (2003) Translational selection and yeast proteome evolution. Genetics 164 (4):1291–1303

Bader GD, Betel D, Hogue CW (2003) BIND: the Biomolecular Interaction Network Database. Nucleic Acids Res 31(1):248–250

Bloom JD, Drummond DA, Arnold FH, Wilke CO (2006) Structural determinants of the rate of protein evolution in yeast. Mol Biol Evol 23(9):1751–1761

Brown KR, Jurisica I (2007) Unequal evolutionary conservation of human protein interactions in interologous networks. Genome Biol 8(5):R95

Brown CJ, Takayama S, Campen AM, Vise P, Marshall TW, Oldfield CJ, Williams CJ, Dunker AK (2002) Evolutionary rate heterogeneity in proteins with long disordered regions. J Mol Evol 55 (1):104–110

Castillo-Davis CI, Hartl DL (2003) Conservation, relocation and duplication in genome evolution. Trends Genet 19(11):593–597

Ceol A, Chatr Aryamontri A, Licata L, Peluso D, Briganti L, Perfetto L, Castagnoli L, Cesareni G (2010) MINT, the molecular interaction database: 2009 update. Nucleic Acids Res 38(Database issue):D532–D539

Chakraborty S, Kahali B, Ghosh TC (2010) Protein complex forming ability is favored over the features of interacting partners in determining the evolutionary rates of proteins in the yeast protein-protein interaction networks. BMC Syst Biol 4:155

Choi SS, Vallender EJ, Lahn BT (2006) Systematically assessing the influence of 3-dimensional structural context on the molecular evolution of mammalian proteomes. Mol Biol Evol 23(11): 2131–2133

Dickerson RE (1971) The structures of cytochrome c and the rates of molecular evolution. J Mol Evol 1(1):26–45

Dosztanyi Z, Chen J, Dunker AK, Simon I, Tompa P (2006) Disorder and sequence repeats in hub proteins and their implications for network evolution. J Proteome Res 5(11):2985–2995

Drummond DA, Bloom JD, Adami C, Wilke CO, Arnold FH (2005) Why highly expressed proteins evolve slowly? Proc Natl Acad Sci USA 102(40):14338–14343

Drummond DA, Raval A, Wilke CO (2006) A single determinant dominates the rate of yeast protein evolution. Mol Biol Evol 23(2):327–337

Dunker AK, Brown CJ, Lawson JD, Iakoucheva LM, Obradovic Z (2002) Intrinsic disorder and protein function. Biochemistry 41(21):6573–6582

Dunker AK, Cortese MS, Romero P, Iakoucheva LM, Uversky VN (2005) Flexible nets. The roles of intrinsic disorder in protein interaction networks. FEBS J 272(20):5129–5148

Duret L, Mouchiroud D (2000) Determinants of substitution rates in mammalian genes: expression pattern affects selection intensity but not mutation rate. Mol Biol Evol 17(1):68–74

Dyson HJ, Wright PE (2002) Coupling of folding and binding for unstructured proteins. Curr Opin Struct Biol 12(1):54–60

Dziembowski A, Seraphin B (2004) Recent developments in the analysis of protein complexes. FEBS Lett 556(1–3):1–6

Feldman I, Rzhetsky A, Vitkup D (2008) Network properties of genes harboring inherited disease mutations. Proc Natl Acad Sci USA 105(11):4323–4328

Fraser HB (2005) Modularity and evolutionary constraint on proteins. Nat Genet 37(4):351–352

Fraser HB, Hirsh AE (2004) Evolutionary rate depends on number of protein-protein interactions independently of gene expression level. BMC Evol Biol 4:13

Fraser HB, Hirsh AE, Steinmetz LM, Scharfe C, Feldman MW (2002) Evolutionary rate in the protein interaction network. Science 296(5568):750–752

Fraser HB, Wall DP, Hirsh AE (2003) A simple dependence between protein evolution rate and the number of protein-protein interactions. BMC Evol Biol 3:11

Gandhi TK, Zhong J, Mathivanan S, Karthick L, Chandrika KN, Mohan SS, Sharma S, Pinkert S, Nagaraju S, Periaswamy B, Mishra G, Nandakumar K, Shen B, Deshpande N, Nayak R, Sarker M, Boeke JD, Parmigiani G, Schultz J, Bader JS, Pandey A (2006) Analysis of the human protein interactome and comparison with yeast, worm and fly interaction datasets. Nat Genet 38 (3):285–293

Giot L, Bader JS, Brouwer C, Chaudhuri A, Kuang B, Li Y, Hao YL, Ooi CE, Godwin B, Vitols E, Vijayadamodar G, Pochart P, Machineni H, Welsh M, Kong Y, Zerhusen B, Malcolm R, Varrone Z, Collis A, Minto M, Burgess S, McDaniel L, Stimpson E, Spriggs F, Williams J, Neurath K, Ioime N, Agee M, Voss E, Furtak K, Renzulli R, Aanensen N, Carrolla S, Bickelhaupt E, Lazovatsky Y, DaSilva A, Zhong J, Stanyon CA, Finley RL Jr, White KP, Braverman M, Jarvie T, Gold S, Leach M, Knight J, Shimkets RA, McKenna MP, Chant J,

Rothberg JM (2003) A protein interaction map of Drosophila melanogaster. Science 302 (5651):1727–1736

Goh KL, Cusick ME, Valle D, Childs B, Vidal M, Barabasi AL (2007) The human disease network. Proc Natl Acad Sci USA 104(21):8685–8690

Hahn MW, Kern AD (2005) Comparative genomics of centrality and essentiality in three eukaryotic protein-interaction networks. Mol Biol Evol 22(4):803–806

Han JD, Bertin N, Hao T, Goldberg DS, Berriz GF, Zhang LV, Dupuy D, Walhout AJ, Cusick ME, Roth FP, Vidal M (2004) Evidence for dynamically organized modularity in the yeast protein-protein interaction network. Nature 430(6995):88–93

Hastings KE (1996) Strong evolutionary conservation of broadly expressed protein isoforms in the troponin I gene family and other vertebrate gene families. J Mol Evol 42(6):631–640

He X, Zhang J (2006) Toward a molecular understanding of pleiotropy. Genetics 173(4): 1885–1891

Hegyi H, Schad E, Tompa P (2007) Structural disorder promotes assembly of protein complexes. BMC Struct Biol 7:65

Hirsh AE, Fraser HB (2001) Protein dispensability and rate of evolution. Nature 411(6841): 1046–1049

Hurst LD, Smith NG (1999) Do essential genes evolve slowly? Curr Biol 9(14):747–750

Ito T, Chiba T, Ozawa R, Yoshida M, Hattori M, Sakaki Y (2001) A comprehensive two-hybrid analysis to explore the yeast protein interactome. Proc Natl Acad Sci USA 98(8):4569–4574

Jeong H, Mason SP, Barabási AL, Oltvai ZN (2001) Lethality and centrality in protein networks. Nature 411(6833):41–42

Jordan IK, Rogozin IB, Wolf YI, Koonin EV (2002) Essential genes are more evolutionarily conserved than are nonessential genes in bacteria. Genome Res 12(6):962–968

Kahali B, Ahmad S, Ghosh TC (2009) Exploring the evolutionary rate differences of party hub and date hub proteins in *Saccharomyces cerevisiae* protein-protein interaction network. Gene 429 (1–2):18–22

Kar G, Gursoy A, Keskin O (2009) Human cancer protein-protein interaction network: a structural perspective. PLoS Comput Biol 5(12):e1000601

Kim PM, Lu LJ, Xia Y, Gerstein MB (2006) Relating three-dimensional structures to protein networks provides evolutionary insights. Science 314(5807):1938–1941

Kim PM, Sboner A, Xia Y, Gerstein M (2008) The role of disorder in interaction networks: a structural analysis. Mol Syst Biol 4:179

Kimura M (1983) The neutral theory of molecular evolution. Cambridge University Press, Cambridge

Li S, Armstrong CM, Bertin N, Ge H, Milstein S, Boxem M, Vidalain PO, Han JD, Chesneau A, Hao T, Goldberg DS, Li N, Martinez M, Rual JF, Lamesch P, Xu L, Tewari M, Wong SL, Zhang LV, Berriz GF, Jacotot L, Vaglio P, Reboul J, Hirozane-Kishikawa T, Li Q, Gabel HW, Elewa A, Baumgartner B, Rose DJ, Yu H, Bosak S, Sequerra R, Fraser A, Mango SE, Saxton WM, Strome S, Van Den Heuvel S, Piano F, Vandenhaute J, Sardet C, Gerstein M, Doucette-Stamm L, Gunsalus KC, Harper JW, Cusick ME, Roth FP, Hill DE, Vidal M (2004) A map of the interactome network of the metazoan C. elegans. Science 303(5657):540–543

Makino T, Gojobori T (2006) The evolutionary rate of a protein is influenced by features of the interacting partners. Mol Biol Evol 23(4):784–789

Makino T, Gojobori T (2007) Evolution of protein-protein interaction network. Genome Dyn 3: 13–29

Makino T, Suzuki Y, Gojobori T (2006) Differential evolutionary rates of duplicated genes in protein interaction network. Gene 385:57–63

Manna B, Bhattacharya T, Kahali B, Ghosh TC (2009) Evolutionary constraints on hub and non-hub proteins in human protein interaction network: insight from protein connectivity and intrinsic disorder. Gene 434(1–2):50–55

Mintseris J, Weng ZP (2005) Structure, function, and evolution of transient and obligate protein-protein interactions. Proc Natl Acad Sci USA 102(31):10930–10935

Nembaware V, Crum K, Kelso J, Seoighe C (2002) Impact of the presence of paralogs on sequence divergence in a set of mouse-human orthologs. Genome Res 12:1370–1376

Pal C, Papp B, Hurst LD (2001) Does the recombination rate affect the efficiency of purifying selection? The yeast genome provides a partial answer. Mol Biol Evol 18(12):2323–2326

Pal C, Papp B, Lercher MJ (2006) An integrated view of protein evolution. Nat Rev Genet 7(5):337–348

Patil A, Nakamura H (2006) Disordered domains and high surface charge confer hubs with the ability to interact with multiple proteins in interaction networks. FEBS Lett 580(8):2041–2045

Podder S, Ghosh TC (2010) Exploring the differences in evolutionary rates between monogenic and polygenic disease genes in human. Mol Biol Evol 27(4):934–941

Podder S, Mukhopadhyay P, Ghosh TC (2009) Multifunctionality dominantly determines the rate of human housekeeping and tissue specific interacting protein evolution. Gene 439(1–2):11–16

Prasad TSK, Goel R, Kandasamy K, Keerthikumar S, Kumar S, Mathivanan S, Telikicherla D, Raju R, Shafreen B, Venugopal A, Balakrishnan L, Marimuthu A, Banerjee S, Somanathan DS, Sebastian A, Rani S, Ray S, Harrys Kishore CJ, Kanth S, Ahmed M, Kashyap MK, Mohmood R, Ramachandra YL, Krishna V, Rahiman BA, Mohan S, Ranganathan P, Ramabadran S, Chaerkady R, Pandey A (2009) Human protein reference database-2009 update. Nucleic Acids Res 37(Database issue):D767–D772

Rain JC, Selig L, De Reuse H, Battaglia V, Reverdy C, Simon S, Lenzen G, Petel F, Wojcik J, Schächter V, Chemama Y, Labigne A, Legrain P (2001) The protein-protein interaction map of Helicobacter pylori. Nature 409(6817):211–215

Raman K (2010) Construction and analysis of protein–protein interaction networks. Autom Exp 2(1):2

Rhodes DR, Tomlins SA, Varambally S, Mahavisno V, Barrette T, Kalyana-Sundaram S, Ghosh D, Pandey A, Chinnaiyan AM (2005) Probabilistic model of the human protein-protein interaction network. Nat Biotechnol 23(8):951–959

Rocha EP, Danchin A (2004) An analysis of determinants of amino acids substitution rates in bacterial proteins. Mol Biol Evol 21(1):108–116

Salathe M, Ackermann M, Bonhoeffer S (2006) The effect of multifunctionality on the rate of evolution in yeast. Mol Biol Evol 23(4):721–722

Salwinski L, Miller CS, Smith AJ, Pettit FK, Bowie JU, Eisenberg D (2004) The database of interacting proteins: 2004 update. Nucleic Acids Res 32(Database issue):D449–D451

Sam L, Liu Y, Li J, Friedman C, Lussier YA (2007) Discovery of protein interaction networks shared by diseases. Pac Symp Biocomput 12:76–87

Singh GP, Ganapathi M, Dash D (2007) Role of intrinsic disorder in transient interactions of hub proteins. Proteins 66(4):761–765

Stelzl U, Worm U, Lalowski M, Haenig C, Brembeck FH, Goehler H, Stroedicke M, Zenkner M, Schoenherr A, Koeppen S, Timm J, Mintzlaff S, Abraham C, Bock N, Kietzmann S, Goedde A, Toksöz E, Droege A, Krobitsch S, Korn B, Birchmeier W, Lehrach H, Wanker EE (2005) A human protein-protein interaction network: a resource for annotating the proteome. Cell 122 (6):957–968

Subramanian S, Kumar S (2004) Gene expression intensity shapes evolutionary rates of the proteins encoded by the vertebrate genome. Genetics 168(1):373–381

Teichmann SA (2002) The constraints protein-protein interactions place on sequence divergence. J Mol Biol 324(3):399–407

Tseng YY, Liang J (2006) Estimation of amino acid residue substitution rates at local spatial regions and application in protein function inference: a Bayesian Monte Carlo approach. Mol Biol Evol 23(2):421–436

Uetz P, Giot L, Cagney G, Mansfield TA, Judson RS, Knight JR, Lockshon D, Narayan V, Srinivasan M, Pochart P, Qureshi-Emili A, Li Y, Godwin B, Conover D, Kalbfleisch T, Vijayadamodar G, Yang M, Johnston M, Fields S, Rothberg JM (2000) A comprehensive analysis of protein-protein interactions in Saccharomyces cerevisiae. Nature 403(6770): 623–627

Wall DP, Hirsh AE, Fraser HB, Kumm J, Giaever G, Eisen MB, Feldman MW (2005) Functional genomic analysis of the rates of protein evolution. Proc Natl Acad Sci USA 102(15): 5483–5488

Warringer J, Blomberg A (2006) Evolutionary constraints on yeast protein size. BMC Evol Biol 6:61

Watts DJ, Strogatz SH (1998) Collective dynamics of "small-world" networks. Nature 393(6684): 440–442

Winter EE, Goodstadt L, Ponting CP (2004) Elevated rates of protein secretion, evolution, and disease among tissue-specific genes. Genome Res 14(1):54–61

Wong P, Althammer S, Hildebrand A, Kirschner A, Pagel P, Geissler B, Smialowski P, Blöchl F, Oesterheld M, Schmidt T, Strack N, Theis FJ, Ruepp A, Frishman D (2008) An evolutionary and structural characterization of mammalian protein complex organization. BMC Genomics 9:629

Xu J, Li Y (2006) Discovering disease genes by topological features in human protein-protein interaction network. Bioinformatics 22(22):2800–2805

Yang J, Gu Z, Li WH (2003) Rate of protein evolution versus fitness effect of gene deletion. Mol Biol Evol 20(5):772–774

Yeates TO, Beeby M (2006) Biochemistry. Proteins in a small world. Science 314(5807): 1882–1883

Zhang J, He X (2005) Significant impact of protein dispensability on the instantaneous rate of protein evolution. Mol Biol Evol 22(4):1147–1155

Zhang L, Li WH (2004) Mammalian housekeeping genes evolve more slowly than tissue-specific genes. Mol Biol Evol 21(2):236–239

Zhou T, Drummond DA, Wilke CO (2008) Contact density affects protein evolutionary rate from bacteria to animals. J Mol Evol 66(4):395–404

Chapter 5
Integration of Evolutionary Biology Concepts for Functional Annotation and Automation of Complex Research in Evolution: The Multi-Agent Software System DAGOBAH

Philippe Gouret, Julien Paganini, Jacques Dainat, Dorra Louati, Elodie Darbo, Pierre Pontarotti, and Anthony Levasseur

Abstract Various strategies have been proposed for predicting protein function. They are derived from the classical homology-based approaches and emerging alternative approaches taking into account gene history in the framework of phylogenetic comparative methods. The growing numbers of available genome sequences and data require bioinformatics tools, in which methodological approaches are set according to the biological issues to be addressed. Much effort has already been devoted to integrating evolutionary biology into bioinformatics tools; e.g., homology-based functional annotation has been successfully integrated in a pipeline-assisted method. In addition, new concepts based on correlation of evolutionary events are emerging. For example, two independent events (e.g., systematic loss of specific genes) that happen repetitively can therefore be functionally linked. However, correlated gene profiles, also called "contextual annotation," makes use of different bioinformatics resources based on multi-agent development. In this chapter, we describe evolutionary concepts and bioinformatics approaches proposed for future functional inference.

P. Gouret • J. Paganini • J. Dainat • E. Darbo • P. Pontarotti
UMR6632, Evolutionary Biology and Modeling, Université de Provence, 3 place Victor Hugo, 13331 Marseille, France
e-mail: philippe.gouret@univ-provence.fr

D. Louati
UMR6632, Evolutionary Biology and Modeling, Université de Provence, 3 place Victor Hugo, 13331 Marseille, France

(LAMSIN-IRD) ENIT, Ecole Nationale d'Ingénieurs de Tunis BP 37, Le Belvèdere 1002-Tunis, Tunisia

A. Levasseur
INRA, UMR1163 de Biotechnologie des Champignons Filamenteux, IFR86-BAIM, Universités de Provence et de la Méditerranée, ESIL, 163 avenue de Luminy, CP 925, 13288 Marseille Cedex 09, France

Universités Aix-Marseille 1 et 2, UMR1163, 163 avenue de Luminy, CP925, 13288 Marseille Cedex 09, France

P. Pontarotti (ed.), *Evolutionary Biology – Concepts, Biodiversity, Macroevolution and Genome Evolution*, DOI 10.1007/978-3-642-20763-1_5,

5.1 Functional Annotation Strategies: Current and Future Approaches

5.1.1 Homology-Based Functional Annotation

Eisen was the first to conceptually rationalize phylogenetic methods to improve the accuracy of functional predictions. In 1998, he proposed a phylogenetic prediction of gene function and compared it to similarity-based functional prediction methods (Eisen 1998). In this work, all known functions on a phylogenetic tree were overlaid. The prediction task could then be split into two steps. In the first step, the tree could be used to decipher orthology and paralogy relationships. Most of the reports based on evolutionary biology methods used ortholog information to transfer functional annotation (see Gouret et al. 2005 and Danchin et al. 2007). Functional assignment could be performed for uncharacterized proteins only if the function of an ortholog was known (and if a similar function was evidenced for all characterized orthologs). Ideally, functional inference should be carried out for experimentally validated orthologs. Bibliographic analysis indicates that orthologs are more likely to keep a similar function than paralogs (e.g., Collette et al. 2003). Theoretically, after duplication, one of the copies is lost, or both duplicates undergo subfunctionalization, or one of the duplicates evolves toward a new function (Force et al. 1999). However, Studer has challenged this assumption, as orthologs and paralogs could have comparable mechanisms of divergence (Studer and Robinson-Rechavi 2009). Different and more complex fates of duplicates could also be evidenced (for a review, see Levasseur and Pontarotti 2011).

In the second step, parsimony reconstruction or alternative reconstructive propagation methods could be used to assign functions of uncharacterized genes by identifying the evolutionary scenario that requires the fewest functional changes over time. Inference of ancestral state on phylogenetic tree requires that character mapping be accurate. Uncertainty about trees and mapping is therefore counterbalanced by introducing Bayesian statistical methods, taking into account this inherent error parameter (Ronquist 2004).

To the best of our knowledge, the first report using both approaches was integrated in the work of Engelhardt et al. (2005). The authors constructed a model of molecular function evolution to infer function in a phylogenetic tree. The model takes into account evidence of varying quality and computes a posterior probability for every possible molecular function for each protein in the phylogeny. Different hypotheses were included in the strategy, i.e., each molecular function may evolve from any other function, and a protein's function may evolve more rapidly after duplication events than after speciation events (Engelhardt et al. 2005). Branch length and duplication are integrated in the methodological approach. In brief, methods may be summarized as propagating functional information from leaves to the root of the phylogeny and then propagating back out to the leaves of the phylogeny, based on the probabilistic model of function evolution.

Homology-based functional annotation is summarized in Fig. 5.1.

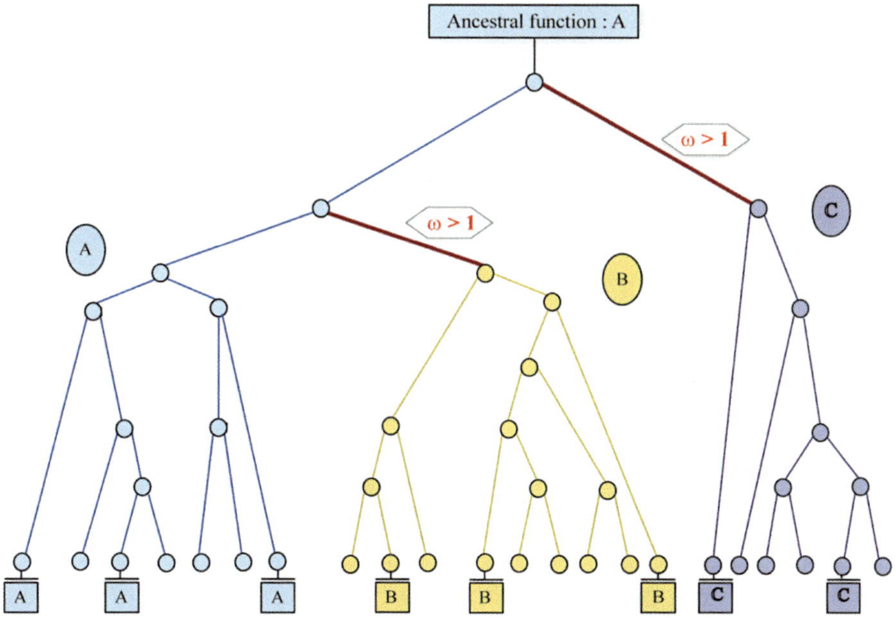

Fig. 5.1 Homology-based functional annotation. Functionally annotated leaves are labeled, respectively, as function A (*blue*), B (*yellow*), and C (*dark blue*). Putative function of non-annotated leaves is inferred after ancestral reconstruction based on propagation of functional information from leaves to the root of the phylogeny. Red branches: evolutionary and functional shift (using $\omega = dN/dS > 1$, i.e., Darwinian selection). (Adapted from Levasseur and Pontarotti 2008)

5.1.2 *Strengthening Functional Annotation: Integration of Correlative Approaches*

Functional prediction using "contextual information" is tricky because of (i) technical difficulty in detecting occurrence profiling and (ii) statistical methods required to correlate and infer function accurately. Co-occurrence and correlated gene profiles could result from phylogenetic inheritance among closely related species. Alternatively, co-occurrence could also result from individual adaptive functions, for instance when genes appear or are lost independently in several distinct lineages (Barker and Pagel 2005). Thus the probability of functional linkage between genes is proportional to the number of multiple independent phylogenetic events. A simplified example of co-occurrence and functional links is depicted in Fig. 5.2. Unlike the overall counting of presence or absence of genes, phylogenetic methods enable us to investigate ancestral states and decipher independent multiple evolutionary events.

Different methods for occurrence profiling have already been proposed, mainly on the basis of the parsimony principle and maximum likelihood (ML).

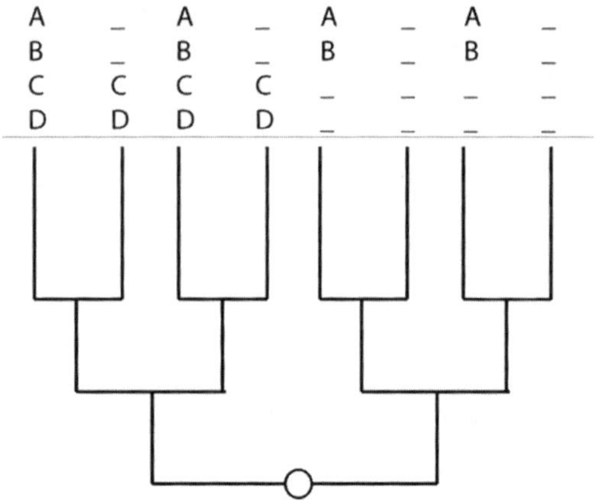

Fig. 5.2 Co-occurrence and functional link. Example of the need for comparative phylogenetic methods. Presence/absence of genes (A, B, C, D) is reported on the leaves of the phylogenetic tree. Here, multiple independent phylogenetic events of gain/loss of gene pairs (i.e., four independent events for genes A and B) are opposed to the apparent correlation arising from shared inheritance of gene pairs loss (resulting from one ancient event for genes C and D). The different steps can be summarized as follows: (i) detection of event: A is lost, (ii) convergence detection: A is lost several times, (iii) co-convergence detection: A and B are lost together several times. Subsequently, statistical tests are carried out. The function of non-annotated genes could be deduced from the correlated annotated genes

As described in the work of Barker and Pagel (2005) and Barker et al. (2007), a common pattern of presence and absence across a range of distinct genomes could be integrated as a method for detecting functionally linked proteins. Thus correlated gains and losses of genes on a phylogenetic tree of species could improve the detection of functionally linked pairs of proteins, compared with the original across-species methods from Pellegrini et al. (1999). Several phylogenetic methods were compared in their work to evaluate the accuracy of their method. Methods were based on either Dollo parsimony (Farris 1977) or ML, including a general model, but also using a constrained model in which the rate of gain of genes is not estimated from the data, but set at a low value. The fixed value of the ML should model gene content evolution better, by preventing the modeling of multiple gains of the same gene in different parts of the phylogeny. In the parsimony case, the reconstructed ancestral states could be very uncertain and parsimony could be applied when rates of changes are rather low. Note that parsimony intervals are proposed to account for the uncertainty of the parsimony methods. For instance, Zhou et al. proposed a dynamic programming algorithm to calculate such parsimony intervals. The best 100 suboptimal ancestral states were determined, and the authors compared the number of correlated events, while allowing for the degree of suboptimality of the reconstructions (Zhou et al. 2006). By contrast, ML accounts for the branch length and uncertainty of topology in the tree, and the estimate of the likelihood values is an independent parameter (i.e., corresponding to all ancestral state possibilities). The authors conclude that all the phylogenetic methods except unconstrained ML achieved higher specificity than the across-species approach (ML model being capable of greater accuracy and sensitivity than a Dollo parsimony-based approach) (Barker et al. 2007).

5.1.3 Toward Reliable Global Functional Annotation: The Need for Bioinformatics

Bioinformatics has unlocked vast amounts of genomic data and developed software applications based on increasingly powerful mathematical algorithms – which themselves produce large volumes of results –, but the amounts of data involved simply cannot be interpreted with any real depth using statistical correlations. We therefore need to develop smart software systems able to support researchers in their efforts, which means systems automatically handling the major routine component of their *in silico* research protocols, and helping analysts interpret the huge volumes of results generated. Such smart software systems could ease the most burdensome part of the workload, leaving researchers to channel their energy into the "sharp end" of their research.

In early 2002, evolutionary biologists were handling vast quantities of biological data made available through the Internet, and running an array of software tools based on probabilistic algorithms working on these data or on data derived from other mathematical tools. The models associated with these tools were all task-specific – sequence similarity, gene prediction, phylogenetic tree-building, and so on. However, they never integrated a large number of concepts employed in biological knowledge and reasoning into a single, integrative software solution. Hence individually, they were unable to answer complex questions posed by biologists or to verify their hypotheses. Consequently, we had to automatically chain mathematical computations through what bioinformaticians call pipelines.

According to the functional annotation strategies described above, homology and correlative approaches were integrated into specific bioinformatics platforms.

A bioinformatics strategy designed for homology-based functional annotation was first implemented by creating FIGENIX (Gouret et al. 2005). FIGENIX is a Java (java.sun.com) platform that automates simple pipeline schemes, such as basic phylogenetic tree-building from a protein sequence by (i) similarity searching against protein databases, (ii) simple filtering, (iii) alignment, and (iv) tree computation. Mathematical tool chaining, through this first version of FIGENIX or any of the pipeline systems available at the time, was unable to completely automate a process: this meant that biologists still had to intervene between computation phases to verify, correct, and synthesize data output from the mathematical tools and guide the workflow to the relevant part of the pipeline. The only way to resolve this automation issue was to introduce an expert system (with Prolog language; Warren et al. 1977) into FIGENIX to model a part of biologists' knowledge and thus act as a human scientist as and when necessary. By introducing specific logical rules in the expert system, a pipeline was created and was dedicated to gene predictions *via* an approach combining ab initio predictions and homology through a lab method. Tested against a known benchmark, the pipeline clearly proved successful. A complex phylogeny pipeline with 50 steps and a lot of expertise modeling was designed. The first version was stabilized in late 2003, and has since enabled the laboratory and its collaborators to produce thousands of phylogenic trees from protein queries. These trees form the basis of our evolutionary research. This pipeline, along with others, was intensively used on laboratory projects, generating several published papers (Danchin et al. 2004, 2006, 2007; Paillisson et al. 2007; Levasseur et al. 2006, 2010). It continued to undergo improvements and enhancements, with upgrades including automatic detection of orthologs in the final process-synthesized tree by online recovery of functional data associated with these orthologs (GO (Ashburner et al. 2000), MGI (www.informatics.jax.org), NCBI (www.ncbi.nlm.nih.gov)), and EST integration (Balandraud et al. 2005). Part of the software developed, called PhyloPattern, emerged as a crucial independent component (Gouret et al. 2009). The aim of this tool was to reproduce human reading of phylogenetic trees, i.e., phylogenetic tree annotation and pattern recognition. Inside the phylogeny pipeline, this tool is used to detect incongruence or isolate specific subtrees, from which biases are then corrected. PhyloPattern now makes it possible to detect events in the history of species, genes, or any other characteristic (from domain to function and further), as well as highlighting artifacts in the phylogenetic trees. We are continuing to improve PhyloPattern as a free open-source JAVA/Prolog API.

5.2 From Pipelines to Multi-Agent Strategies

In 2005, it became clear that the "pipeline approach," even with the controlling expertise introduced, remained limited to computation processes. In addition, functional annotation using the correlative approaches strategy required flexible and more sophisticated data processing architecture. Computation processes are essential, but are not really able to resolve complex tasks of interest to the laboratory, such as automatically highlighting genetic events in the human genome and detecting convergences and co-convergences among these events. Any solution to these issues needs to be driven by expertise through parallel and more "intelligent" processes than the rigid, deterministic pipelines. We also note that the "pipeline approach" does not extend to establishing an explicitly described semantic universe that would allow accurate meta descriptions of data. It thus remains impossible to raise the abstraction level of software tasks, and interfacing them with other software systems is not natural.

Integration of correlated gene profiles for functional annotation requires a three-step process: (i) specific detection of all evolutionary events, (ii) correlation using phylogenetic comparative methods leading to a compelling statistical results, and (iii) deducing the function of non-annotated genes from the correlated annotated genes.

5.3 Technical System Specifications

Accordingly, a new software system was conceived and is able to implement complete automation of actual full research via bottom-up (from biological data) strategies specified by the laboratory, rather than "just" complex computation workflows. We opted for the following research strategy: (i) working from known or computed features to find evidence for generating new hypotheses, (ii) attempting to verify hypotheses to transform them into features, (iii) correlating verified features to deduce new features, and so on. A set of characteristic specifications was drawn up:

- The treatments had to be flexible, modular, and parallelized.
- The strategies for identifying and verifying the facts had to be led by expertise.
- Communication with external software systems (online databases, web services) should systematically gather the relevant results produced by these platforms, such as Ensembl (Hubbard et al. 2009), NCBI, String (Szklarczyk et al. 2011), and ArrayExpress (Parkinson et al. 2011).
- The results had to be placed in an accurately described semantic universe that was not redundant but interfaced with data from external systems.
- Some modules had to work together and communicate directly, while others, such as modules for intelligent correlations of events, had to work in stand-alone mode directly on the mass of results produced by the full set of modules.
- The modules had also be able to work at different times.
- The system had to be resistant to failure; as such, very costly computational treatments should have to be run only once.

5.4 Technical State of the Art

The field of biology now has a number of software tools, approaches, standards, and publications that could be recycled for our needs. The type of system targeted here required establishing an integrated data model, placed between structured biological data (e.g., genomic databases) or unstructured data (publications) located inside or outside the laboratory, and the research strategies desired by laboratory researchers. Software systems clearly have to work with large-scale data banks, but what is most important now is to work with different kinds of data, many of which are not a direct representation of biological objects but are more abstract concepts.

We could therefore rule out relational database management systems, which are not powerful enough or flexible enough to describe semantics in biology. Some recently developed software tools such as the alignment expert system ALEXSYS (Aniba et al. 2009) are based on the UIMA framework (http://sourceforge.net/projects/uima-framework/), which offers a powerful architecture and is well-suited to the introduction of a virtual model on unstructured data, i.e., building meta-information from artifacts such as scientific publications (also see DiscoveryLink

(Hass et al. 2001) or BioMOBY (Wilkinson and Links 2002)). We are more focused on trying to directly model actual genomics or evolutionary concepts. Also, the UIMA approach is only "object-oriented," and we believe that this kind of modeling architecture is not rich enough to integrate the complexity of biological paradigms, especially compared with approaches based on mathematical first-order logic ontology techniques such as Description Logic (http://dl.kr.org/). The W3C-standardized OWL language (http://www.w3.org/TR/owl-ref/) is an XML representation of DL. Initially applied to the semantic web, it is fast becoming a standard for ontology modeling. In DL, relations between classes are not limited to aggregation or inheritance links but can be formalized with logical formulae. However, we note that DL does not integrate concepts of inductive, temporal, or fuzzy logic, which in the long term could direct the natural extension of our systems.

Biology now has many ontologies (e.g., NCI Cancer Ontology: http://www.mindswap.org/2003/CancerOntology/). Some are defined in OWL but to our knowledge, none computationally exploit the descriptive capacity of description logic (DL). This situation is surely set to change. We note the existence of relational ontology (Smith et al. 2005), placed between "object" modeling and DL modeling, which attempts to standardize relations in biological ontologies. This point will be revisited below. There appears to be a continuing dichotomy between the activity of defining ontologies, considered as vocabularies by many biologists, and the establishment of DL-based software and databases within and between laboratories or institutes. We believe that this dichotomy is an error, as it has very adverse repercussions, such as poor software systems and bad interoperability.

As stated above, to fully automate *in silico* research strategies, the type of system we are targeting has to be less rigid and deterministic than pipelines. A natural candidate solution would be multi-agent systems. In bioinformatics, these systems are used essentially to model and simulate biological networks (reactive agents), although they are also used to parallelize mathematical computations through agents with very fine granularity. They are rarely employed for building integrative applications where "smart" agents work with biological information. Nevertheless, like the FIPA institute (http://www.fipa.org/), we are convinced that this kind of architecture built from cognitive agents (with large granularity) communicating inside an ontological semantic universe can be applied to bioinformatics automation. The JADE software framework (http://jade.tilab.com/) is a Java implementation of FIPA specifications. At our lab, we used JADE to develop a first prototype multi-agent system named CASSIOPE (Rascol et al. 2009), dedicated to highlighting conserved synteny.

Recently, eHive emerged from EBI as a new workflow system (Severin et al. 2010). It is built as a multi-agent "blackboard" architecture. Here, the blackboard, i.e., the communication area between agents, is reduced to chaining rules between agents. Thus the tasks produced by the system are driven by predefined functional relations between agents and not by the autonomous interpretation, by agents, of the data resulting from other agents' work. The Ehive blackboard database has a rigid structure with no data modeling. Also, agents' source code is written with the Perl language, which albeit very widely used in bioinformatics remains very poor in expertise and knowledge modeling.

As stated earlier, we are seeking to deploy expertise-driven research strategies, which means that all agents need to be built with expert-system architectures. Rule engines do exist – one example is Jess (www.jessrules.com) – but it would be preferable to write our own engine in Prolog language to reap the benefit of tools we developed previously, especially PhyloPattern. After years of hands-on experience, we can confirm that the Prolog language is very well-suited to bioinformatics. Its benefits for the target system include: (i) a natural capacity to generate all the solutions for a question, (ii) easy and native manipulations of lists and tree structures, which are intensively used in bioinformatics data, (iii) development of expert systems in backward- and/or forward-chaining mode (verification and/or production of facts), (iv) formalisms (e.g., ontological relations) representable directly in the language's syntax, (v) brevity and simplicity of knowledge descriptions, and (vi) interpreted language that strengthens the experimental aspect of certain developments.

5.5 System Architecture

Our system was called DAGOBAH. It is shaped as a multi-agent software (see Fig. 5.3), with a voluntarily hybrid model summing of a model called "Belief Desire Intention" with a model called "Blackboard" (Ferber 1995). The BDI model is suitable for cognitive agents with high granularity and therefore high "intelligence." In the BDI model, agents have a plan formed for our purposes by logical rules. This highly flexible rule system is used by each agent to implement a specific strategy, but can also be used as a traditional expert system to produce high-level facts deduced from simpler facts. For example, an agent capable of sifting through actions to detect several equally probable genetic events from a phylogenetic tree will be able to retain only one event, through a set of logical rules associated with a set of criteria.

The semantics for one rule is defined as follows:

- $Action_1 \ldots Action_k$
  $ConditionFact_1 \ldots ConditionFact_n \rightarrow ConclusionFact_1 \ldots ConclusionFact_m$
  $ToBeRemovedFact_1 \ldots ToBeRemovedFact_z$

The meaning is "if all condition facts (n) are known by the agent ($\subset$ Belief) and if at least one of the conclusion facts (m) is not present and if the agent is capable of achieving all actions (k) ($\subset$ Intention) successfully, then all conclusions (m) ($\subset$ Desire) are considered truthful, and all indicated facts (z) are removed from the agent's knowledge."

Here is an example rule, used in the DAGOBAH agent dedicated to searching for domain architecture events. We suppose that for a specific protein with the domain architecture A-B-C, DAGOBAH detects an event that produced the B-C part of the architecture by analyzing the phylogenetic tree of domain B, and we suppose

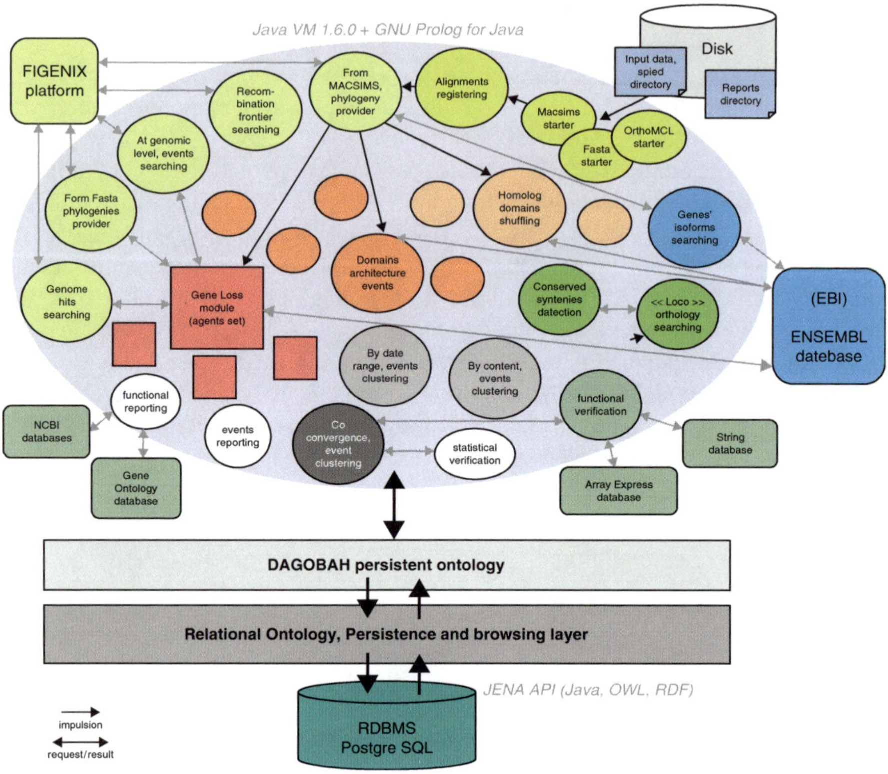

Fig. 5.3 DAGOBAH multi-agents system architecture. All agents (*disks*) or modules (*squares*) (set of agents) that compose DAGOBAH are contained in the large blue ovoid. Around it are displayed the external software systems interacting with the agents by the network. At the bottom of the scheme is shown the ontological database, containing the biological results produced and shared by the agents

that DAGOBAH hesitates between identifying the event as a shuffling or a gain. A simple rule, if it is applicable, allows DAGOBAH to definitely assert there is a gain (see Fig. 5.4):

. *verify_similarity_of_signal_between(P1, P2, [B, C])*
event_found_under_ancestral_node(N),
apomorphic_chosen_protein(P1, [A, B, C]), $\rightarrow$ *gain_event_found(N, P1, [C])*
plesiomorphic_chosen_protein(P2, [A, B])
event_found_under_ancestral_node(N)

The "Blackboard" model introduces an area of information shared by agents, i.e., any important result produced by an agent is placed on the blackboard. The blackboard architectural model chosen in DAGOBAH is defined as a persistent ontology (an ontological database) representing the semantic universe in which the agents work. These results are used by other agents, unless they are forced to

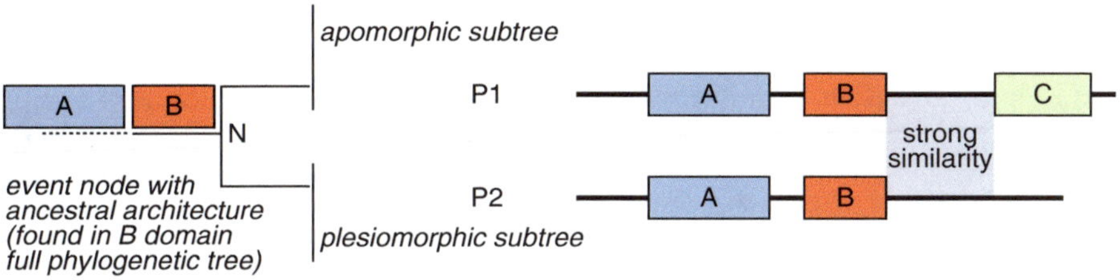

Fig. 5.4 A virtual example for a domains architecture event. Here again event is confirmed because the genomic signal between domains B and C on the apomorphic sequence is strongly conserved after domain B on the plesiomorphic sequence

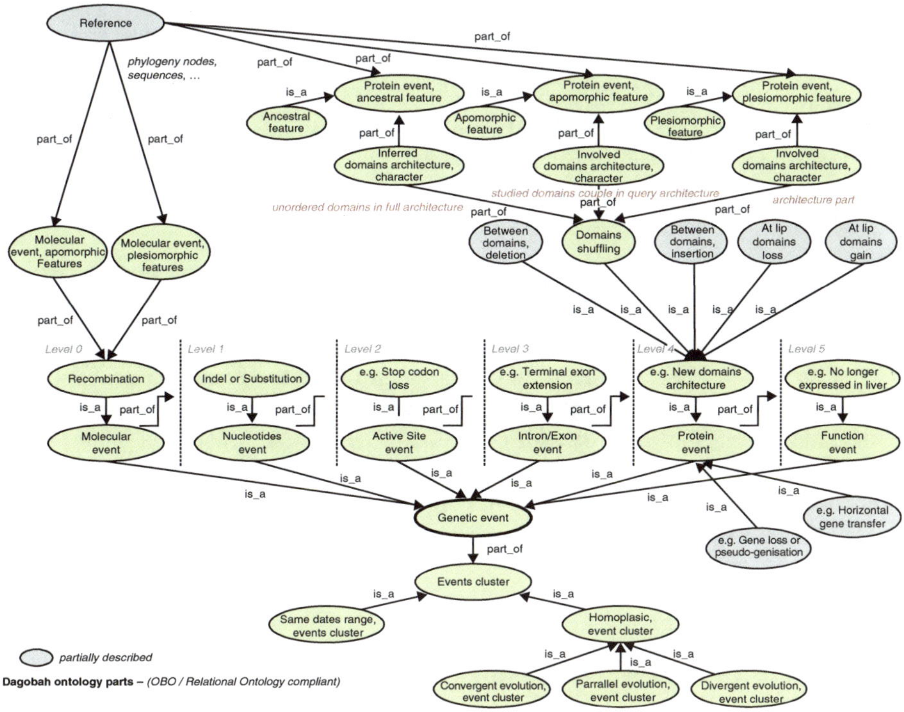

Fig. 5.5 The core of DAGOBAH ontology. Some genetic event classes laid out by their reading level are presented. As an example, we give all the classes participating in a nonhomologous domain shuffling event, induced by a recombination event. Clustering classes are also displayed with their inheritance relationships

explicitly and systematically exchange them. Figure 5.5 illustrates the main parts of the DAGOBAH ontology. Genetic event classes are grouped by reading level. For example, a recombination event can be described at a "protein" level if we are talking about domains involved in recombination, but also at a "molecular" level if we are talking about the position of the recombination on a chromosomal region. Ancestral, apomorphic, and plesiomorphic features associated with an event are always explicitly expressed. This model is also particularly well-suited to studying automatic correlations of genetic events, and is able to correlate several events detected by DAGOBAH and temporally localized between speciation event pairs. For example, DAGOBAH may find that two genes A and B are lost twice "together" for two different lineages, which could prove very interesting in a functional perspective. In this case, if the "function" of gene A is known and the "function" of B is not, we can assume that the B gene may be involved in the "same" function as A. "By Dates" event clusters and homoplasic event clusters are the sources of a co-convergent event clustering process in DAGOBAH. For example, a "convergent evolution event cluster" is produced for events that have the same apomorphic feature objects.

The DAGOBAH ontological database must not have redundancy vs. external databases (like Ensembl; Hubbard et al. 2009). Consequently, we only model, by classes and relations, those concepts associated with specific laboratory research themes, and references were kept only to biological data or results held in external databases. The current DAGOBAH ontology adopts the Relational Ontology standard, although in the future we will probably abandon this standard so as to fully exploit the capabilities of Description Logic.

5.6 DAGOBAH Functionalities and Summarized Strategies

As described in Fig. 5.2, the strategies used in DAGOBAH can be conceptually subdivided into these different steps: (i) detection of evolutionary events, i.e., gain or loss of genes, shuffling, etc. (ii) detection of convergence between one or more gene pairs, (iii) detection of co-convergence between linked genes, (iv) search for functionally annotated gene and infer the function of correlated non-annotated gene. These four steps can be considered as forming the core of the phylogenetic comparative methods.

5.7 Detection of Events (New Architecture Appearance)

The current DAGOBAH version offers a broad panel of functions, ranging from automatic detection of genetic events to homologous domain shuffling, nonhomologous domain shuffling, insertion, deletion, gain and loss, plus gene losses and pseudogenization, and on to horizontal gene transfer and duplications (compilation on gene and species trees). A simplified summary of DAGOBAH's general strategy for event detection is:

1. Use "domain-annotated" protein alignments built from a query protein to outsource phylogeny trees building (domain trees and protein trees) to the FIGENIX platform.
2. Automatically read these trees with PhyloPattern to highlight possible events.

3. Seek to verify and clarify the putative events at a genomic level.

For new protein domain architecture events, actual examples of putative events in trees are given in the PhyloPattern publication. For this kind of event, a dedicated DAGOBAH agent studies each consecutive domain pair in the query protein architecture to investigate whether the association is the result of an event. Ideally, it finds an event's phylogenetic pattern (see Fig. 5.4) on each domain phylogenetic tree, which strengthens the event hypothesis.

The full confirmation of the event is achieved at genomic level by searching for an alignment break position between two DNA segments – one associated with the most representative apomorphic sequence and the other associated with the most representative plesiomorphic sequence. DNA segments are extracted between the domains involved (see Fig. 5.6). The most representative apomorphic sequence is chosen as the one nearest the parent node (the agent uses neighbor joining for branch lengths), while the most representative plesiomorphic sequence is chosen as the one whose domain architecture is closest to the ancestral node architecture (Dollo, Sankoff, and Mirkin parsimony algorithms (Sankoff 1975; Farris 1977; Mirkin et al. 2003) are integrated into PhyloPattern and used by the agent to infer ancestral domain architectures). If several plesiomorphic sequences share the same architecture comparison "score," the agent chooses a sequence from the nearest species in the species tree.

Gene losses and pseudogenization are studied by a set of agents in DAGOBAH, which form a module named GeneLoss. It starts the study by searching for missing species in the biggest ortholog group of the query protein tree. Each species is then studied by independent agents.

Describing the strategy in schematic terms, agents set out to determine whether the species is really missing, whether a new gene should be annotated, or whether there are some mutations or indels that can explain a pseudogenization process.

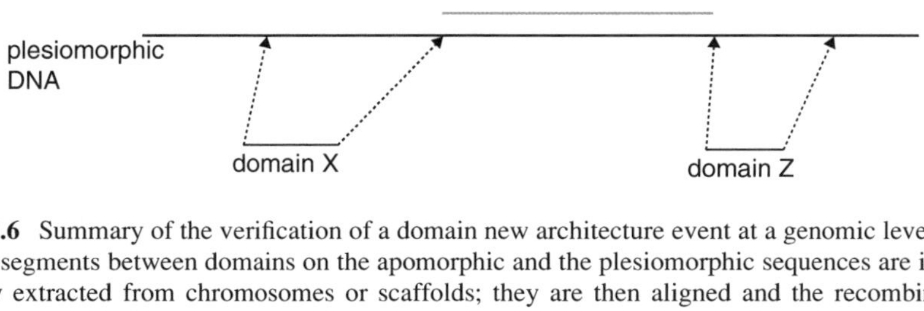

Fig. 5.6 Summary of the verification of a domain new architecture event at a genomic level. The DNA segments between domains on the apomorphic and the plesiomorphic sequences are intelligently extracted from chromosomes or scaffolds; they are then aligned and the recombination point is searched for as an alignment break

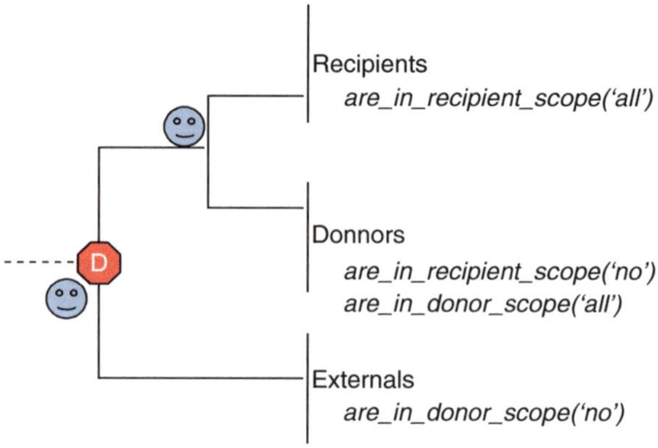

Fig. 5.7 A pattern to detect horizontal gene transfers from a phylogenetic gene tree. This means a duplication node, because the subtree does not have to match the species tree. The "donor" subtree must contain only species of a specific scope, and not from the "recipient" scope and *vice versa*

Full complex GeneLoss module strategy and results will be published separately at a later date.

Horizontal gene transfer events are detected from the query protein tree. A recipient species scope and a donor species scope are defined so as to orient the search. The dedicated agent uses PhyloPattern to annotate each internal node of the tree with two tags: are_in_recipient_scope_species and are_in_donor_scope_species, which can take three values: "no" if no species of a subtree falls in a scope, "some" if some species of a subtree fall in a scope, or "all" if all the species of a subtree fall in a scope. Then, *via* PhyloPattern, the agent applies a specific phylogenetic pattern (see Fig. 5.7) that directly gives the branch with potential HGT events.

The expert idea behind this pattern is to search the gene tree to find recipient species closer to donor species than other species that are normally placed between the recipient and donor species in the species tree.

5.8 Convergence and Co-Convergence Detection

Another important function in DAGOBAH is event convergence and co-convergence detection as conceptually described in the correlative approaches described above. Convergence identification is easy to obtain from the DAGOBAH ontological database, as a dedicated agent groups events into homoplasic convergent clusters. For example, two events are in the same convergent cluster if they have the same apomorphic character. The definition of an apomorphic character can easily be user-defined as a Prolog "ontological" pattern. The clustering mechanism is independent of the pattern definition. Co-convergence detection is a more complex task. It starts by homoplasic clustering, after which an agent produces date range clustering. Inside DAGOBAH, events are dated with tuples:

[TaxidSpeciationBefore, NumberOfDuplicationsBefore, NumberOfDuplicationsAfter, TaxidSpeciationAfter]

This tuple is determined by taking the nearest speciation event (SBE) before the event (E) on its parent branch. NumberOfDuplicationsBefore equals the number of duplication events on the branch between SBE and E. TaxidSpeciationBefore is the common parent taxid of all species in the SBE subtree. The same approach is then reapplied for the next speciation event. Date range clustering is also "user-defined" through date range patterns. Two events whose dates fit the same date pattern are pooled in the same date range cluster.

Co-convergence clusters are built with a hierarchical clustering method. A minimum co-convergent cluster is formed by four events: Eh1, Eh2, Eh1′, Eh2′. Eh1 and Eh1′ have to be in the same homoplasic cluster, while Eh2 and Eh2′ have to be in another homoplasic cluster. Eh1 and Eh2 have to be in the same date range cluster, while Eh1′ and Eh2′ have to be in another date range cluster.

We can model this basic cluster as a square:

--- Eh1, Eh2,

--- Eh1′, Eh2′

The clusters can be rectangular, if they come from more date clusters than homoplasic clusters (shape 1) or the opposite (shape 2). The hierarchical clustering method enables us to build the biggest possible clusters, and implies the definition of a distance method between two clusters. Our distance method favors clusters with shape 1 rather than shape 2.

Once the biggest clusters are determined, the agents seek to verify them, both statistically, *via* the Pagel method (Pagel 1994), and functionally, using the String database (Szklarczyk et al. 2011) to see whether proteins associated with events in the same homoplasic cluster belong to the same protein interactions network, and using the ArrayExpress database (Parkinson et al. 2011) to see whether proteins associated with events in the same homoplasic cluster concern the same expression experiments.

In conclusion, DAGOBAH is designed to exploit the modern functional annotation strategies and specially the evolutionary-based biology concepts. In addition, it could be addressed to various general biological questions such as searches of conserved synteny regions from a given region associated to a species to another target species.

All public results produced by DAGOBAH are openly available on the IODA Web site (http://ioda.univ-provence.fr/).

Acknowledgments This research was supported by the contract MIE (Maladies Infectieuses Emergentes-Programme Interdisciplinaire, CNRS) and ANR EvolHHuPro (ANR-07-BLAN-0054-01).

References

Aniba MR, Siguenza S, Friedrich A, Plewniak F, Poch O, Marchler-Bauer A, Thompson JD (2009) Knowledge-based expert systems and a proof-of-concept case study for multiple sequence alignment construction and analysis. Brief Bioinform 10:11–23

Ashburner M, Ball CA, Blake JA, Botstein D, Butler H, Cherry JM, Davis AP, Dolinski K, Dwight SS, Eppig JT, Harris MA, Hill DP, Issel-Tarver L, Kasarskis A, Lewis S, Matese JC, Richardson JE, Ringwald M, Rubin GM, Sherlock G (2000) Gene ontology: tool for the unification of biology. The gene ontology consortium. Nat Genet 25:25–29

Balandraud N, Gouret P, Danchin EG, Blanc M, Zinn D, Roudier J, Pontarotti P (2005) A rigorous method for multigenic families' functional annotation: the peptidyl arginine deiminase (PADs) proteins family example. BMC Genomics 6:153

Barker D, Pagel M (2005) Predicting functional gene links from phylogenetic-statistical analyses of whole genomes. PLoS Comput Biol 1:e3

Barker D, Meade A, Pagel M (2007) Constrained models of evolution lead to improved prediction of functional linkage from correlated gain and loss of genes. Bioinformatics 23:14–20

Collette Y, Gilles A, Pontarotti P, Olive D (2003) A co-evolution perspective of the TNFSF and TNFRSF families in the immune system. Trends Immunol 24:387–394

Danchin E, Vitiello V, Vienne A, Richard O, Gouret P, McDermott MF, Pontarotti P (2004) The major histocompatibility complex origin. Immunol Rev 198:216–232

Danchin EG, Gouret P, Pontarotti P (2006) Eleven ancestral gene families lost in mammals and vertebrates while otherwise universally conserved in animals. BMC Evol Biol 6:5

Danchin EG, Levasseur A, Rascol VL, Gouret P, Pontarotti P (2007) The use of evolutionary biology concepts for genome annotation. J Exp Zool B Mol Dev Evol 308:26–36

Eisen JA (1998) Phylogenomics: improving functional predictions for uncharacterized genes by evolutionary analysis. Genome Res 8:163–167

Engelhardt BE, Jordan MI, Muratore KE, Brenner SE (2005) Protein molecular function prediction by Bayesian phylogenomics. PLoS Comput Biol 1:e45

Farris JS (1977) Phylogenetic analysis under Dollo's law. Syst Zool 26:77–88

Ferber J (1995) Les systèmes multi-agents. InterEdition, Paris

Force A, Lynch M, Pickett FB, Amores A, Yan YL, Postlethwait J (1999) Preservation of duplicate genes by complementary, degenerative mutations. Genetics 151:1531–1545

Gouret P, Vitiello V, Balandraud N, Gilles A, Pontarotti P, Danchin EG (2005) FIGENIX: intelligent automation of genomic annotation: expertise integration in a new software platform. BMC Bioinform 6:198

Gouret P, Thompson JD, Pontarotti P (2009) PhyloPattern: regular expressions to identify complex patterns in phylogenetic trees. BMC Bioinform 19 10:298

Haas LM, Schwarz, Kodali P, Kotlar E, Rice JE, Swope WC (2001) DiscoveryLink: A system for integrated access to life sciences data sources. IBMSJ 40:489–511.

Hubbard TJ, Aken BL, Ayling S, Ballester B, Beal K, Bragin E, Brent S, Chen Y, Clapham P, Clarke L, Coates G, Fairley S, Fitzgerald S, Fernandez-Banet J, Gordon L, Graf S, Haider S, Hammond M, Holland R, Howe K, Jenkinson A, Johnson N, Kahari A, Keefe D, Keenan S, Kinsella R, Kokocinski F, Kulesha E, Lawson D, Longden I, Megy K, Meidl P, Overduin B, Parker A, Pritchard B, Rios D, Schuster M, Slater G, Smedley D, Spooner W, Spudich G, Trevanion S, Vilella A, Vogel J, White S, Wilder S, Zadissa A, Birney E, Cunningham F, Curwen V, Durbin R, Fernandez-Suarez XM, Herrero J, Kasprzyk A, Proctor G, Smith J, Searle S, Flicek P (2009) Ensembl. Nucleic Acids Res 37:D690–D697

Levasseur A, Pontarotti P (2008) An overview of evolutionary biology concepts for functional annotation: advances and challenges. In: Pontarotti P (ed) Evolutionary biology from concept to application. Springer, Berlin, pp 209–215

Levasseur A, Pontarotti P (2011) The role of duplications in the evolution of genomes highlights the need for evolutionary-based approaches in comparative genomics. Biol Direct 6:11

Levasseur A, Gouret P, Lesage-Meessen L, Asther M, Asther M, Record E, Pontarotti P (2006) Tracking the connection between evolutionary and functional shifts using the fungal lipase/feruloyl esterase a family. BMC Evol Biol 6:92

Levasseur A, Saloheimo M, Navarro D, Andberg M, Pontarotti P, Kruus K, Record E (2010) Exploring laccase-like multicopper oxidase genes from the ascomycete trichoderma reesei: a functional, phylogenetic and evolutionary study. BMC Biochem 11:32

Mirkin BG, Fenner TI, Galperin MY, Koonin EV (2003) Algorithms for computing parsimonious evolutionary scenarios for genome evolution, the last universal common ancestor and dominance of horizontal gene transfer in the evolution of prokaryotes. BMC Evol Biol 3:2

Pagel M (1994) Detecting correlated evolution on phylogenies: a general method for the comparative analysis of discrete characters. Proc R Soc Lond B 255:37–45

Paillisson A, Levasseur A, Gouret P, Callebaut I, Bontoux M, Pontarotti P, Monget P (2007) Bromodomain testis-specific protein is expressed in mouse oocyte and evolves faster than its ubiquitously expressed paralogs BRD2, -3, and -4. Genomics 89:215–223

Parkinson H, Sarkans U, Kolesnikov N, Abeygunawardena N, Burdett T, Dylag M, Emam I, Farne A, Hastings E, Holloway E, Kurbatova N, Lukk M, Malone J, Mani R, Pilicheva E, Rustici G, Sharma A, Williams E, Adamusiak T, Brandizi M, Sklyar N, Brazma A (2011) ArrayExpress update–an archive of microarray and high-throughput sequencing-based functional genomics experiments. Nucleic Acids Res 39:D1002–D1004

Pellegrini M, Marcotte EM, Thompson MJ, Eisenberg D, Yeates TO (1999) Assigning protein functions by comparative genome analysis: protein phylogenetic profiles. Proc Natl Acad Sci USA 96:4285–4288

Rascol VL, Levasseur A, Chabrol O, Grusea S, Gouret P, Danchin EG, Pontarotti P (2009) CASSIOPE: an expert system for conserved regions searches. BMC Bioinform 10:284

Ronquist F (2004) Bayesian inference of character evolution. Trends Ecol Evol 19:475–481

Sankoff D (1975) Minimal mutation trees of sequences. SIAM J Appl Math 28:35–42

Severin J, Beal K, Vilella AJ, Fitzgerald S, Schuster M, Gordon L, Ureta-Vidal A, Flicek P, Herrero J (2010) eHive: an artificial intelligence workflow system for genomic analysis. BMC Bioinform 11:240

Smith B, Ceusters W, Klagges B, Köhler J, Kumar A, Lomax J, Mungall C, Neuhaus F, Rector AL, Rosse C (2005) Relations in biomedical ontologies. Genome Biol 6:R46

Studer RA, Robinson-Rechavi M (2009) How confident can we be that orthologs are similar, but paralogs differ? Trends Genet 25:210–216

Szklarczyk D, Franceschini A, Kuhn M, Simonovic M, Roth A, Minguez P, Doerks T, Stark M, Muller J, Bork P, Jensen LJ, von Mering C (2011) The STRING database in 2011: functional interaction networks of proteins, globally integrated and scored. Nucleic Acids Res 39: D561–D568

Warren DH, Pereira LM, Pereira F (1977) Prolog - the language and its implementation compared with Lisp. Proceedings of the 1977 symposium on artificial intelligence and programming languages

Wilkinson MD, Links M (2002) BioMOBY: an open source biological web services proposal. Brief Bioinform 3:331–341

Zhou Y, Wang R, Li L, Xia XF, Sun Z (2006) Inferring functional linkages between proteins from evolutionary scenarios. J Mol Biol 359:1150–1159

Part II
Biodiversity and Evolution

Chapter 6
A New Animal Model for Merging Ecology and Evolution

Gabriele Procaccini, Ornella Affinito, Francesco Toscano, and Paolo Sordino

Abstract The theory of evolution has recently been in turmoil, with great interest in applying empirical information from EvoDevo, genomics, and ecology into the framework of quantitative genetic studies of evolution. *Ciona* is a small genus of sea squirts within the class Ascidiacea of the subphylum Tunicata, the sister group of vertebrates, a phylogenetic position that has contributed to fuel the interest in studying development and evolution in ascidians. *Ciona* species display several traits of evolutionary interest, e.g., conservative anatomy, high genetic polymorphism, cryptic speciation, metapopulation structure and invasive behavior. Some of these aspects may depend on the ecology of these marine animals, which display a great ecophysiological tolerance and unpredictable colonization capabilities. In addition, natural populations show the occurrence of spontaneous mutations with phylomimicking phenotypes. Here we review some key features of this talented marine organism that promise to provide insights in specific aspects of the expanded evolutionary biology.

6.1 The Model System

The ascidian genus *Ciona* (Chordata, Tunicata) has attracted the interest of biologists for over a century because of the ecological importance and the key position in the evolutionary path leading to vertebrates. *Ciona intestinalis* (L. 1767) is a marine invertebrate that lives on shallow hard bottoms, where it can represent a major component of coastal benthic ecosystems. The life cycle is characterized by a short pelagic stage with a chordate-like tadpole larva that, upon settlement and metamorphosis, loses the chordate body plan and generates an invertebrate form. To increase the rate of successful reproduction, *Ciona* is hermaphrodite like all

G. Procaccini • O. Affinito • F. Toscano • P. Sordino
Stazione Zoologica Anton Dohrn, Villa Comunale, 80121 Naples, Italy
e-mail: gabriele.procaccini@szn.it; ornella.affinito@szn.it; francesco.toscano@szn.it; paolo.sordino@szn.it

P. Pontarotti (ed.), *Evolutionary Biology – Concepts, Biodiversity, Macroevolution and Genome Evolution*, DOI 10.1007/978-3-642-20763-1_6,

tunicates, and like many ascidians, it spawns large numbers of gametes on a daily basis. Moreover, due to the high tolerance of *C. intestinalis* to ecophysiological parameters, such as euthrophic conditions following anthropization, this opportunistic species is considered a sentinel of environmental conditions in coastal biotopes. These biological traits may contribute to the cosmopolitan, and sometimes invasive, distribution of *intestinalis* in temperate to sub-boreal waters of both hemispheres. Phylogenetic and comparative genomic studies have demonstrated that the Linnean taxon is a complex of at least four (*C. intestinalis* spA-D) morphologically cryptic but genetically distinct sibling species of disjoint geographical distribution (see Sect. 6.2; Suzuki et al. 2005; Caputi et al. 2007; Nydam and Harrison 2007; Zhan et al. 2010) (Fig. 6.1). *C. intestinalis* spA is the cryptic species for which the genome sequence is available (Dehal et al. 2002).

Evolutionary developmental biology is providing strong evidence that embryonic processes are conserved between basal chordates and vertebrates. Therefore, it is not surprising that this animal occupies a prominent role in several branches of biology (Satoh 1994). A unique blend of advantageous features for experimental analysis has made it a powerful model organism for genetics and genomics of chordates due to ease of manipulability. Its genome carries fewer genes and less genetic redundancy than in vertebrates (Dehal et al. 2002; Small et al. 2007a). In addition, the application of bioinformatics and experimental approaches to the *Ciona* system, such as means for manipulating gene expression, provides reverse and forward genetics tools for studying the genetic basis of cellular and developmental processes common to all chordates (Hendrickson et al. 2004; Sasakura et al. 2007; Sordino et al. 2008; Veeman et al. 2008).

Besides advancements in culturing and formal genetics, yet research with *C. intestinalis* is largely based on sampling in nature. Indeed, growing *Ciona* strains is labor intensive, and sporadic loss of genotypes occurs due to yet suboptimal

Fig. 6.1 Sympatric individuals of *Ciona intestinalis* spA and spB from North European coasts. Photo by L. Caputi

culturing conditions (Kano 2007; Cirino et al. 2002; Joly et al. 2007). Simple, efficient, and rapid method for long-term storage of *Ciona* sperm in liquid nitrogen can greatly facilitate strain management and experimental design. Obligation to sample natural populations for laboratory needs has prompted better understanding of genetic diversity in the wild as well as of *Ciona* population biology *sensu lato*, also in the light of preserving natural resources.

However, knowledge of classical genetics is still fragmentary. Progress in genetic studies is revealing a high degree of genetic polymorphisms at individual and population scales (Dehal et al. 2002; Boffelli et al. 2004). Therefore, to understand the complex genetic structure observed between natural populations at different geographical scales (Schmidtke and Engel 1980; Kano et al. 2001; Small et al. 2007b; Sordino et al. 2008; Zhan et al. 2010Caputi et al. Personal Communication), it is important to address genetic polymorphism and gene flow within and among distant and close populations by means of unlinked markers of nuclear, mitochondrial, and ribosomal origins (see Sect. 6.3). This allows to resolve historic phylogeographical patterns of *C. intestinalis* populations and to determine current patterns of allelic flow between genotypes. Resolution of the genetic structure of populations is also a prerequisite for undertaking analysis of environmental selection across gradients of physical conditions within specific areas.

What ecological and evolutionary forces generate and maintain variation? Which and how many loci are involved? Which is their effect? Answering these questions may help to elucidate the particular evolutionary and demographic history of natural populations, identifying new genomic regions and candidate genes of evolutionary significance (Stinchcombe and Hoekstra 2007). The increasing knowledge concerning *C. intestinalis* population biology makes this species an ideal tool for studying microevolutionary processes. The advent of next-generation genomic technologies in *C. intestinalis* research provides a useful instrument to study environmental impact and adaptation of this invasive species by population genomics studies, as well as to understand the involvement of epigenetic mechanisms in the control of ascidian development. Genomic technologies can be used to analyze the dynamic distribution of epigenetic marks such as cytosine methylation, posttranslational modification of histone tails, and nucleosome composition, at distinct stages of development and in different environmental settings. The resulting spatiotemporal information will provide detailed insights into how the ascidian transcriptome is controlled by epigenetic processes, and what is the effect of environmental cues on the chromatin landscape. We suggest that combining ecology, population genetics and genomics, and the study of spontaneous mutations in *Ciona* may aid in understanding which types of microevolutionary mechanisms and factors do generate variation (see Sect. 6.5).

6.2 Distribution and Ecology

Ciona intestinalis has always been considered as a species with cosmopolitan distribution, and broad environmental tolerance to unstable or fluctuating environments (Hoshino and Nishikawa 1985; Therriault and Herborg 2008a, b).

Nevertheless, the recent finding of cryptic species within the so-called *C. intestinalis*, has questioned this general assumption. Two *C. intestinalis* cryptic species are more widely distributed and have a mostly disjoint distribution, which overlaps only in few areas. *C. intestinalis* spA, in fact, lives in temperate seas around the globe (Mediterranean Sea, south European and South American Atlantic coasts, and Pacific Ocean), while *C. intestinalis* spB is found in North Atlantic waters (Caputi et al. 2007; Zhan et al. 2010). They coexist in the English Channel, where they can theoretically hybridize (Caputi et al. 2007; Nydam and Harrison 2011a, b). A third and a fourth cryptic species, defined as spC and spD, have been described only for one location in the Mediterranean Sea and the Black Sea, respectively, within the distribution range of spA (Nydam and Harrison 2007, 2010; Zhan et al. 2010) (Fig. 6.2).

Specific studies on life cycle and life history of *C. intestinalis* spA are lacking. As a matter of fact, most published information deals with the biology and ecology of *C. intestinalis* spB (Suzuki et al. 2005; Caputi et al. 2007), referring to Scandinavian-Danish (Svane and Havenhand 1993; Petersen and Svane 1995; Petersen et al. 1995; Petersen 2007) and eastern Canadian populations (Ramsay et al. 2008, 2009). It has been shown that recruitant waves of spB larvae settle close to or on adults individuals, forming multigenerational clusters (Havenhand and Svane 1991), and that larval settlement is enhanced by the formation of a fouling canopy on a bare substratum that is subsequently monopolized by ascidian clusters (Schmidt 1983). Thus, the environment into which *C. intestinalis* settles is strongly affected by the occurrence and density of conspecific individuals (Marshall and Keough 2003).

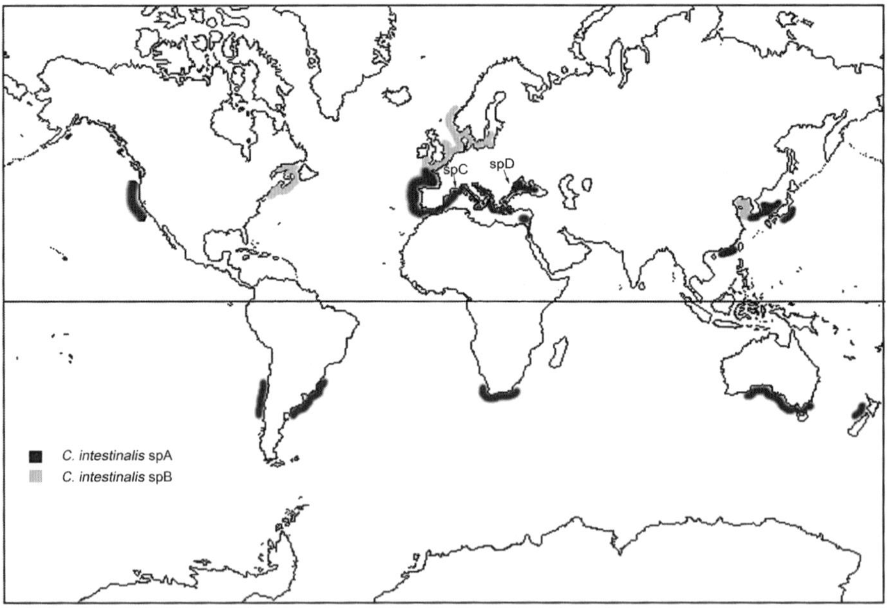

Fig. 6.2 Worldwide distribution of *Ciona intestinalis* cryptic species

Recruitment is also driven by the seasonal fluctuations of density that *C. intestinalis* experiences in the different sites the taxon colonizes, as a function of the year and particularly season (Lambert and Lambert 1998). Moreover, predation does not seem to be a selective factor for recruitment as *C. intestinalis* lives in environments where the number of predators is reduced (Petersen and Svane 1995; de Oliveira Marins et al. 2009).

According to known distribution, spB is more adapted to cold water, as also reflected in its response to the unfavorable seasons. *C. intestinalis* spB populations, in fact, experience a reduction in density of individuals in winter (Petersen and Svane 1995), while spA populations completely disappear in late summer and winter, reoccurring when the environmental conditions are again favorable in spring and autumn as shown for the southern California harbors (Pérès 1952; Sabbadin 1958; Lambert and Lambert 1998). Due to the lack of specific information, many questions related to the ecology and distribution of spA are still open. For example, how T° - S ‰ seasonal variations influence the reoccurrence of the species in the benthic community, being the taxon offspring sensible to the quality of the environment (Marshall and Keough 2003) even if the species tolerates moderate excess of nitrogen, reduced water clarity, low oxygen levels, and periodic algal blooms (Carman et al. 2007). *C. intestinalis* inhabits harbors and confined natural environments where thanks to its capacity to tolerate a wide range of temperature and salinity, produces viable gametes in different environmental conditions, which are an important extrinsic ecological forcing influencing its life history, in particular the occurrence timing (Dybern 1965; Lambert and Lambert 1998; Carver et al. 2003). Tolerance to salinity and temperature also varies ontogenetically, embryos and larvae being less tolerant to high temperatures, and young post-metamorphic ascidians more resistant than adults to low temperatures. The favorable temperature range is between 15°C and 20°C, with a salinity value of 35‰ (Marin et al. 1987). *C. intestinalis* has a remarkable filter suspension feeder capacity with a high efficiency in terms of particles range in relation to temperature which affects the clearance rates (Petersen 2007). Examining stomach and gut contents of *C. intestinalis*, it is possible to distinguish an amorphous matrix made of particulate matter rich of pelagic and benthic diatoms (Fig. 6.3). The species represents also a useful sentinel organism being able to tolerate oil pollution, industrial and radioactive wastes into marine environments concentrated in its tunic or filtered from the water column. Recently, the invasive behavior of *Ciona* species has been reported throughout the globe (e.g., Canada, Argentina, South Africa, New Zealand; reviewed in Zhan et al. 2010). In most newly introduced areas, *C. intestinalis* is a very competitive species that rapidly covers nearly 100% of the available substratum excluding native species in a short period of time (Ramsay et al. 2008). The ecological impact caused by the introduction of nonindigenous ascidian species in natural coastal environments has raised a growing concern in recent years (Whitlatch et al. 1995; Lambert 2001). Nonindigenous species cause different types of influences, as the reduction of biodiversity, the alteration of ecosystems, and the decline of native species (Everett 2000; Pimentel et al. 2000). The ecological impact also reflects on the coastal human activities,

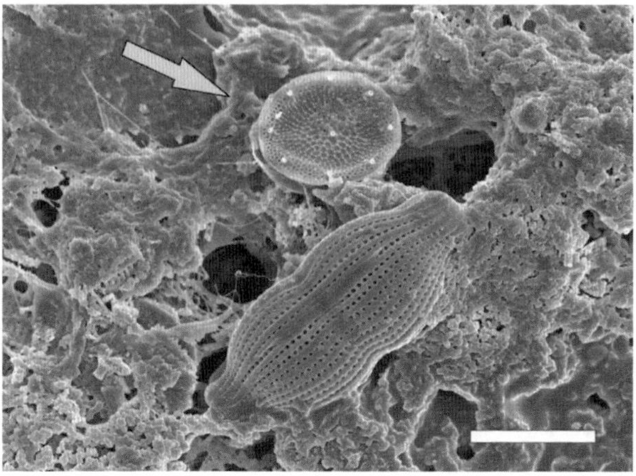

Fig. 6.3 Stomach content of *Ciona intestinalis* spA. Note particulate matter embedding fragments and recognizable pelagic diatoms, such as one belonging to the genus *Thalassiosira* sp. (*light gray arrow*), and an unidentified species. Scale bar 5 μm

with severe economic consequences. For this reason, understanding the evolutionary and ecological causes responsible for the rapid spread of invasive species represents one of the main challenges for conservation biologists and environmental managers.

6.3 Population Genetics and Phylogeography

Studies of phylogeography structure, genetic diversity, and evolutionary changes can help in understanding the potential for colonization and establishment of alien species, the geographic patterns of invasion and range expansion, and the potential for evolutionary responses to novel environments, giving important insights in the definition of management practices (Everett 2000; Holland 2000; Pimentel et al. 2000). Understanding population dynamics and spread potential of invasive species means also determining the degree of population connectivity and investigating factors driving genetic exchange at various geographical scales. Population connectivity in aquatic invasive species can be influenced by water currents, natural and human-mediated pathways of propagule dispersal, species' life histories, and by variation in environmental and community composition across geographical scales (Lee 2002; Darling and Folino-Rorem 2009; Jesse et al. 2009; Goldstien et al. 2010; Sorte et al. 2010).

In *C. intestinalis*, allelic DNA polymorphism across the entire genome is very high (1.2%, based on Single Nucleotide Polymorphisms – SNPs, and insertion/deletions; Dehal et al. 2002), including non-synonymous SNP frequency in important developmental genes such as *Hox5* and *Hox10* (0.27% and 1.89% in spA and spB, respectively) (Caputi et al. 2008). The average genome-wide SNP heterozygosity in *C. savignyi* is even higher, reaching 4.5% (Small et al. 2007b). Because of that, it has been argued that *C. savignyi* exhibits the highest levels of SNP variation and structural polymorphism among multicellular organisms (Small et al. 2007b).

In *Ciona* populations, high levels of genetic diversity may derive from several factors, among which the mutation–selection balance between reappearing phenotypes and their elimination by natural selection, the reproductive strategies, the geographical connectivity, and the association with the large effective population size. From here on we focus our attention on population genetic structure and population connectivity at both large and regional geographical scales of *C. intestinalis* spA.

6.3.1 Genetic Diversity

A high level of polymorphism has been demonstrated at various enzyme (Schmidtke and Engel 1980) and microsatellite loci (Procaccini et al. 2000; Andreakis et al. 2007; Zhan et al. 2010). In allozymes, average heterozygosity was 0.319 and the average number of alleles per locus was 2.6 (Schmidtke and Engel 1980). For microsatellites, allelic richness is higher, ranging from 5.1 to 5.9 (Zhan et al. 2010). The high degree of polymorphism observed in *C. intestinalis* populations determines extremely high values of expected heterozygosity (HE) in all analyzed populations, with values ranging from 0.65–0.69 on eight populations distributed worldwide (Zhan et al. 2010), to 0.73 on three populations in the Gulf of Naples (Tyrrhenian Sea) (Sordino et al. 2008) (Table 6.1). These values are higher than those recorded in other solitary tunicates such as *Styela clava* (HE = 0.48–0.63, Dupont et al. 2009, 2010; Goldstien et al. 2010). Small et al. (2007b) suggested that extreme polymorphism in *Ciona* is probably associated with large effective population size rather than an elevated mutation rate.

Despite the high expected heterozygosis, populations show a relevant heterozygote deficit, which determines a significant deviation from Hardy–Weinberg equilibrium (HWE) ($P < 0.001$; Sordino et al. 2008; Zhan et al. 2010). Similar or higher departure ratio from HWE was previously reported in other ascidians including the solitary *S. clava* (Dupont et al. 2010) and the colonial *Botryllus schlosseri* (Ben-Shlomo et al. 2006). The departure resulting from heterozygote

Table 6.1 Genetic diversity at microsatellite loci for *Ciona intestinalis* spA and spB

Pop	*n*	Na	H_O	H_E	F_{IS}	F_{ST}	Reference
spA							
3	26–52	6.8–9.3	0.42–0.48	0.65–0.69	0.30–0.42	0.03–0.05	Zhan et al. (2010)
3	20	6.25–7.83	0.38–0.50	0.56–0.71	0.20–0.38	–	Sordino et al. (2008)
2	20	7.58–7.83	0.57–0.71	0.66–0.71	–	0.18	Caputi et al. (2007)
spB							
9	21–49	8.9–15	0.31–0.63	0.80–0.871	0.28–0.63	0.001–0.15	Zhan et al. (2010)
2	20	6.08–7.17	0.56–0.60	0.63–0.65	–	0.24–0.81	Caputi et al. (2007)

Pop number of populations, *n* number of individuals, *Na* number of alleles; H_O observed heterozygosity, H_E expected heterozygosity, F_{IS} inbreeding coefficient within individuals relative to the subpopulations, F_{ST} inbreeding coefficient within subpopulations relative to the total deficit might be explained in three different ways. First, recurrent inbreeding. Different studies suggested that *C. intestinalis* larvae generally settle very close to the adult individuals (Petersen and Svane 1995; Howes et al. 2007), increasing the possibility of breeding with related individuals. Additionally, *C. intestinalis* self-sterility is not complete; about 15–20% of individuals can self-fertilize (Rosati and Santis 1978; Kawamura et al. 1987). Second, subpopulation structure. The possibility that a temporal and/or spatial Wahlund effect can be another cause for the massive heterozygote deficit, is suggested by the enhanced subpopulation structure, the high connectivity among population, which are highly dynamic in their fluctuations, and the metapopulation structure shown at small geographic scale. Third, the presence of null alleles. Null alleles are very likely to be present in *C. intestinalis*, due to the highly polymorphic genome, and have been recognized to be one important cause for heterozygote deficiency in many marine species (e.g., Hedgecock et al. 2004; Zhan et al. 2007).

6.3.2 Genetic Structure

Despite high genetic polymorphism of populations, geographic isolation and theoretical short-distance dispersal of larvae, an impressive level of genetic homogeneity is revealed by microsatellites and mtDNA analyses. In the Mediterranean basin, *C. intestinalis* spA shows a single mitochondrial COI haplotype and low degree of microsatellite-based differentiation (Caputi et al. 2007; Caputi et al. personal communication), suggesting a recent bottleneck event followed by rapid recolonization by a dominant mitochondrial haplotype with high dispersal capacity (Fig. 6.4). In comparison, spB mitochondrial COI shows higher geographical structure, suggestive of fixed populations (Caputi et al. 2007; Zhan et al. 2010). From this point of view, the Mediterranean Sea seems to act as a metapopulation formed by genetically homogeneous clusters of spA individuals occupying fragmented habitats in completely different environmental conditions (Caputi et al. 2007). Another theoretical factor linked to the observed lack of sharp genetic structure is the absence of barriers to gene flow among populations that may dilute the effects of genetic bottlenecks and decrease the genetic differentiation among populations (Nei et al. 1975). Nevertheless, we hypothesize that the short larval dispersal and ecological preference for enclosed habitats of *C. intestinalis* act as barriers to natural patterns of gene flow.

Global Fst values, ranging from 0.0353 to 0.0543, are dramatically lower than the average values (0.2354) recorded for populations of the same areas, suggesting that gene flow among populations at a large geographical distance is more active than gene flow among close-by localities (Zhan et al. 2010). These results are in agreement with preliminary observations showing high levels of migration between continents (Caputi et al. personal communication).

How can we explain the lower gene flow at regional scale, compared to global scale, coupled with the absence of genetic structure and low theoretical dispersal of

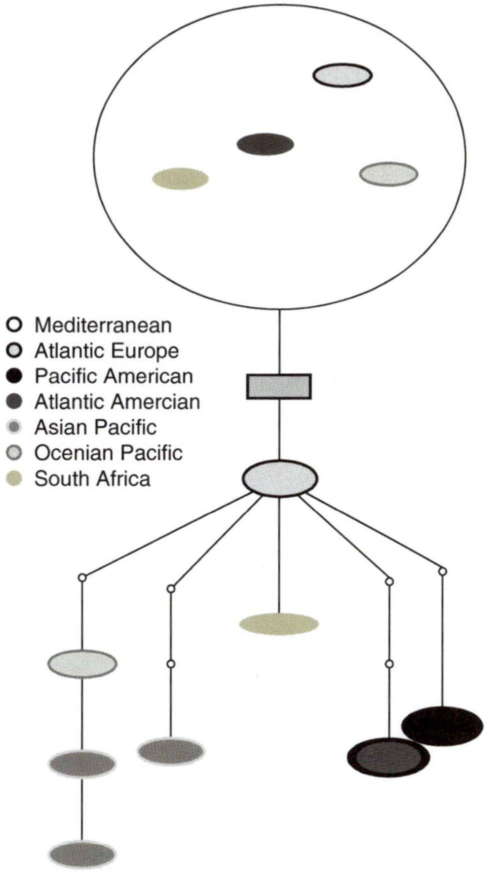

Fig. 6.4 Nested Clade Analysis of *C. intestinalis* spA populations performed on the COI statistical parsimony cladogram. Modified from Caputi et al. personal communication

larvae? In our opinion there are two possible reasons. The most obvious explanation is that the geographic structure of *C. intestinalis* spA does not reflect only natural dispersal patterns but results also from propagation vectors of anthropogenic origin. The second is that local populations can be adapted to specific environmental conditions and isolated from local gene flow, due to their persistence in enclosed environments such as lagoons and marinas.

Larvae can be entrapped in water currents, which may offer an effective method of long-distance dispersal in "open" environments lacking of physical barriers (Palumbi 1992). By the way, this dispersal method appears to have a limited range and is not expected to be sufficient to homogenize genetic variation at an intercontinental scale (Dupont et al. 2010). Generally, human-mediated vectors such as ballast water exchange, vessel hull fouling, and aquaculture trading have been considered responsible for such a genetic pattern and the recent range expansion of *C. intestinalis* (Carver et al. 2003; Lambert 2007). Looking at the isolation of populations living in enclosed environments, no significant differentiation was detected between open-shore and lagoon sites in the Gulf of Naples (Tyrrhenian

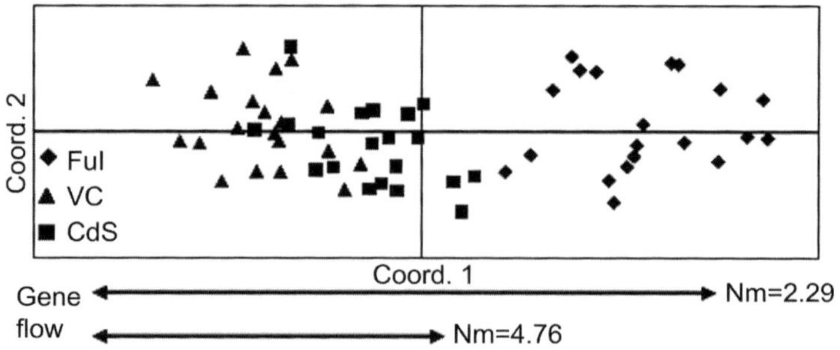

Fig. 6.5 Principal coordinate analysis showing significant differentiation between isolated (Ful) and open-sea (VC/CdS) populations. (Modified from Sordino et al. 2008)

Sea, Italy). The only exception is represented by the innermost population of the Fusaro lagoon (Sordino et al. 2008), which shows genotypic proportions deviating significantly from HWE expectations at several loci. This population occupies specific microhabitats and it oscillates between periodical extinction events from the upper layers and subsequent recolonization with the likely contribution of deeper refugees (Riisgard et al. 1998). These environmental constraints coupled with nonrandom mating, or eventually selfing, offer the reproductive assurance of selected mating combinations and an increased probability that locally adapted genotypes are able to persist (Fields and Johnston 2005). In general, we can assume that local adaptation and physical barriers experienced by populations living in confined environments can add an important variable in the interpretation of neutral patterns of gene flow in *C. intestinalis* (Fig. 6.5).

6.4 Naturally Occurring Mutations

One of the oldest and unsolved problems in evolutionary biology is which mutations generate evolutionarily relevant phenotypic variation. To understand the genetic basis of evolutionary change in nature, we need to address questions about the relative importance of coding *vs.* regulatory DNA in morphological variation, the origin of phenotypic traits by either natural selection or genetic drift, and how the ecology mediates such processes. Because the genetic variation responsible for phenotypic variation is but a subset of the genetic variation in general, it is essential to understand the evolutionary processes underlying this variation if we are to identify genes that cause morphological transitions. Furthermore, it is clear that allelic variation affects responses to key environmental stressors.

Several traits make ascidians suitable for studying the role of developmental genes in natural variation. Transparent and externally fertilized embryos develop by bilaterally symmetrical cleavage, according to a determined pattern, with a defined and well-characterized cell lineage and segregation of developmental fate of cells until the end of gastrulation. Larvae are valuable for understanding the elements of structure, developmental processes, and possibly also the evolution of the vertebrate body plan, including organogenesis of the chordate nervous system.

Natural variation and neutral population genetics can be used for the characterization of phylogenetically relevant mutations in *C. intestinalis*, under the assumption that the accumulation and distribution of mutant phenotypic classes is mostly influenced by historical and environmental factors (i.e., effective population size, genetic variability, reproductive strategies, geographical barriers, environmental harshness). It has been previously shown that up to 20% individuals from natural populations are heterozygote carriers of developmental mutations that segregate in a way typical of recessive alleles (Hendrickson et al. 2004; Sordino et al. 2008). The percentage of heterozygote carriers varies from 13.4% to 19.5% among populations, as well as the relative contribution of lethal and nonlethal mutant phenotypes that may be sharply different (20–60% and 40–80%, respectively). It has been argued that these phenotypes contribute to local adaptation but do not represent potentially independent evolutionary lineages, as suggested by the low morphological complexity of the species, which is adapted to an ecological scenario featuring low intraspecific competition and stressful environmental conditions (Sordino et al. 2008).

Changes in terms of presence, position, size, and structure of the larval brain sensory organs have occurred repeatedly in ascidian phylogeny (Jeffery 2004). In *C. intestinalis*, mutations that affect sensory organ phenotype are of special interest to uncover the changes in regulatory gene networks that underlie morphological diversity. Phylomimicking mutations, as they are called, have been also termed "hopeful monsters" by Richard Goldschmidt, who proposed that they were generated through alterations of development (Goldschmidt 1940). Which are the molecular networks controlling organ specification and how morphogenetic structures evolved in the chordate lineage? Which seemingly different mechanisms drove the remodelling of the larval body plan in ascidians, as seen in the independent loss of tail and brain sensory organs in Molgulid and Clavelinid species? Mutations that affect development of the *Ciona* embryo in a similar way might have played analogous roles during evolution, but the degree to which these mutations are identical to those upon which ascidian evolution depends is unknown. The molecular identity of *Ciona* mutation can be disclosed by positional cloning via bulked segregant analysis (BSA) with AFLP reactions and the support of genetic and physical linkage maps and the *C. intestinalis* genomic sequence (Dehal et al. 2002; Kano et al. 2006; Kano 2007; Veeman et al. 2008). Then, the orthologue of the genes mutated in *Ciona* can be isolated from specific ascidian lineages, and compared at the levels of sequence, expression, function, and regulation in order to address their implication in phenotypic radiation. At the population level, the molecular identity of specific mutations that underlie evolutionary shift can be correlated with the performance in natural environments, e.g., fitness of mutated sensory organs compared with the wild type, and with the rules of population genetics that allow their repeated fixation under conditions of natural selection.

The combination of marker-assisted analysis of population structure and screening of phenotypic classes is central not only in shaping models in developmental biology but also in evolutionary genetics of populations. The comparison between frequencies of natural variation and population genetic patterning represents a unique opportunity to quantify the importance of biogeographic and anthropogenic factors in affecting mutation–selection dynamics (Sordino et al. 2008).

6.5 Ciona Intestinalis as a Model for the Ecology and Genetics of Adaptation

Assessing the dynamics and timescales of phenotypic and genetic polymorphism in different environments requires the study of the dynamics of natural populations through extensive field work. *C. intestinalis* represents an ideal species for studies in the wild (see Sect. 6.2). Here we suggest that *Ciona* species can become a model organism for deciphering the evolutionary potential of invasive species in the newly colonized habitats and, more in general, for understanding the microevolution of natural populations in response to changing environmental conditions. This is particularly important if we consider the continuous and increasing impact of human activities on natural environments. One of the main challenges in evolutionary biology is the understanding of evolutionary processes in natural populations and their relationship with environmental conditions and environmental quality. In a recent review, Charmantier and Garant (2005) discuss three important points, which should be taken into consideration for evaluating the evolutionary potential of a species in the natural environment. First, unfavorable environmental conditions determine lower rate of evolution although the rapid increasing of human disturbance on natural environments makes it difficult to derive general trends for all the species. Second, different genotypes can have different levels of interaction with the environment, which reflects on a different evolutionary potential in stressful conditions. Third, authors stress that environments can be variable in space and time, making that the evolutionary potential of a single population can vary when environmental conditions change. In this case, the potential for microevolution is constrained by either a lack of heritable variation in unfavorable environments, or by a reduced strength of selection in favorable environments (Wilson et al. 2006).

In species which are displaced from their natural habitat and move to colonize new habitats, the change of environmental conditions to which they must adapt can be orders of magnitude faster than what would occur in the same natural habitat. Successful invasive species show to possess high evolutionary potential, which is theoretically proportional to the amount of additive genetic variation present (Fisher 1930). Nevertheless, very little is known about the role of genetic diversity in process of adaptation to new environments.

C. intestinalis seems to possess very high genetic polymorphism (see Sect. 6.3), which could be at the basis of its invasive success, in particular in some areas (Ramsay et al. 2008). Also, the existence of a significant number of natural mutants seems to confirm the high evolutionary potential of the species. Nevertheless, the life cycle of the species, the ecological requirements which seem to favor its life in isolated and enclosed environments, with low water quality, make the scenario more complicated. The deficit in heterozygosis encountered in most of the natural populations analyzed so far (Sordino et al. 2008; Zhan et al. 2010) seems to counteract the positive effect toward rapid evolution provided by the high genetic polymorphism.

Until now, no studies have specifically focused on the relationship between genetic polymorphism and adaptive potential in *C. intestinalis*, taking into consideration the life cycle of the species and its ecological requirements. Recently, we started a multidisciplinary research program which aims to study population ecology of the adult and larval stages, population genetics of natural populations, and evolutionary significance of natural mutants, in relation to the biotic and abiotic component of a coastal lagoon. The available molecular tools existing for this species (see Sect. 6.1) makes it an excellent model for this type of integrative studies. Actually, being *C. savignyi* and the two cryptic species *C. intestinalis* spA and spB separated by comparable genetic distance (13–15%), they altogether offer a unique experimental model for approaching levels of evolutionary divergence in developmental programs among sister taxa. Our research aims at gaining insights in population structure and diversity, to relate with temporal and spatial population dynamics; to correlate population genetic structure and the occurrence of abnormal phenotypes as an important focus for understanding selective forces that shape natural finite populations, and to gain insights into the embryological and evolutionary mechanisms that generate animal diversity. We also hope to give insights in the understanding of the evolutionary potential of invasive species in general. Whether a new mutation can contribute to morphological and, ultimately, taxonomic radiation, entails the possibility that the frequency of this genetic change increases in the population, leading progressively to reproductive isolation and fixation of the change. As future perspective, we are considering the possibility to apply a genome scan approach using gene-linked polymorphic markers in order to identify genes under selection to be related to both the presence of mutants and the population genetic patterns observed.

References

Andreakis N, Caputi L, Sordino P (2007) Characterization of highly polymorphic nuclear microsatellite loci from the ascidian *Ciona intestinalis*. Mol Ecol Notes 7:610–612

Ben-Shlomo R, Paz G, Rinkevich B (2006) Postglacialperiod and recent invasions shape the population genetics of botryllid ascidians along european atlantic coasts. Ecosystems 9:1118–1127

Boffelli D, Weer CV, Weng L, Lewis KD, Shoukry MI, Pachter L, Keys DN, Rubin EM (2004) Intraspecies sequence comparisons for annotating genomes. Genome Res 14:2406–2411

Caputi L, Andreakis N, Mastrototaro F, Cirino P, Vassillo SP, Sordino P (2007) Cryptic speciation in a model invertebrate chordate. Proc Natl Acad Sci USA 104:9364–9369

Caputi L, Andreakis N, Affinito O, Vassillo M, Procaccini G, Sordino P (submitted) Recent expansion and global divergence of *Ciona intestinalis* sp. A, a strong marine competitor

Caputi L, Borra M, Andreakis N, Biffali E, Sordino P (2008) SNPs and Hox gene mapping in *Ciona intestinalis*. BMC Genomics 9:39–50

Carman MR, Bullard SG, Donnelly JP (2007) Water quality, nitrogen pollution, and ascidian diversity in coastal waters of southern Massachusetts, USA. J Exp Mar Biol Ecol 342:175–178

Carver CE, Chisholm A, Mallet AL (2003) Strategies to mitigate the impact of *Ciona intestinalis* (L.) biofouling on shellfish production. J Shellfish Res 22:621–631

Charmantier A, Garant D (2005) Environmental quality and evolutionary potential: lessons from wild populations. Proc R Soc Lond B 272:1415–1425

Cirino P, Toscano A, Caramiello D et al (2002) Laboratory culture of the ascidian *Ciona intestinalis* (L.): a model system for molecular developmental biology research. Mar Mod Elec Rec. http://www.mbl.edu/html/BB/MMER/CIR/CirCon.html

Darling JA, Folino-Rorem NC (2009) Genetic analysis across different spatial scales reveals multiple dispersal mechanisms for the invasive hydrozoan *Cordylophora* in the Great Lakes. Mol Ecol 18:4827–4840

Dehal P, Satou Y, Campbell RK et al (2002) The draft genome of *Ciona intestinalis*: insights into chordate and vertebrate origins. Science 298:2157–2167

Dupont L, Viard F, Dowell MJ, Wood C, Bishop JDD (2009) Fine- and regional-scale genetic structure of the exotic ascidian *Styela clava* (Tunicata) in southwest England, 50 years after its introduction. Mol Ecol 18:442–453

Dupont L, Viard F, Davis MH, Nishikawa T, Bishop JDD (2010) Pathways of spread of the introduced ascidian *Styela clava* (Tunicata) in Northern Europe, as revealed by microsatellite markers. Biol Invasions 12:2707–2721

Dybern BI (1965) The life cycle of *Ciona intestinalis* (L.) f. *typica* in relation to the environmental temperature. Oikos 16:109–131

Everett RA (2000) Patterns and pathways of biological invasions. Trends Ecol Evol 15:177–178

Fields S, Johnston M (2005) Whither model organism research? Science 307:1885–1886

Fisher RA (1930) The genetical theory of natural selection. Oxford University Press, Oxford

Goldschmidt R (1940) The material basis of evolution. Yale University Press, New Haven

Goldstien SJ, Schiel DR, Gemmell NJ (2010) Regional connectivity and coastal expansion: differentiating pre-border and post-border vectors for the invasive tunicate *Styela clava*. Mol Ecol 19:874–885

Havenhand JN, Svane I (1991) Roles of hydrodynamics and larval behaviour in determining spatial aggregation in the tunicate *Ciona intestinalis*. Mar Ecol Prog Ser 68:271–276

Hedgecock D, Li G, Hubert S, Bucklin K, Ribes V (2004) Widespread null alleles and poor cross-species amplification of microsatellite DNA loci cloned from the Pacific oyster, *Crassostrea gigas*. J Shellfish Res 23:379–385

Hendrickson C, Christiaen L, Deschet K et al (2004) Culture of adult ascidians and ascidian genetics. Methods Cell Biol 74:143–170

Holland BS (2000) Genetics of marine bioinvasions. Hydrobiologia 420:63–71

Hoshino Z, Nishikawa T (1985) Taxonomic studies of *Ciona intestinalis* (L.) and its allies. Publ Seto Mar Biol Lab 30:61–79

Howes S, Herbinger CM, Darnell P, Vercaemer B (2007) Spatial and temporal patterns of recruitment of the tunicate *Ciona intestinalis* on a mussel farm in Nova Scotia, Canada. J Exp Mar Biol Ecol 342:85–92

Jeffery WR (2004) Evolution and development of brain sensory organs in molgulid ascidians. Evol Dev 6:170–179

Jesse R, Pfenninger M, Fratini S et al (2009) Disjunct distribution of the Mediterranean freshwater crab *Potamon fluviatile* - Natural expansion or human introduction? Biol Invasions 11: 2209–2221

Joly JS, Kano S, Matsuoka T, Auger H, Hirayama K et al (2007) Culture of Ciona intestinalis in closed systems. Dev Dyn 236:1832–1840

Kano S (2007) Initial stage of genetic mapping in *Ciona intestinalis*. Dev Dyn 236:1768–1781

Kano S, Chiba S, Satoh N (2001) Genetic relatedness and variability in inbred and wild populations of the solitary ascidian *Ciona intestinalis* revealed by arbitrarily primed polymerase chain reaction. Mar Biotechnol 3:58–67

Kano S, Satoh N, Sordino P (2006) Primary genetic linkage maps of the ascidian, *Ciona intestinalis*. Zoolog Sci 23:31–39

Kawamura K, Fujita H, Nakauchi M (1987) Cytological characterization of self incompatibility in gametes of the ascidian, *Ciona intestinalis*. Dev Growth Diff 29:627–642

Lambert G (2001) A global overview of ascidian introductions and their possible impact on the endemic fauna. In: Sawada H, Yokosawa H, Lambert CC (eds) The biology of ascidians. Springer, Tokyo

Lambert G (2007) Invasive sea squirts: a growing global problem. J Exp Mar Biol Ecol 342:3–4

Lambert CC, Lambert G (1998) Non indigenous ascidians in southern California harbors and marinas. Mar Biol 130:675–688

Lee CE (2002) Evolutionary genetics of invasive species. Trends Ecol Evol 17:386–391

Marin MG, Bressan M, Beghi L, Brunetti R (1987) Thermo-aline tolerance of *Ciona intestinalis* (L., 1767) at different developmental stages. Cah Biol Mar 28:47–57

Marshall DJ, Keough MJ (2003) Effects of settler and density on early post-settlement survival of *Ciona intestinalis* in the field. Mar Ecol Prog Ser 259:139–144

Nei M, Maruyama T, Chakraborty R (1975) Bottleneck effect and genetic-variability in populations. Evolution 29:1–10

Nydam ML, Harrison R (2007) Genealogical relationships within and among shallow-water *Ciona* species (Ascidiacea). Mar Biol 151:1839–1847

Nydam ML, Harrison RG (2010) Polymorphism and divergence within the ascidian genus *Ciona*. Mol Phylogenet Evol 56:718–726

Nydam ML, Harrison RG (2011a) Introgression despite substantial divergence in a broadcast spawning marine invertebrate. Evolution 65:429–442

Nydam ML, Harrison RG (2011b) Reproductive protein evolution in two cryptic species of marine chordate. BMC Evol Biol 11:18

Oliveira Marins de F, Silva Oliveira da C, Viera Maciel NM, Skinner LF (2009) Reinclusion of *Ciona intestinalis* (Ascidiacea: Cionidae) in Brazil –a methodological view. JMBA2 – Biodivers Rec. 1–5

Palumbi SR (1992) Marine speciation on a small planet. Trends Ecol Evol 7:114–118

Pérès JM (1952) Recherches sur le cycle sexuel de «*Ciona intestinalis* (L.)». Arch Anat Microsc Morphol Exp 41:153–183

Petersen JK (2007) Ascidian suspension feeding. J Exp Mar Biol Ecol 342:127–137

Petersen JK, Svane I (1995) Larval dispersal in the ascidian *Ciona intestinalis* (L.). Evidence for a closed population. J Exp Mar Biol Ecol 186:89–102

Petersen JK, Schou O, Thor P (1995) Growth and energetics in the ascidian *Ciona intestinalis*. Mar Ecol Prog Ser 120:175–184

Pimentel D, Lach L, Zuniga R, Morrison D (2000) Environmental and economic costs of non-indigenous species in the United States. Bioscience 50:53–65

Procaccini G, Pischetola M, Di Lauro R (2000) Isolation and characterization of microsatellite loci in the ascidian *Ciona intestinalis* (L.). Mol Ecol 9:1924–1926

Ramsay A, Davidson J, Landry T, Arsenault G (2008) Process of invasiveness among exotic tunicates in Prince Edward Island, Canada. Biol Invasions 10:1311–1316

Ramsay A, Davidson J, Bourque D, Stryhn H (2009) Recruitment patterns and population development of the invasive ascidian *Ciona intestinalis* in Prince Edward Island, Canada. Aquat Invasions 4:169–176

Riisgard HU, Jensen AS, Jürgensen C (1998) Hydrography, near-bottom currents and grazing impact of the filter-feeding ascidian *Ciona intestinalis* in a Danish fjord. Ophelia 49:1–16

Rosati F, Santis RD (1978) Studies on fertilization in the ascidians. I. Self-sterility and specific recognition between gametes of *Ciona intestinalis*. Exp Cell Res 112:111–119

Sabbadin A (1958) Il ciclo biologico di *Ciona intestinalis* (L.), *Molgula manhattensis* (De Kay) e *Styela plicata* (Lesuer) nella laguna veneta. Arch Oceanogr Limnol 11:1–28

Sasakura Y, Oogai Y, Matsuoka T, Satoh N, Awazu S (2007) Transposon mediated transgenesis in a marine invertebrate chordate: *Ciona intestinalis*. Genome Biol 8(Suppl 1):S3

Satoh N (1994) Developmental biology of ascidians. Cambridge University Press, New York

Schmidt GH (1983) The hydroid *Tubularia larynx* causing "bloom" of the ascidians *Ciona intestinalis* and *Ascidiella aspersa*. Mar Ecol Prog Ser 12:103–105

Schmidtke J, Engel W (1980) Gene diversity in tunicate populations. Biochem Genet 18:503–508

Small KS, Brudno M, Hill MM, Sidow A (2007a) A haplome alignment and reference sequence of the highly polymorphic *Ciona savignyi* genome. Genome Biol 8:R41

Small KS, Brudno M, Hill MM, Sidow A (2007b) Extreme genomic variation in a natural population. Proc Natl Acad Sci USA 104:5698–5703

Sordino P, Andreakis N, Brown ER et al (2008) Natural variation of model mutant phenotypes in *Ciona intestinalis*. PLoS One 3:e2344

Sorte CJ, Williams SL, Carlton JT (2010) Marine range shifts and species introductions: comparative spread rates and community impacts. Global Ecol Biogeogr 19:303–316

Stinchcombe JR, Hoekstra HE (2007) Combining population genomics and quantitative genetics: finding the genes underlying ecologically important traits. Heredity 100:158–170

Suzuki M, Nishikawa T, Bird A (2005) Genomic approaches reveal unexpected genetic divergence within *Ciona intestinalis*. J Mol Evol 61:627–635

Svane I, Havenhand JN (1993) Spawning and dispersal in *Ciona intestinalis* (L.). P.S.Z.N.I. Mar Ecol 14:53–66

Therriault TW, Herborg L (2008a) A qualitative biological risk assessment for vase tunicate *Ciona intestinalis* in Canadian waters: using expert knowledge. ICES J Mar Sci 65:781–787

Therriault TW, Herborg L (2008b) Predicting the potential distribution of the vase tunicate *Ciona intestinalis* in Canadian waters: informing a risk assessment. ICES J Mar Sci 65:788–794

Veeman MT, Nakatani Y, Hendrickson C, Ericson V, Lin C, Smith WC (2008) *chongmague* reveals an essential role for laminin-mediated boundary formation in chordate convergence and extension movements. Development 135:33–41

Whitlatch RB, Osman RW, Frese A (1995) The ecology of two introduced marine ascidians and their effects of epifaunal organisms in Long Island Sound. In: Balcom N (ed) Proceedings of the Northeast Conference on Non-Indigenous Aquatic Nuisance Species: Reg Conf, pp 29–48

Wilson AJ, Pemberton JM, Pilkington JG et al (2006) Environmental coupling of selection and heritability limits evolution. PLoS Biol 4:e216

Zhan A, Bao Z, Hui M et al (2007) Inheritance pattern of EST-SSRs in self-fertilized larvae of the bay scallop *Argopecten irradians*. Ann Zoolog Fennici 44:259–268

Zhan A, Macisaac HJ, Cristescu ME (2010) Invasion genetics of the *Ciona intestinalis* species complex: from regional endemism to global homogeneity. Mol Ecol 19:4678–4694

Chapter 7
Rapid Evolution of Simple Microbial Communities in the Laboratory

Margie Kinnersley, Jared W. Wenger, Gavin Sherlock, and Frank R. Rosenzweig

Abstract Classical models predict that asexual populations evolve in simple unstructured environments by clonal replacement, yet laboratory evolutionary studies have uncovered persistent polymorphism, driven either by frequency-dependent selection or mutualistic interactions. We have studied the evolution of microbes in simple unstructured environments as a way to illuminate the evolution of biodiversity. We sought to understand how complexity arises in an *Escherichia coli* population founded by a single clone and propagated under glucose limitation for >770 generations. When coevolved clones are cultured separately, their transcriptional profiles differ from their common ancestor in ways that are consistent with our understanding of how *E. coli* adapts to glucose limitation. A majority of the 180 differentially expressed genes shared between coevolved clones is controlled by the global regulators RpoS, Crp, and CpxR. Clone-specific expression differences include upregulation of genes whose products scavenge overflow metabolites such as acetate, enabling cross-feeding. Unexpectedly, we find that when coevolved clones are cultured together, the community expression profile more closely resembles that of minority clones cultured in isolation rather than that of the majority clone cultured in isolation. We attribute this to habitat modification and regulatory feedback arising from consumption of overflow metabolites by niche specialists. Targeted and whole-genome sequencing reveal *acs*, *glpR*, and *rpoS* mutations in the founder that likely predispose evolution of niche specialists. Several mutations bringing about specialization are compensatory rather than gain-of-function. Biocomplexity can therefore arise on a single limiting resource if consumption of that resource results in the creation of others that are differentially

M. Kinnersley • F.R. Rosenzweig
Division of Biological Sciences, University of Montana, 32 Campus Dr., Missoula, MT 59812, USA
e-mail: Frank.Rosenzweig@mso.umt.edu

J.W. Wenger • G. Sherlock
Department of Genetics, Stanford University School of Medicine, Stanford, CA 94305, USA

P. Pontarotti (ed.), *Evolutionary Biology – Concepts, Biodiversity, Macroevolution and Genome Evolution*, DOI 10.1007/978-3-642-20763-1_7,

accessible to adaptive mutants. These observations highlight the interplay of founder genotype, biotic environment, regulatory mutations, and compensatory changes in the adaptive evolution of asexual species and somatic cells.

7.1 Enacting Evolutionary Plays in the Laboratory

Biologists have long sought to understand mechanistically how adaptive genetic variation arises and persists. Experimental studies using model organisms such as *Drosophila* (Dobzhansky and Wright 1947; Dobzhansky and Spassky 1947; Wright and Dobzhansky 1946) and *C. elegans* (Estes et al. 2004; Estes and Lynch 2003; Denver et al. 2005) transformed the search for such mechanisms from a retrospective to a prospective endeavor. However, long generation times, sexual recombination, and logistical constraints on lab population size make multicellular eukaryotes imperfectly suited to study the tempo, trajectory and mechanisms by which evolution occurs in asexual species and in the somatic cells of sexual organisms. In such cells, new genetic variation is limited by the rate of mutation supply and, in bacteria, also by the incidence of horizontal gene transfer. Fortunately, evolution in asexual species and somatic cells can be studied using microbial models, such as the yeast *Saccharomyces cerevisiae* and the bacterium, *Escherichia coli* (Zeyl 2006; Rozen and Lenski 2000). Early microbial studies led to two generalizations concerning the emergence and persistence of genetic variation in large, asexual populations. First, over *ecological time* and in the absence of spatial structure and differential predation, competition for the same limiting resource selects for one fittest variant, an insight that came to be known as the "competitive exclusion principle" (Gause 1934; Hardin 1960). Second, over *evolutionary time* variation arising by mutation is subject to "periodic selection," leading to a succession of genotypes each more fit than its immediate predecessor ("clonal replacement") (Muller 1932; Atwood et al. 1951; Novick and Szilard 1950). These generalizations led to the expectation that large, clonal populations evolving under resource limitation should exhibit limited genetic variation.

Classical models of asexual evolution posit that in simple environments, evolving asexual populations' complexity should be transient and limited in scope. Experimental evidence now suggests otherwise. Multiple genotypes that arise from a single founder clone can coexist over evolutionary time; in other words, out of one comes many (*e unum pluribus*). This phenomenon has been documented in spatially and temporally unstructured chemostats (Helling et al. 1987; Rosenzweig et al. 1994), in temporally structured batch cultures (Spencer et al. 2007; Friesen et al. 2004; Le Gac et al. 2008; Turner et al. 1996; Rozen and Lenski 2000), and in spatially structured microcosms (Rainey and Travisano 1998). In each setting, the emergence and persistence of polymorphism in the absence of sexual recombination seems to require that cohabitants exploit alternative ecological opportunities (i.e., unoccupied niche space), and/or accept trade-offs between being a specialist and a generalist (as reviewed in Rainey et al. 2000), also see Zhong et al. 2004). Alternatively, new adaptive genotypes may arise at such a rapid rate that there is always clonal interference (see below). In serial dilution batch culture, multiple growth parameters may be selected upon (reviewed in Rosenzweig and Sherlock 2009). Clones may arise that differ with respect to lag time, maximum specific growth rate, or enhanced capacity to survive and/or reproduce as cultures transition into stationary phase. Periodic changes in population density and nutrient levels may bring balancing selection to bear on these different phenotypes, especially if antagonistic pleiotropy precludes evolution of one fittest genotype incorporating all of these advantageous traits. In spatially structured environments, selection may favor mutants better adapted to particular regions or better able to colonize microhabitats formed at the boundaries between such regions. In continuous nutrient-limited environments (e.g., chemostats), theory (Monod 1942; Kubitschek 1970) predicts that selection will favor clones better able to scavenge the limiting resource or more efficiently convert that resource to progeny. Ultimately, the outcome of the "evolutionary play" in any of these "ecological theaters" will depend on founder genotype, the complexity of genetic pathways that lead to different adaptive strategies, as well as the propensity of key steps along those pathways to undergo mutation and to act pleiotropically.

7.2 Alternative Dénouements to the Evolutionary Play

Increasing evidence points to the possibility of not one, but three possible evolutionary outcomes when asexual microbes evolve in simple environments: *clonal replacement*, as described above, *clonal interference*, where fixation of a single fittest clone is deferred because independent beneficial mutations arise in multiple, independent clones that are in competition with one another (Gerrish and Lenski 1998; de Visser and Rozen 2006; Kao and Sherlock 2008), and what we propose to call *clonal reinforcement,* where the presence of one genotype actually favors the emergence and persistence of other genotypes by virtue of cooperative (mutualistic) interactions, as described below. The last two outcomes, clonal interference and clonal reinforcement, involve the coexistence of multiple genotypes in the same environment, i.e., the existence of stable polymorphisms. Because mutualism is recognized to be a pervasive force in evolution (Thompson 2005), and because mutualism has been at least as important as competition in bringing about biological innovation (Boucher 1985), we are experimentally investigating the genetic and environmental factors that determine how population genetic complexity arises and persists as microbes evolve in the laboratory.

7.3 *E unibus plurum*: Genetic Bases for the Emergence of Stable Polymorphism in Clonal Populations

Only recently have we begun to discover the mechanisms by which complexity in the form of balanced polymorphisms originate in large, asexual populations, in effect how out of one comes many (*e unibus plurum*). In serial batch culture, differences in the activity of the global regulator RpoS help explain coexistence of two *E. coli* isolates with different propensities to survive extended stationary phase (Gerrish and Lenski 1998), although the precise genetic basis for these differences remains obscure. In a spatially structured microcosm founded by a single clone of *Pseudomonas fluorescens*, a methylesterase structural mutant arose and persisted because the resulting change in exopolysaccharide production enabled the mutant to colonize the air–broth interface (Rainey et al. 2000; Bantinaki et al. 2007). Finally, in glucose-limited chemostats, polymorphic *E. coli* populations repeatedly evolved, in part owing to local regulatory mutations that alter expression of a single operon (*acs-actP-yjcH*) (Treves et al. 1998). When adaptive clones from one such population were grown in monoculture, strain-specific differences in expression of ca. 20% of identifiable proteins suggested the presence of other mutations with highly pleiotropic effects (Kurlandzka et al. 1991). Thus, regardless of experimental system, uncertainty remains as to whether either regulatory or structural mutations consistently deliver greater fitness increments, which category of mutation better explains the maintenance of diversity, and whether one type is more likely to precede the other in an evolutionary sequence leading to balanced polymorphism.

Theoretical considerations have led some to argue that the major phenotypic changes underlying adaptive radiation are more likely due to regulatory than to structural mutations (Carroll 2005; Wray 2007 and references therein), This argument is based on the perception that changes in coding sequences more likely exert large pleiotropic effects than changes in the expression of those sequences, in particular changes that arise from the mutation of *cis*-regulatory elements affecting single genes. In effect, this type of regulatory mutation enables selection more easily to "tinker" (*sensu* Jacob 1977), as it provides a mechanism to alter functionality in one process while still preserving the role of pleiotropic genes in others (Rozen et al. 2009). Also, and this fact too often goes unappreciated, a discrete *cis*-regulatory mutation preserves the capacity to restore the ancestral pattern of expression via compensatory or back mutations. The proposition that regulatory mutations play a greater role in adaptive diversification has been criticized on empirical and theoretical grounds by Hoekstra and Coyne who point out the vastly greater number of examples where adaptation is attributable to structural rather regulatory mutations, as well as the facts that *cis*-acting elements offer much smaller targets for mutation than ORFs, and that in many species, pleiotropic effects arising from structural mutations may be buffered by gene duplication (Hoekstra and Coyne 2007).

7.4 *Saccharomyces cerevisiae* as Dramatis Personae: Complexity in a Simple, Constant Environment Can Arise from Clonal Interference

Kao and Sherlock (2008) investigated population dynamics in another microbial model, *Saccharomyces cerevisiae*, and sought to determine whether clonal replacement or clonal interference occurred as cells evolved under glucose limitation.

They evolved eight populations of yeast in glucose-limited chemostats for 440 generations, beginning each experiment with population sizes between 10^8 and 10^9. Populations consisted of equal parts of otherwise isogenic yeast expressing 1 of 3 fluorescent proteins, red, yellow and green; the abundance of each of these subpopulations during the experimental evolution was monitored using FACS. In each population, there was clear and unequivocal evidence of clonal interference, with subpopulations expanding and contracting, as presumably adaptive clones carrying one or another of the fluorescent markers expanded during the evolution. An example population, which was characterized further, is shown in Fig. 7.1.

As can be seen in Fig. 7.1, not only were they able to determine that clonal interference was occurring, they could also identify when adaptive clones increased in frequency, based on the expansion of one of the subpopulations. Using live cell FACS, different colored subpopulations were isolated at the generation times indicated, when the subpopulation had reached a maximum. Kao and Sherlock then tested seven clones from those subpopulations in pair-wise competitions with the progenitor strain to identify adaptive clones, and selected clones M1 through

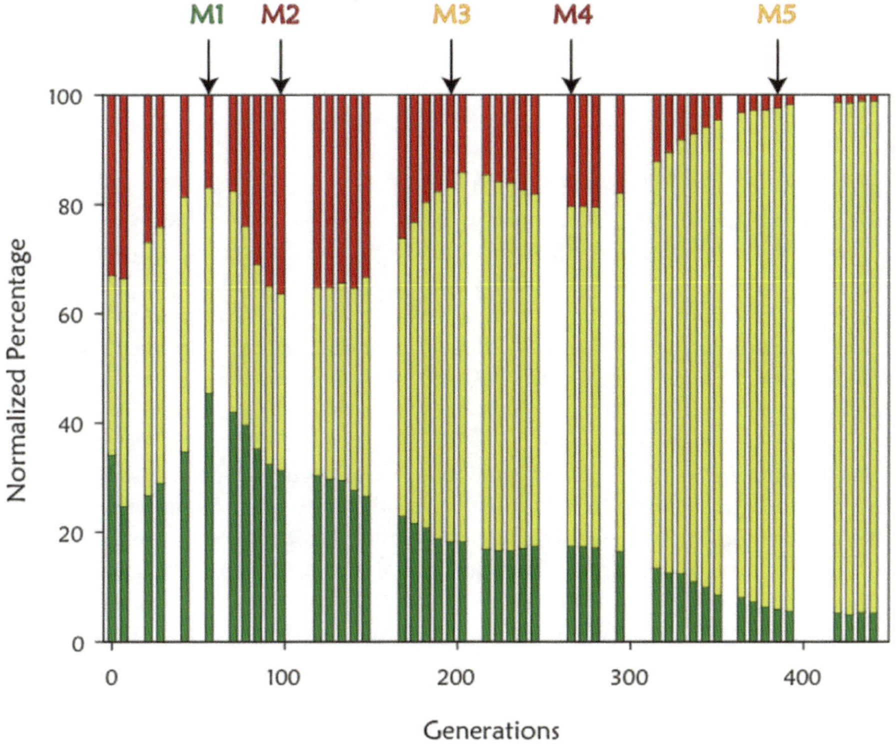

Fig. 7.1 Clonal interference in yeast cultured in a glucose-limited chemostat. Population dynamics of the evolution experiment from which adaptive clones marked M1–M5 were isolated and characterized ($N \approx 10^9$ cells). The times when putative adaptive subpopulations reach their maxima are marked with *arrows* (Kao and Sherlock 2008)

M5 as the most adaptive from each of those subpopulations. High-throughput sequencing of M1 through M5 revealed different mutations in a transcriptional repressor, *MTH1* in M1-M3, and independent amplification of the hexose transporters (*HXT6/7*) in M4 and M5. M3 through M5 also contain additional mutations in the Ras-cAMP signaling pathway, which have subsequently been shown to be adaptive (Kvitek and Sherlock 2011). In addition, there also exists reciprocal sign epistasis between the *MTH1* and the *HXT* mutations, thus these two mutations lead to mutually exclusive paths on the adaptive landscape.

7.5 *Escherichia coli* as Dramatis Personae: Complexity in a Simple, Constant Environment Can Arise from Clonal Reinforcement

Helling et al. (1987) addressed the issue of the relative importance of structural versus regulatory change in a seminal work demonstrating how diversity arises in *E. coli*. Experimental populations were founded using a single clone (JA122), which was then evolved in aerobic, glucose-limited chemostats at a constant dilution rate ($D = 0.2\ h^{-1}$) and constant temperature (30°C). From fluctuations in a neutral marker (phage T5 resistance), they inferred that adaptive mutations occurred about every 50–100 generations. After 765 generations, four strains in one such population could be distinguished on the basis of colony size and ampicillin sensitivity. Three of these phenotypes were shown to stably coexist in reconstruction experiments, wherein the majority clone strain, CV103, was followed in rank order of abundance by CV116 and CV101 (Rosenzweig et al. 1994). Each strain exhibited a characteristic pattern of protein expression, as determined by 2D protein gel electrophoresis, when grown in glucose-limited chemostat monoculture; as a group, evolved clones significantly differed from their common ancestor at ~160 expressed proteins of ~700 that could be resolved (Kurlandzka et al. 1991).

Relative to the common ancestor JA122, all evolved clones demonstrated enhanced uptake of the glucose analogue ^{14}C-α-methylglucoside (αMG), and CV103 accumulated significantly more α-MG than any other clone (Helling et al. 1987), even though its yield and maximum specific growth rate, μ_{max}, were less than that of the other adaptive clones. The equilibrium glucose concentration (the amount detectable in a culture of actively dividing cells at steady state) was an order of magnitude less in CV103 than in CV101 chemostats and less than half that observed for CV116 (Table 7.1). Unlike CV101 and CV116, however, CV103 left excreted metabolizable carbon in the chemostat, effectively creating metabolic niches conducive to the evolution of cross-feeding (Table 7.1). The other strains filled those niches, efficiently scavenging overflow metabolites near to or below detection limit (Rosenzweig et al. 1994). Acetate-scavenging strains were subsequently observed in 6 out of 12 independent evolutionary populations founded by cells of similar genetic background grown under similar conditions (Treves et al. 1998).

Table 7.1 Co-evolved *E. coli* and their common ancestor are phenotypically differentiated

Strain	Characteristics	Specific growth rate (h^{-1})	Relative growth yield	Glucose uptake (μmol αMG/min/g)	Equilibrium [glucose] (nmol/mL)	Equilibrium [acetate] (nmol/mL)
JA122	F^- *thi1, lacY1, araD139, gdh supE44, hss1*; lysogenic for λ; pBR322Δ5	0.44 ± 0.01	1.14 ± 0.02	1.19 ± 0.09	1.84 ± 0.48	194 ± 20
CV101	Derivative of JA122; isolated after 773 generations, AmpR	0.50 ± 0.02	1.11 ± 0.02	1.66 ± 0.06	0.88 ± 0.31	0 ± 0
CV103	As CV101, but independent isolate, which forms small colonies on TA, Amp^R	0.40 ± 0.01	0.81 ± 0.04	2.46 ± 0.16	0.07 ± 0.03	252 ± 70
CV115	Derivative of JA122, isolated after 773 generations, lacks plasmid	0.55 ± 0.02	1.11 ± 0.02	ND	ND	ND
CV116	As CV115 but forms small colonies on TA	0.60 ± 0.01	1.20 ± 0.03	1.61 ± 0.11	0.19 ± 0.05	40 ± 25

Adapted from Kinnersley et al. (2009)

The Helling et al. experiments exemplify how adaptive evolution occurs in the context of niche diversification, in essence, how biodiversity builds upon itself through the cooperative use of a limiting resource. Because cross-feeding interactions evolve repeatedly (Treves et al. 1998) and in diverse ways (even via cannibalism! Rozen et al. 2009), because these interacting populations can be taken apart and reassembled (Rosenzweig et al. 1994), and because reassembled populations can persist for hundreds of generations (Helling et al. 1987), *E. coli* experimental evolution offers an ideal system in which to investigate conditions that favor emergence of metabolic partnerships.

7.6 A Closer Look at How Mutualism Evolves in the Laboratory

As noted, chemostat theory predicts that under continuous nutrient limitation, selection will favor clones that better scavenge the limiting resource or more efficiently convert that resource to progeny (Dykhuizen and Dean 1994). Theory also predicts that under resource limitation, one adaptive clone may create ecological opportunity for another if its metabolic activity produces a substrate or substrates that support growth by other clones. This prediction has been demonstrated empirically: *E. coli* populations founded by a single clone and evolved under glucose limitation repeatedly give rise to interdependent, polyclonal communities supported by cross-feeding of overflow metabolites.

We have recently sought to better understand the mechanistic basis for this phenomenon (Kinnersley et al. 2009). Specifically, we evaluated global gene expression of community members grown both in isolation and as a group under evolutionary conditions. We observed ~180 genes significantly altered in their expression (Fig. 7.2), and attributed many shared increases and decreases to shared mutations in the stationary phase sigma factor, *rpoS* and the maltose operon operator, *mglO*. Expression differences that distinguish isolates occur mainly in the majority clone, CV103 (Fig. 7.3). Many of these genes are either regulated, or are predicted to be regulated by the cAMP receptor protein (CRP) and/or the global stress regulator CpxR. Of particular significance, targeted sequencing uncovered in the founder, JA122 regulatory mutations in *acs* and *glpR* not present in the standard *E. coli* wild-type, K12 MG1655. As the products of these genes are involved in acetate and glycerol catabolism, we hypothesize that these mutations predispose the system to evolve cross-feeding. Among adaptive clones, we discovered shared mutations in *rpoS* and *mglO*, and mutations that distinguish clones from one another at $_{p}$*acs, malT*, and *glpK*. Remarkably, the "community" expression profile is similar to the monoculture profiles of sub-dominant clones, suggesting that biochemical interactions among clones may alter CRP-CpxR regulation (see Fig. 7.4). Mechanistically, we speculate that CV103 experiences feedback inhibition when cultured alone, but not in coculture, because its mutualist partners consume the inhibitory molecules. Altogether, our results suggest that both *cis*- and *trans*-regulatory changes underlie adaptive diversification in a simple, unstructured, resource-limited environment, and that founder genotype and chemical interactions among clones not only facilitate coevolution, but also strongly impact their respective patterns of gene expression.

7.7 Sequencing Reveals Surprising Genetic Complexity in an Evolved Mutualism

Sequencing of 13 candidate loci led to the discovery of the genetic changes described above, as well as a silent substitution in glycerol kinase (*glpK)* in glycerol scavenging clone CV116 (Kinnersley et al. 2009). Relative to the ancestral strain, we observed few mutations in the four evolved clones, virtually all of them in *cis*- or *trans*-acting regulatory sequences. This observation was consistent with prior estimates that no more than eight adaptive mutations had occurred in the experimental population (inferred by scoring periodic selection) (Helling et al. 1987), as well as speculations that these were responsible for dramatic changes in protein (Kurlandzka et al. 1991) and RNA expression (Kinnersley et al. 2009).

To determine whether he had a true picture of genetic diversity in our consortium, we have undertaken whole-genome sequencing. Single-end, 36 bp Illumina sequencing reads (with qualities) were used to identify SNPs resulting from the

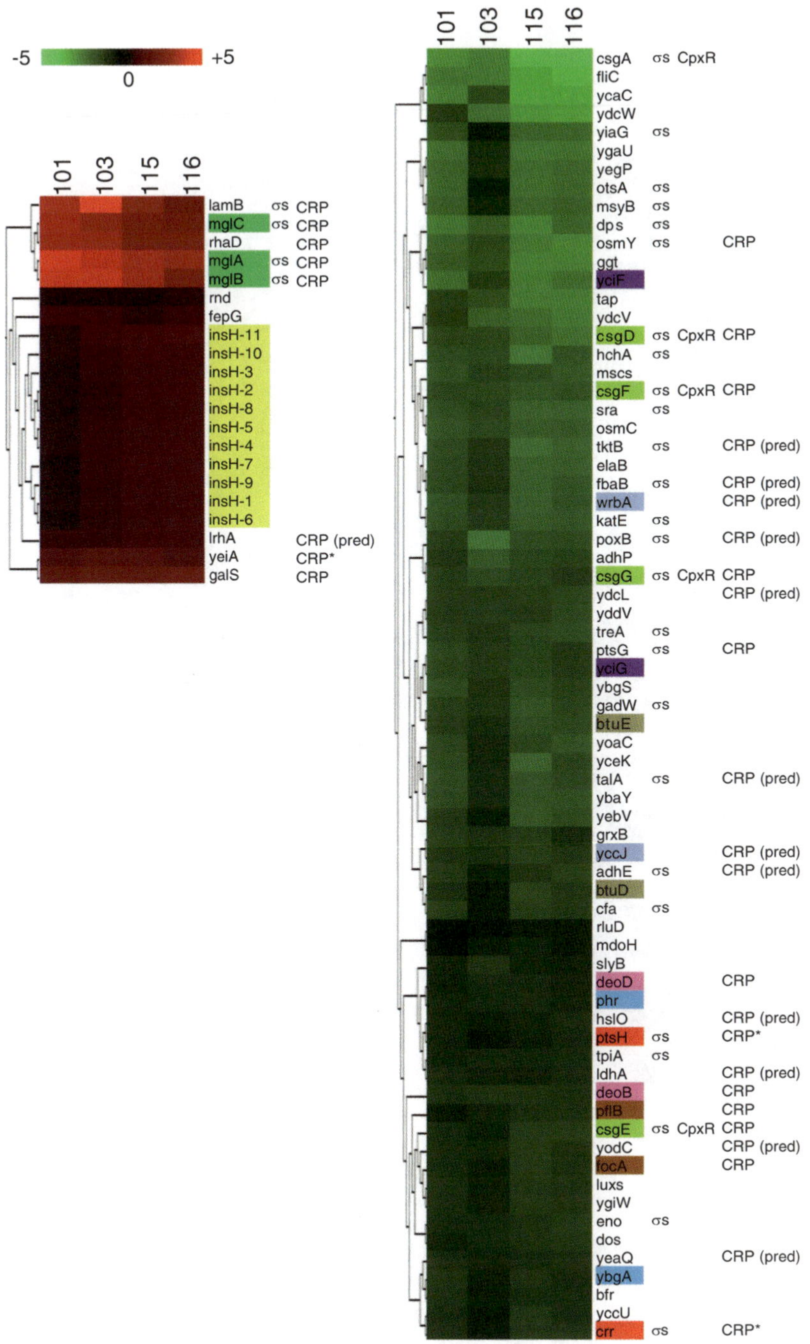

Fig. 7.2 Expression profiling of coevolved *E. coli* clones in chemostat monoculture. Relative to their common ancestor, adaptive clones share many changes in gene expression, most of which are under the control of global regulators RpoS, CpxR, and CRP (2-class SAM, Kinnersley et al. 2009)

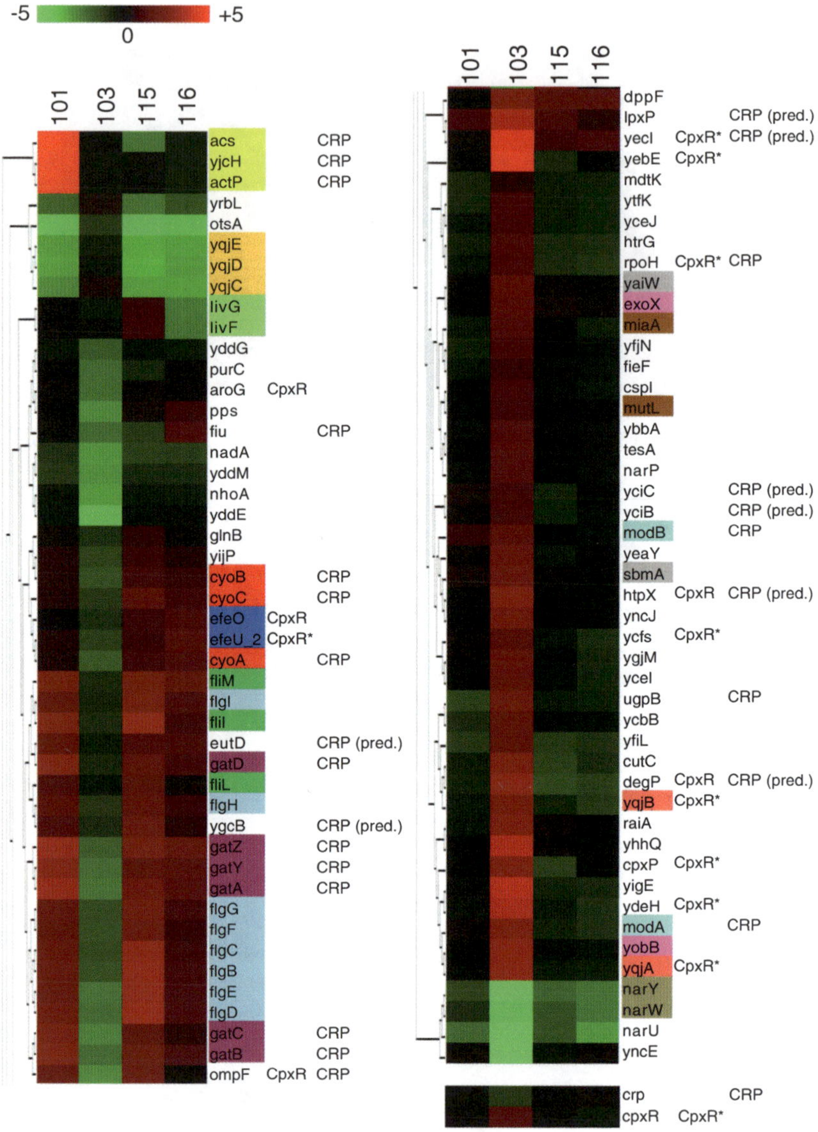

Fig. 7.3 Transcriptional profiling of coevolved *E. coli* clones in isolation reveals that most clone-specific differences occur in clone, CV103 (4-class SAM, Kinnersley et al. 2009). A majority of these differences are in genes under the control of CRP and CpxR, the latter of which can undergo nonenzymatic phosphorylation by acetyl phosphate

evolution. Contrary to expectation, whole-genome sequences of these clones have revealed unexpectedly high levels of polymorphism. In fact, hundreds of single nucleotide polymorphisms distinguish mutualist clones from their common ancestor, and the majority of these SNPs are in coding sequences. Remarkable as this

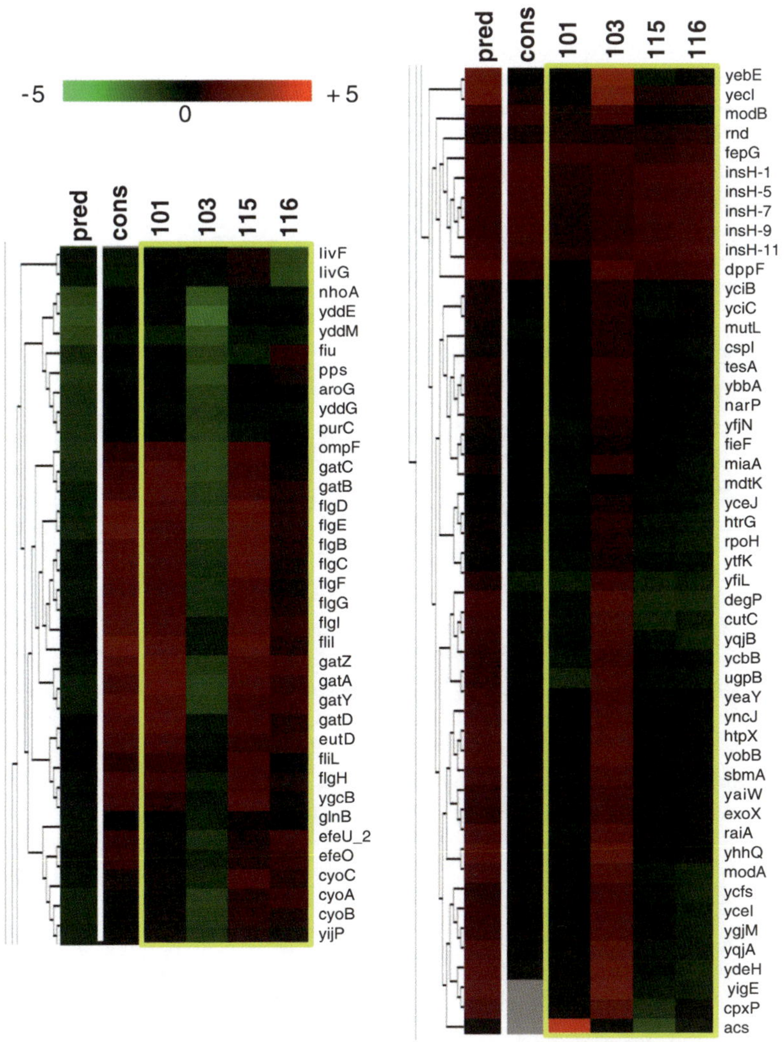

Fig. 7.4 The transcription profile of the coevolved community resembles that of minority clones, not that of the dominant clone. We hypothesize that consumption of acetate and glycerol by minority clones relieves feedback effects by these compounds on the dominant clone (Kinnersley et al. 2009)

may seem, it is likely that these data *underestimate* total genetic diversity in the mutualist population, both because as tools improve there are likely more variants that were identified from the data (false negatives), and because we only sequenced individual clones. We already know from array-Comparative Genome Hybridization (aCGH) data that the dominant clone CV103 can be distinguished from its partners by a 27-gene deletion that consists of multiple genes in anaerobic metabolism and nitrogen metabolism (Kinnersley et al. 2009). Thus, much exciting work lies ahead in determining which of these mutations are beneficial, which are neutral, and which are deleterious but have hitchhiked with selectively favored alleles. Given that so many of these mutations are in coding regions and that many are non-synonymous, it is easy to imagine that many impact fitness. It is tempting to speculate that even in a simple, unstructured environment the rate at which beneficial mutations accumulate is much higher than previously imagined. Indeed, this speculation finds theoretical support in recent work by Sniegowski and Gerrish (2010).

7.8 Prospects for Future Study

The two studies we have highlighted illustrate two alternative mechanisms by which population genetic complexity can be maintained in a simple, resource-limited environment: *clonal interference* (Kao and Sherlock 2008) and *clonal reinforcement* (Kinnersley et al. 2009). Clearly, the classical model (Muller 1932) of *clonal replacement* (periodic selection) imperfectly explains how adaptive evolution occurs in asexual species and somatic cells, even under the simplest possible experimental conditions. Thus, we anticipate that this venerable work will soon be relegated to use as a heuristic device and/or null model. Urgently needed are investigations aimed at defining the ecological and genetic boundary conditions within which either clonal interference or clonal reinforcement act alone or in concert with periodic selection to maintain population genetic diversity. Defining these boundaries will depend on answers to the following general questions. First, just how clonal is a clonal population? The discovery that most mutations are neutral or mildly deleterious suggests that the rate of mutation accumulation in asexual populations may approximate the rate of mutation supply, especially under slow-growth conditions (e.g., chemostats) where modest decrements to maximum specific growth rate μ_{max}, may not come under selection. Second, for a given type of selection, how much does founder genotype, and/or the genotype of early arising adaptive mutants, constrain evolutionary "dénouement?" Repeated evolution of cross-feeding in the Helling et al. *E. coli* strains (Treves et al. 1998) suggests that possible adaptive solutions are partly constrained by the ancestral strain background; moreover, apparent reciprocal sign epistasis between adaptive mutations in yeast (Kvitek and Sherlock 2011) suggests that the evolution of certain adaptive mutations may preclude the acquisition of others in the same genetic background. Finally, while it is theoretically and intuitively obvious how spatial and temporal heterogeneity support evolution of population genetic complexity, the roles played by physical factors such as temperature or irradiation, and chemical factors such as limiting resource type are not. With regard to the latter, we speculate that in the case of microbes, simple, non-fermentable substrates such as acetate or lactate may be less likely to support mutualism because the types of secondary metabolites produced are few and their abundance limited. The same reasoning can be applied to predicting the effect of other limiting resources, e.g., phosphorus, nitrogen and sulfur, where we anticipate that inorganic substrates would poorly support metabolic cooperation among clones, as compared to organic substrates. We eagerly look forward to answers to these questions, as they have far-reaching implications for better understanding both the early evolution of life on our planet, as well as for better predicting the outcomes of chronic infectious disease and cancer, both of which are examples of evolution in action.

Acknowledgments The authors gratefully acknowledge fruitful discussions with Dan Kvitek, Evgueny Kroll, and Carla Boulianne-Larsen, and financial support from NIH-NHGRI (HG003328-01) and NASA (NNX07AJ28G) to GS and FR, respectively.

References

Atwood KC, Schneider LK, Ryan FJ (1951) Periodic selection in *Escherichia coli*. Proc Natl Acad Sci USA 37:146–155

Bantinaki E, Kassen R, Knight CG, Robinson Z, Spiers AJ, Rainey PB (2007) Adaptive divergence in experimental populations of *Pseudomonas fluorescens*. III. Mutational origins of wrinkly spreader diversity. Genetics 176:441–453

Boucher D (1985) The biology of mutualism: ecology and evolution. Croom Helm, London

Carroll SB (2005) Evolution at two levels: on genes and form. PLoS Biol 3:e245

de Visser JAGM, Rozen DE (2006) Clonal interference and the periodic selection of new beneficial mutations in *Escherichia coli*. Genetics 172:2093–2100

Denver DR, Morris K, Streelman JT, Kim SK, Lynch M, Thomas WK (2005) The transcriptional consequences of mutation and natural selection in *Caenorhabditis elegans*. Nat Genet 37: 544–548

Dobzhansky T, Spassky B (1947) Evolutionary changes in laboratory cultures of *Drosophila pseudoobscura*. Evolution 1:191–216

Dobzhansky T, Wright S (1947) Genetics of natural populations. Xv. rate of diffusion of a mutant gene through a population of *Drosophila pseudoobscura*. Genetics 32:303–324

Dykhuizen DE, Dean AM (1994) Predicted fitness changes along an environmental gradient. Evol Ecol 8:541

Estes S, Lynch M (2003) Rapid fitness recovery in mutationally degraded lines of *Caenorhabditis elegans*. Evolution Int J Org Evolution 57:1022–1030

Estes S, Phillips PC, Denver DR, Thomas WK, Lynch M (2004) Mutation accumulation in populations of varying size: the distribution of mutational effects for fitness correlates in *Caenorhabditis elegans*. Genetics 166:1269–1279

Friesen ML, Saxer G, Travisano M, Doebeli M (2004) Experimental evidence for sympatric ecological diversification due to frequency-dependent competition in *Escherichia coli*. Evolution Int J Org Evolution 58:245–260

Gause GF (1934) Experimental analysis of Vito Volterra's mathematical theory of the struggle for existence. Science 79:16–17

Gerrish PJ, Lenski RE (1998) The fate of competing beneficial mutations in an asexual population. Genetica 102–103:127–144

Hardin G (1960) The competitive exclusion principle. Science 131:1292–1297

Helling RB, Vargas CN, Adams J (1987) Evolution of *Escherichia coli* during growth in a constant environment. Genetics 116:349–358

Hoekstra HE, Coyne JA (2007) The locus of evolution: evo devo and the genetics of adaptation. Evolution Int J Org Evolution 61:995–1016

Jacob F (1977) Evolution and tinkering. Science 196:1161–1166

Kao KC, Sherlock G (2008) Molecular characterization of clonal interference during adaptive evolution in asexual populations of *Saccharomyces cerevisiae*. Nat Genet 40:1499–1504

Kinnersley MA, Holben WE, Rosenzweig F (2009) *E unibus plurum*: genomic analysis of an experimentally evolved polymorphism in *Escherichia coli*. PLoS Genet 5:e1000713

Kvitek DJ, Sherlock G (2011) Reciprocal sign epistasis between frequently experimentally evolved adaptive mutations causes a rugged fitness landscape. PloS Genetics 2011 April; 7 (4): e100256 PMCID: PMC3084205

Kubitschek HE (1970) Introduction to research with continuous cultures. Prentice-Hall, Englewood Cliffs

Kurlandzka A, Rosenzweig RF, Adams J (1991) Identification of adaptive changes in an evolving population of *Escherichia coli*: the role of changes with regulatory and highly pleiotropic effects. Mol Biol Evol 8:261–281

Le Gac M, Brazas MD, Bertrand M, Tyerman JG, Spencer CC, Hancock REW, Doebeli M (2008) Metabolic changes associated with adaptive diversification in *Escherichia coli*. Genetics 178:1049–1060

Monod J (1942) Recherche sur la croissance des cultures bactériennes. Hermann et Cie, Paris

Muller HJ (1932) Some genetic aspects of sex. Am Nat 66:118–138

Novick A, Szilard L (1950) Experiments with the chemostat on spontaneous mutations of bacteria. Proc Natl Acad Sci USA 36:708–719

Rainey PB, Travisano M (1998) Adaptive radiation in a heterogeneous environment. Nature 394:69–72

Rainey PB, Buckling A, Kassen R, Travisano M (2000) The emergence and maintenance of diversity: insights from experimental bacterial populations. Trends Ecol Evol 15:243–247

Rosenzweig F, Sherlock G (2009) Through a glass, clearly: experimental evolution as a window on genome evolution. In: Garland T, Rose M (eds) Experimental evolution: applications and methods. University of California Press, Berkeley, pp 353–388

Rosenzweig RF, Sharp RR, Treves DS, Adams J (1994) Microbial evolution in a simple unstructured environment: genetic differentiation in *Escherichia coli*. Genetics 137:903–917

Rozen DE, Lenski RE (2000) Long-term experimental evolution in *Escherichia coli*, VIII, dynamics of a balanced polymorphism. Am Nat 155:24–35

Rozen DE, Philippe N, de Visser JA, Lenski RE, Schneider D (2009) Death and cannibalism in a seasonal environment facilitate bacterial coexistence. Ecol Lett 12:34–44

Sniegowski PD, Gerrish PJ (2010) Beneficial mutations and the dynamics of adaptation in asexual populations. Philos Trans R Soc Lond B Biol Sci 365:1255–1263

Spencer CC, Bertrand M, Travisano M, Doebeli M (2007) Adaptive diversification in genes that regulate resource use in *Escherichia coli*. PLoS Genet 3:e15

Thompson J (2005) The geographic mosaic of coevolution. University of Chicago Press, Chicago

Treves DS, Manning S, Adams J (1998) Repeated evolution of an acetate crossfeeding polymorphism in long-term populations of *Escherichia coli*. Mol Biol Evol 15:789–797

Turner PE, Souza V, Lenski RE (1996) Tests of ecological mechanisms promoting the stable coexistence of two bacterial genotypes. Ecology 77:2119

Wray GA (2007) The evolutionary significance of cis-regulatory mutations. Nat Rev Genet 8: 206–216

Wright S, Dobzhansky T (1946) Genetics of natural populations. Xii. Experimental reproduction of some of the changes caused by natural selection in certain populations of *Drosophila pseudoobscura*. Genetics 31:125–156

Zeyl C (2006) Experimental evolution with yeast. FEMS Yeast Res 6:685–691

Zhong S, Khodursky A, Dykhuizen DE, Dean AM (2004) Evolutionary genomics of ecological specialization. Proc Natl Acad Sci USA 101:11719–11724

Chapter 8
Use of Paleontological and Phylogenetic Data in Comparative and Paleobiological Analyses: A Few Recent Developments

Michel Laurin

Abstract Comparative biology has progressed tremendously but unevenly in the last decades, through incorporation of methodological progress in phylogenetics and in statistical methods that incorporate phylogenetic data into statistical analyses of character correlation or evolution. This review presents a few methods of general interest to comparative biologists, such as phylogenetic independent contrasts (PIC) and variance partition with phylogenetic eigenvector regression. In evo-devo, heterochrony detection has usually been done using event pairing, in the last decade. That method uses a topology, but does not exploit branch length information. A recently proposed method based on squared-change parsimony and PIC exploits both topology and branch lengths, and it outperforms event pairing. Molecular evolution can also benefit from a phylogenetic perspective, as shown in recent studies on genome size evolution. In paleobiology, phylogenies are still rarely and often incompletely incorporated in analyses. Recent developments facilitate time-tree compilations and the combination of paleontological and molecular age data, and new branch length transformation methods can help to standardize PIC, to determine if the characters evolved according to a Brownian motion model, and to deal with clades about which no age information is available.

8.1 Introduction

Comparative biology has roots extending at least into the early nineteenth century, through the works of Lamarck (1809), one of the very first evolutionists, and it could probably even be argued that some pre-evolutionary biologists did comparative

M. Laurin
Département Histoire de la Terre, UMR 7207, CNRS/MNHN/UPMC, Centre de Recherches sur la Paleodiversité et les Paléoenvironments, Muséum national d'Histoire naturelle, Bâtiment de Géologie, Case Postale 48, 43 rue Buffon, F-75231 Paris Cedex 05, France
e-mail: michel.laurin@upmc.fr

P. Pontarotti (ed.), *Evolutionary Biology – Concepts, Biodiversity, Macroevolution and Genome Evolution*, DOI 10.1007/978-3-642-20763-1_8,

biology, although in a different theoretical framework (Mayr 1982). The related field of paleobiology can be considered, to an extent, as a special form of comparative biology because one of the most reliable methods of paleobiological inference consists in demonstrating, in extant relatives of extinct taxa, a correlation between an attribute that usually fossilizes (e.g., skeletal characters in vertebrates) and another that is usually not observed in fossils but for which we need to infer the presence or value (such as a behavior or basal metabolic rate). Demonstrating such a correlation is a typical comparative biology problem, and paleobiologists rely extensively on work on extant taxa to draw their inferences (e.g., de Buffrénil and Rage 1993; Canoville and Laurin 2010).

Comparative biology has changed over time, as it incorporated new techniques and conceptual developments. The advent of cladistics (Hennig 1965) and later, of molecular phylogenetics and dating (Zuckerkandl and Pauling 1965) have greatly improved our knowledge of the tree of life, thus greatly facilitating the work of comparative biologists, to the extent that closely related taxa have to be compared to study transformation series. By now, most (but not all; see below) comparative biologists have integrated phylogenetics in their routine work.

Another very important development in comparative biology was the development of statistical methods that accounted for the statistical nonindependence of comparative data. Indeed, standard statistical methods assume that data about each point (terminal taxa, in the context of comparative analyses) are independent of each other. The very existence of the tree of life indicates that for many datasets, this assumption is violated; whether or not this happens depends mainly on the taxonomic sampling and the evolutionary rate of the characters, but empirical work (e.g., Freckleton et al. 2002; Laurin 2004; Cubo et al. 2005) and simulations (e.g., Martins et al. 2002; Laurin 2010a) show that this problem is pervasive. The first statistical method developed to solve this problem was the phylogenetic independent contrasts (Felsenstein 1985; abbreviated as PIC below), a method that has inspired most (Grafen 1989; Martins and Hansen 1997; Pagel 1997) but not all subsequent comparative methods (Gittleman and Kot 1990; Desdevises et al. 2003; Cubo et al. 2008).

This brief review shows that progress in comparative biology depends rather critically (but not exclusively) on the incorporation of phylogenetic data into the analysis, and on the use of comparative methods that adequately use these phylogenetic data. Below, I will first discuss briefly problems that may arise when this is not done. I will then present briefly some recent comparative methods, and show that most of them require phylogenies with estimated branch lengths. Getting these lengths remains difficult, despite recent progress in molecular and paleontological dating, and many authors still do not bother getting these. Thus, this topic deserves a discussion, which will lead into a short digression in a chronic problem in recent molecular dating studies, namely, the underuse of paleontological literature.

Throughout this discussion, I try to emphasize the most important points, but this review nevertheless emphasizes, to an extent, my own modest contributions to these fields, for the simple reasons that this is where my expertise lies, and that my recent work on these topics is scattered in various journals and book chapters. Thus, a review summarizing these recent developments may be useful.

8.2 The Traditional Approach and Why It Is Being Abandoned

The most basic task of comparative biology, namely, showing a correlation between two features (e.g., body size and basal metabolic rate) used to be done using standard statistical methods, such as least-squares linear regressions. The problem with this approach is mostly that the statistical significance of the relationship is not assessed properly because the number of degrees of freedom is overestimated. Indeed, closely related species tend to resemble each other in most characters, so in a data matrix with n taxa, we do not have n independent data points. For instance, a standard simple linear regression is represented by equation (8.1):

$$y = ax + b \tag{8.1}$$

Two constants are estimated (a and b), and we should consequently expect to have n-2 degrees of freedom. In comparative biology, this is not true, but the number of degrees of freedom is difficult to estimate, and most comparative methods modify the data before performing regressions (see below). These problems are expected on the basis of theoretical considerations (Felsenstein 1985), and have been shown to occur in various situations represented by simulation parameters (Purvis et al. 1994; Martins et al. 2002; Laurin 2010a).

The importance of this problem in comparative biology cannot be overemphasized, especially because several studies are still conducted with inadequate comparative methods. Let us consider the classical problem of assessing the presence of evolutionary trends concerning some of the most basic questions about the history of life. For instance, did complexity of organisms increase over time (McShea 1996)? Did body size increase over time, a trend known as the Cope-Depéret rule (Laurin 2004)? Such studies are still being conducted with a great variety of methods, which hampers meaningful comparisons of results and of the reliability of analyses because the same data analyzed by different methods can yield contradictory results, for instance about the presence (Hone et al. 2008) or absence (Butler and Goswami 2008) of a trend of increasing body size in Mesozoic birds. In a recent simulation study attempting to remedy this situation (Laurin 2010a), I have shown that a simple, non-phylogenetic linear regression of body size vs. geological age of origin of terminal taxa, still used recently to assess evolutionary trends (e.g. Hone et al. 2008), has greatly inflated type I error rate, ranging from 0.12 to 0.18, at the 0.05 threshold. However, simple linear regression had good power, and yielded correct regression coefficient (slope) estimates (Laurin 2010a).

Similar problems pervade comparative biology and extend to the assessment of evolution of qualitative (discrete) characters. This applies also, for instance, to the field of evo-devo, in which a rigorous comparative phylogenetic framework is unfortunately still often lacking. A good example of this is provided by the classical work on *Hox* gene expression patterns in developing vertebrate appendage buds. Sordino et al. (1995) showed that the teleost *Danio rerio* lacked the discrete third phase of *Hox* D-10 to D-13 gene expression pattern (Fig. 8.1d, f) that characterizes the tetrapod limb bud (Fig. 8.1c, e), then documented at least in the mouse, but since then demonstrated in other tetrapod taxa, such as in the chick. The territory where that third expression phase is located (Fig. 8.1e) corresponds more or less with the autopod (hand and foot). Furthermore, *Hox* A-11 is expressed at the apex of the fin in *D. rerio* (Fig. 8.1h), but in a territory proximal to the future autopod in tetrapod limb buds (Fig. 8.1g). Therefore, Sordino et al. (1995) concluded that the presence of a third phase of *Hox* gene expression pattern in tetrapod limbs supports the conclusion that the autopod is a neomorph. The problem with this interpretation is that with data only on one actinopterygian (the teleost *Danio rerio*), few tetrapods, and no other taxa, the polarity of the change in *Hox* gene expression patterns could not be established (the condition in the ancestral osteichthyan could not be determined unambiguously). Furthermore, the teleost *Danio rerio* has a diminutive paired fin endoskeleton lacking a metapterygial axis (usually considered to be homologous with the main axis of tetrapod limbs) that is probably reduced from that of the earliest actinopterygians, judging by the more developed fin endoskeleton (with a metapterygial axis) found in more basal actinopterygians, and this raises the possibility that *D. rerio* lost the third *Hox* gene expression phase when its paired

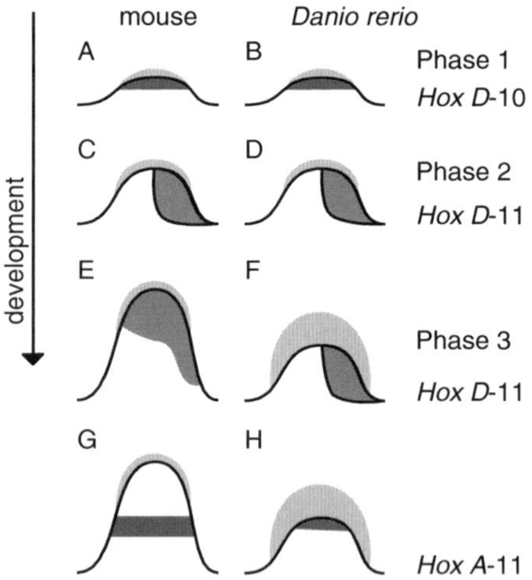

Fig. 8.1 *Hox* gene expression pattern in actinopterygian and tetrapod appendages. *Hox* gene expression pattern in mouse limb buds (*left*) and in fin buds of the teleost *Danio rerio* (*right*). Proximal is below, and cranial is to the left, in all figure parts. The zones of various *Hox* gene expressions are shaded dark gray; the apical ectodermal ridge, in which fin rays (dermal skeleton) develop, is in light gray. Note that in *Danio*, the third expression phase (**f**) is not distinct from the second one (**d**; the same expression pattern prevails), contrary to the pattern displayed by the mouse. (Redrawn from Sordino et al. 1995; modified version from Laurin 2010b)

fin endoskeleton was simplified. I pointed out this problem (Laurin et al. 2000) in a review paper, and raised doubts about these conclusions, but my concerns were ignored. Nevertheless, the subsequent discovery of a tetrapod-like third phase of *Hox* gene expression pattern in the basal actinopterygian *Polyodon spathula* (Davis et al. 2007) that retains a metapterygial axis in its paired fins confirmed my alternative interpretation.

The few examples mentioned above illustrate the need for a rigorous analytical method in comparative or evolutionary biology, if attributes of taxa are compared. Although this contribution focuses on quantitative characters, similar conclusions apply to all kinds of data, from nucleotides to ecology and behavior.

8.3 Modern Comparative Methods

8.3.1 Phylogenetic Independent Contrasts

Felsenstein (1985) laid the foundation for statistical analysis of comparative data by proposing the method of phylogenetic independent contrasts (PIC), a method that will feature prominently in this paper because of its widespread use. Its popularity is shown by the fact that on August 8, 2007, the ISI reported 2,382 citations for the paper that presented it (Felsenstein 1985). This method works by making comparisons between sister-groups (the most closely related taxa on a tree, terminal taxa, or higher taxa, represented by nodes). Thus, for n terminal taxa, if the tree is fully resolved (dichotomous), *n*-1 contrasts can be taken (Fig. 8.2). These contrasts are based on the difference in character value between taxa because despite the phylogenetic relationships of taxa, differences in character value should be statistically independent, if measured between taxa, and if no path linking contrasted taxa overlaps another such path. Thus, the raw (unstandardized) contrast between taxa 1

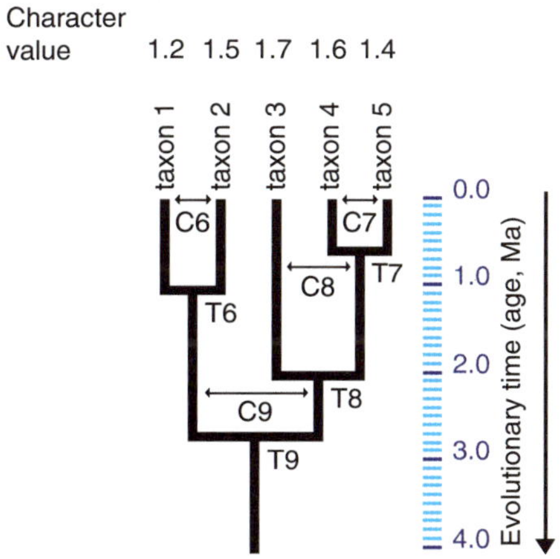

Fig. 8.2 Phylogenetic independent contrasts. Five terminal taxa (1–5) and four higher taxa (T6–9) showing the four phylogenetically independent contrasts that can be taken (C6–C9), each of which corresponding to a higher taxon

and 2 (C6) is $1.5 - 1.2 = 0.3$. However, contrasts should be standardized because they are expected to be greater between distantly related taxa (e.g., C8, between taxon 3 and T7) than between closely related ones (e.g., C7, between taxa 4 and 5), and PIC is based on parametric linear regressions. If characters evolve according to a Brownian motion model (the basic assumption of PIC), variance in the characters is expected to increase linearly with time. Thus, standardization is performed by dividing the raw contrasts by the square root of the path length involved in the contrast. This path length is the sum of lengths of branches connecting the contrasted taxa. For contrasts between terminal taxa, this length is easy to calculate; for instance, for C6, given that the contrasted taxa (1 and 2) are contemporary (extant) and that their last common ancestor dates from 1 Ma, each branch measures 1 Ma, and the path length is 2 Ma. Thus, the standardized contrasts C6 would be $0.3/2^{0.5}$. For contrasts involving higher taxa (e.g., C8, C9), the branches have to be lengthened because the nodal values (of higher taxa) are estimated, not measured; therefore, this introduces error and the variance is expected to be greater. We need not dwell further on how to compute the PIC; the above explanation should suffice for our purpose. Nevertheless, the adequacy of the standardization can be verified using various statistical tests, four of which are available in the PDAP module of Mesquite (Midford et al. 2008). It is important to check that the contrasts are adequately standardized, because otherwise, results will not be reliable. Inadequate standardization may have several causes: the characters may not have evolved according to a Brownian motion model (rather frequent), or errors may be present in the topology, branch lengths (both of which are rather common, if not the rule), or character value measurement. When character data are reliable (fairly common) and the when the phylogeny is considered also reliable (much less common), such tests yield information about the evolutionary model because they can suggest that the characters did not evolve according to a Brownian model. Some methods, such as phylogenetic regression (Grafen 1989) or PGLS (Martins and Hansen 1997; Pagel 1997), can actually yield more detailed data about the probable model of character evolution, if we assume no errors in data measurements and in the phylogeny.

8.3.2 Variance Partition with Phylogenetic Eigenvector Regression (PVR)

Another comparative method, variance partition with PVR, is based on an entirely different principle. It works on values of terminal taxa (not contrasts), but tries to control for phylogenetic effects by using a phylogenetic distance matrix (Fig. 8.3). That matrix simply shows the phylogenetic distances (sum of branch lengths on the path linking the compared taxa). It cannot be used directly; instead, a principal coordinate analysis (a technique related to principal component analysis) is performed to extract coordinates that reflect the distances between taxa. Given the structure of the tree, for *n* terminal taxa, *n*-1 axes are necessary to represent the position of all taxa without distortion, but not all these axes can be used because no

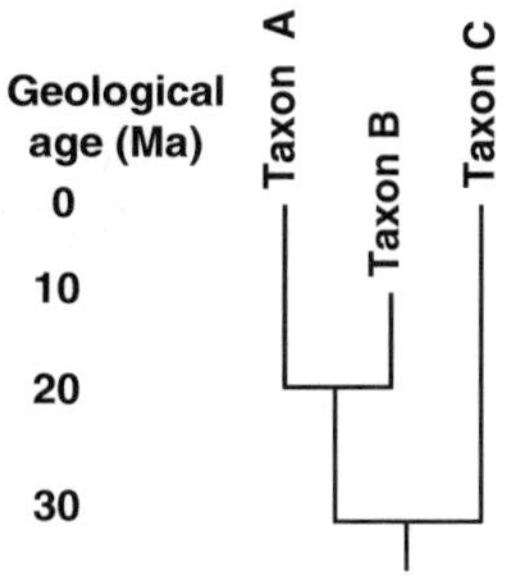

Phylogenetic distance matrix

Taxa	A	B	C
A	-	30	60
B		-	50
C			-

Fig. 8.3 Phylogentic tree, branch lengths, and the corresponding phylogenetic distance matrix (used in variance partition with PVR, among other comparative techniques)

degrees of freedom would be left to compute the statistical significance of regression coefficients. Thus, axes are selected, either using a broken stick model (Diniz-Filho et al. 1998), that selects the first few axes that explain more phylogenetic variance than expected by chance alone, or by regressing the coordinates of these axes against the dependent character, to determine which axes have a significant effect (Desdevises et al. 2003). After that, regressions allow determining the portion of the variance in the dependent character reflecting the independent characters, the phylogenetic signal, and the covariance between both (and some variance remains unexplained). The statistical significance of the effect of the independent characters and of the phylogenetic effects can also be tested.

Regressing the dependent character on the independent characters and on the selected principal coordinate axes representing the phylogeny allows estimating the total explained variance. Partial regressions are then used to establish the portion of variance explained only by the independent characters, only by the phylogeny (the statistical significance of both of these can be established), the portion explained by covariance between the independent characters and the phylogeny (whose statistical significance cannot be assessed), and the residual variance.

There is no need to delve further into the mathematics involved in variance partition with PVR because the purpose of this brief review is to show how all recent comparative methods for continuous (quantitative) characters require a phylogeny with estimated branch lengths.

8.3.3 *The Use of Phylogenies in Evo-Devo*

Phylogenies can also be useful to analyze other types of data (such as discrete data) or to assess other types of problems, such as heterochrony, rather than character correlation. For instance, several recent studies have dealt with how to analyze developmental sequence data (typically based on relative or absolute time data on the position of various events in ontogeny of several species) to detect heterochronies. This problem is complex because there is no universal developmental time metric, and ontogenies can differ drastically between species, by the number of events that they include, by rather extensive differences in sequences, etc. (Jeffery et al. 2002, 2005). Thus, Smith (1997) developed a method called "event pairing" perfected in subsequent studies (e.g., Jeffery et al. 2002, 2005) to circumvent these problems and analyze developmental data on several species simultaneously. The method, initially developed to compare the craniofacial development of marsupials and placentals (Smith 1997), relies on coding the relative time (before, simultaneous, or after) between two events. All events are inserted into a C by C table (where C represents the individual events) and the table gives the relative order between the events listed in the various rows and those listed in the columns. A separate table is made for each taxon, and the data are subsequently treated to see the relative timing (the heterochronies). A full explanation of the method would require considerable developments that are beyond the scope of this chapter (see Smith 1997; Jeffery et al. 2002, 2005), but the point to remember is that analyzing developmental data to detect heterochronies using this method requires a topology (but no branch lengths).

An alternative method using both topology and branch lengths was recently proposed. That method, called the "continuous analysis," relies on squared-change parsimony to infer ancestral (nodal) values and PIC to calculate 95% confidence intervals (CIs) for these ancestral (nodal) values (Germain and Laurin 2009). The method consists in estimating the sequence position (or standardized time, if such data are available) of an event in a given ancestor along with the 95% CI on this value. Then, the observed or inferred sequence position (or standardized time) of the same event in the descendant is compared; if it lies outside the 95% CI of the ancestor, the heterochrony is statistically significant. This method was used to infer the ancestral cranial ossification sequence for urodeles with that of a potential sister-group (the Permo-Carboniferous branchiosaur *Apateon*). This method showed that contrary to previous claims (Schoch and Carroll 2003), *Apateon* was significantly different from the reconstructed ancestral urodele sequence (Germain and Laurin 2009). In any case, the shared similarities turn out to be mostly primitive, as shown by an event pairing analysis (Schoch 2006).

The relevance of the continuous analysis to this contribution is that like PIC and PVR, it uses branch length information whenever it is available, contrary to event-pairing analyses. It is thus not surprising that simulations show that the continuous analysis has a lower Type I error rate, and that it is more powerful (Germain and Laurin 2009).

8.3.4 Phylogenies and Paleontological Data in Paleogenomics

Given the increasing popularity of molecular biology, a brief illustration of how branch length data can contribute to genomics may be relevant. In addition to their widespread use in studies on the evolution of gene expression patterns and of the genes themselves, phylogenies can be used to study genome size (and any other quantitative molecular character) evolution. Thus, Organ et al. (2011) recently took advantage of a correlation between genome size and osteocytic lacuna size to infer the size of genomes of early tetrapods and thus better constrain scenarios on genome size evolution. It has long been known that among extant tetrapods, urodeles have the largest genomes, and that birds have the smallest genomes. However, the polarity of change was difficult to assess from extant taxa alone and at least three main scenarios could explain the observed distribution: (1) the ancestral tetrapod genome was large, as in urodeles, and shrank to various extents in all taxa except for urodeles; (2) the ancestral tetrapod genome had a moderate size, as found in extant placental mammals, and it expanded in amphibians and shrank in birds; or (3) the ancestral tetrapod genome was small, as in birds, and it increased in all other taxa to various extents. Discriminating between the three scenarios with data from extant taxa alone is very difficult because evolutionary trends usually require temporally spread data, as shown by simulations (Laurin 2010a). Thus, the finding that all studied early tetrapods (some amphibians and amniotes) had mid-sized genomes like extant mammals shows that the second scenario is the correct one (Fig. 8.4). This study incorporated a time-calibrated tree at various steps of the analysis, namely, in the assessment of the correlation between genome size and osteocyte lacuna size in extant taxa in which both are known, and in the inference on the evolution of genome size in extant and extinct taxa, using a Bayesian method (Organ et al. 2011).

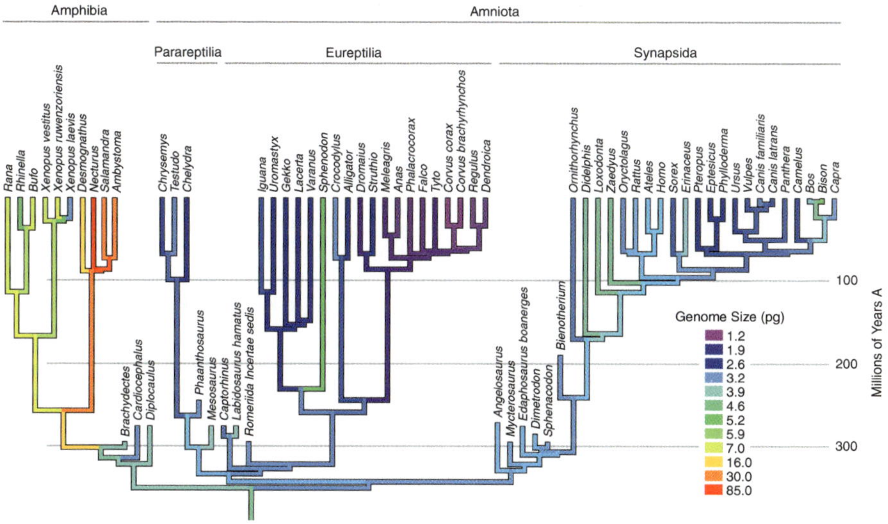

Fig. 8.4 Evolution of genome size in tetrapods based on observed values in extant taxa, and inferred values for extinct taxa based on osteocyte lacuna size. (Reproduced from Organ et al. 2011)

This survey illustrates the usefulness of time-calibrated trees and paleontological data in a wide array of fields in comparative biology. Similar examples could have been taken from evolutionary physiology (Careau et al. 2007), functional morphology (Pouydebat et al. 2008), ecology (Canoville and Laurin 2010), or conservation biology (Faith 1992).

8.3.5 Phylogenies in Paleobiology

Most paleobiological studies have exploited phylogenetic data little, if at all, until recently. This is the case of most studies based on an observed correlation between bone microanatomy and lifestyle (aquatic to terrestrial) to infer the lifestyle of various extinct tetrapods, as was done on extant and extinct snakes (de Buffrénil and Rage 1993). Even the latest studies in this field (e.g., Canoville and Laurin 2010) use the phylogeny only to assess the relationship between bone microanatomy and lifestyle and to build general paleobiological inference models. Thus, the linear discriminant models used by Canoville and Laurin (2010) assess the probability that an extinct taxon was aquatic, amphibious, or terrestrial by using the distribution of quantitative long bone microanatomical characters of taxa of known lifestyle. A graphical representation can also be obtained and shows the relative position of taxa of known lifestyles (along with polygons that encompass all taxa of each given lifestyle) and of the extinct taxa of unknown lifestyle (Fig. 8.5). Ironically, in this particular case, all extinct taxa of unknown lifestyle fit outside the polygons representing the distribution of extant taxa, so the inferences must be viewed with caution, but they can nevertheless be made based on the distance to the centroid and the variance. Thus, the early Permian amniote *Mesosaurus* and Triassic diapsid *Neusticosaurus* were certainly aquatic.

More inference methods are available for quantitative characters, and some of these use classical, well-known statistical methods. Thus, Pouydebat et al. (2008) used multiple linear regression models to infer the grasping behavior of three extinct primate taxa (the Pliocene hominid *Australopithecus afarensis* and the Miocene hominoids *Oreopithecus bambolii* and *Proconsul africanus*) based on the hand proportion, which is itself correlated with frequency of use of various such behaviors. Again, these inferences did not take into consideration the systematic position of these taxa. However, it is possible to incorporate the phylogeny into such inferences, either using the Bayesian method used, among others, by Organ et al. (2011), namely, the program BayesTraits (Pagel and Meade 2006), or by using a program (PhyloPars) originally designed to estimate missing values in comparative datasets (Bruggeman et al. 2009). Both of these methods use character correlation and the systematic position of extinct taxa to make their inferences, which should result in more reliable estimates. Unfortunately, the interface of BayesTraits is not very user-friendly, and PhyloPars works only with quantitative characters.

The phylogeny is often used in paleobiology to infer character history, often based on extant and/or extinct taxa. When extinct taxa are included, the character

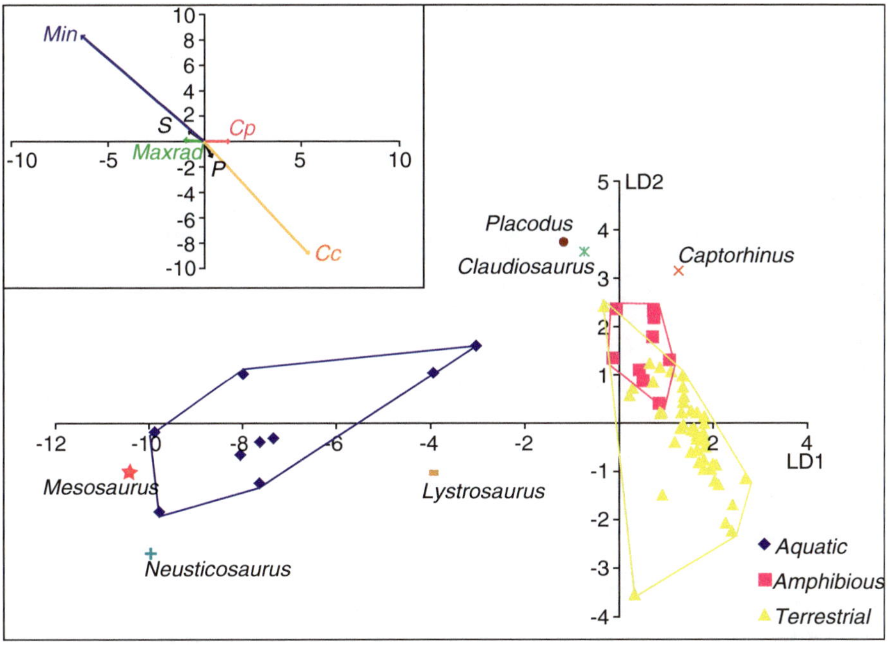

Fig. 8.5 Distribution of taxa of known (in the *polygons*) and of unknown lifestyle, according to their bone microanatomical and body size characters. (Reproduced from Canoville and Laurin 2010)

states of terminal taxa are often inferred before the optimization is carried out. Thus, Canoville and Laurin (2010) used parsimony optimization of the inferred or observed lifestyle of 28 terminal taxa to reconstruct the history of the conquest of land by vertebrates. However, as expected, some parts of the tree are ambiguous. Thus, parsimony indicates that the first amniote was probably amphibious or terrestrial (Fig. 8.6), a conclusion that gives some support to Romer's (1958) suggestion that the first amniotes retained the amphibious lifestyle of their distant ancestors. However, it is easy to use the phylogeny to better investigate this question. Canoville and Laurin (2010) suggested that in this case, the value of the quantitative characters that are used in the linear discriminant inference models can be inferred on the node of interest (here Amniota) and that these inferences could be used to infer the lifestyle. Confidence intervals on all the quantitative characters can also be computed using PIC, allowing a sensitivity analysis of the lifestyle inference. This analysis suggests that the first amniote was amphibious (Canoville and Laurin 2010: supplementary online material 9). Of course, Bayesian methods would allow for a more complete and more rigorous incorporation of various sources of uncertainty into this analysis, but remain impractical because of the software limitations evoked above. Nevertheless, the various phylogeny-informed methods recently developed open exciting perspectives in paleobiology.

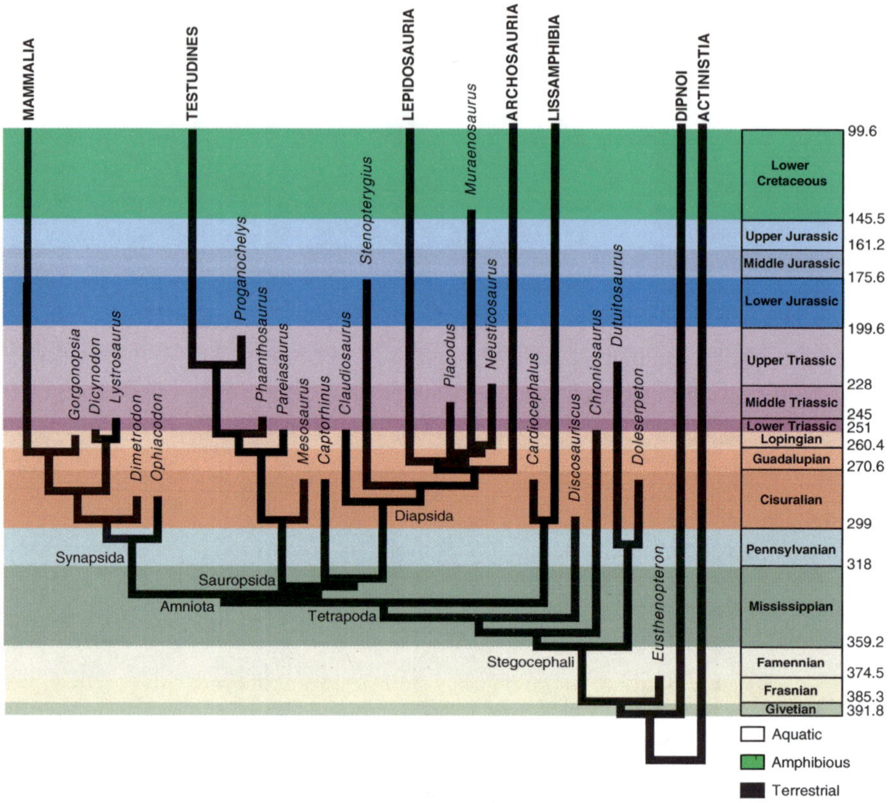

Fig. 8.6 Optimization of the lifestyle (considered as an ordered character) on a phylogeny including extinct taxa whose lifestyle was inferred using microanatomical data and various inference models (especially, linear discriminant models). (Reproduced from Canoville and Laurin 2010)

8.4 Branch Lengths in Comparative Methods

As mentioned above, a frequent problem when applying PIC is that contrasts are not adequately standardized. Data transformation may solve this problem, but when it does not, transforming the branch lengths is the next logical step. Unfortunately, many methods of branch length transformations are not particularly biologically meaningful, and obscure (or discard) the relationship between branch lengths and time. For instance, an exponential or natural log transformation makes subsequent calculation of evolutionary rates difficult. Setting all branches of equal lengths, often done to save time (actually used mostly when no attempt was made to collect branch length data), precludes any meaningful calculation of evolutionary rate. Other methods are more useful and sophisticated, such as Grafen's (1989) rho transform, which consists in a power transformation of branch lengths that distorts the tree to change the relative lengths of terminal and more basal branches. This is useful to adjust the analysis to reflect the amount of phylogenetic signal present in the data; if it is large, the internal branches are rather long; if it is small, internal branches are short, and terminal branches are long. However, when the rho transform needs to be used, the investigator will normally conclude that the analyzed characters did not evolve according to a Brownian motion model. This may, in some cases, represent overinterpretation of the results (see below), especially considering that when Grafen (1989) proposed his method, branch length data were usually unavailable, so he proposed his method to adapt "artificial" branch lengths to better fit the data.

All branch length transformation methods discussed above assume that if divergence time data were used to build the tree, the resulting branch lengths were more or less correct. This may not be so, and it is conceivable that in at least some cases, it is precisely branch lengths that cause the lack of adequate standardization of contrasts because they are wrong (and the characters really evolved according to a Brownian motion model). Two new simple branch length transformation methods were developed by Josse et al. (2006) to facilitate paleontological tree construction (Marjanović and Laurin 2007), and they can be used to check if slight modifications of the initial lengths yield adequate standardization. A few empirical tests on bone microanatomical data demonstrate that in many cases, adequate PIC standardization can be obtained by thus manipulating initial branch lengths (Laurin et al. 2009; Canoville and Laurin 2010). The resulting lengths may remain plausible estimates of evolutionary time (because the exact length of each branch is usually only moderately well-constrained), which allows determination of absolute evolutionary rates of characters. This is essential if evolutionary rates of characters need to be compared between studies in which the taxonomic sample overlaps partly, if at all. This method can test, to an extent, the hypothesis that the characters have evolved according to a Brownian motion model, even if the initial branch lengths were slightly inaccurate. Given that the Brownian model is the simplest model of character evolution (Martins et al. 2002), it should not be discarded in favor of more complex models needlessly.

8.5 Getting Branch Length Data

The last section assumes that branch length data reflecting evolutionary time can be obtained. This is becoming increasingly frequent, but only a minority of comparative studies incorporates such lengths because they are time-consuming to collect, as my recent experience as an editorial board member of the *Journal of Evolutionary Biology* (2008–current) has made me realize. Paleontological data can be used to estimate minimal and (with less precision) maximal divergence dates, but such data are typically scattered in the literature, although a few recent compilations (Benton 1993; Benton and Donoghue 2007; Marjanović and Laurin 2007) ease this process. Molecular divergence dates are becoming increasingly common and their reliability is improving as more taxa and genes are being sequenced, and as the analytical methods become more sophisticated (Sanderson 1997, 2002; Thorne and Kishino 2002). Contrary to the earliest attempts at molecular dating that assumed a single evolutionary rate over the whole tree (Zuckerkandl and Pauling 1965), all modern molecular dating methods rely on a "relaxed molecular clock" that allows for each branch to have its own rate. Recent compilations of molecular timetrees (Hedges and Kumar 2009) should prove invaluable for comparative biologists, especially for taxa in which the fossil record is patchy or inexistent (as in taxa that lack mineralized body parts, such as nematodes, or those in which the morphology is not very informative, as in many eubacteria).

However, combining paleontological and molecular ages is not straightforward. The fossil record provides mostly minimal divergence dates that usually underestimate the actual divergence dates by an unknown amount, whereas molecular ages represent attempts at dating the actual divergences. For reasons explained below, molecular ages are usually older (sometimes much older) than paleontological dates and may often tend to be overestimated, for at least two reasons. First, a methodological factor tends to inflate such ages because molecular divergence age is an "asymetrically bounded random variate" (Rodríguez-Trelles et al. 2002), and random variations around the actual age scale divisively forward (to the present) but multiplicatively backward (to the past). Therefore, the arithmetic means of such age estimates (which are used by molecular dating software to estimate the actual divergence date from several genes and/or portions of the evolutionary tree) are upwardly biased. Second, most molecular phylogeneticists tend to enforce a single (or very few) maximal age constraint in a tree, but several minimal age constraints (e.g., Roelants et al. 2007). This also creates an upward bias, as recently shown empirically (Marjanović and Laurin 2007). This second factor is especially important because the greatest part in the variance in molecular age estimates appears to result from calibration choice, rather than the algorithm used to analyze the molecular sequence data (Marjanović and Laurin 2007).

Thus, it may not be appropriate to simply mix paleontological and molecular divergence dates into a timetree because they appear often fundamentally different. For this reason, Laurin et al. (2009) suggested to use the lower bound (minimal age) of the 95% CI of molecular ages along with paleontological ages to compile an initial timetree. These ages are entered as real (paleontological data) or virtual (molecular age data) taxa into the tree. In most cases, this timetree will underestimate true ages, but if it fails to adequately standardize PICs, the stratigraphic tools (Josse et al. 2006) can be used to lengthen the branches while keeping all time constraints at their minimal age. The result is to push back in time some or all nodes. This procedure can be repeated a few times to check if such transformations adequately standardize PICs while retaining plausible branch lengths. Given that the uncertainty on actual divergence dates is substantial (the 95% CI often represents the data ± 10–50%), there is usually ample room to perform such branch-length manipulations. Thus, in the case of lissamphibians, the initial time-calibrated tree compiled by Laurin et al. (2009) implied a divergence data between urodeles and anurans near the Permo/Triassic boundary (251 Ma), which is barely

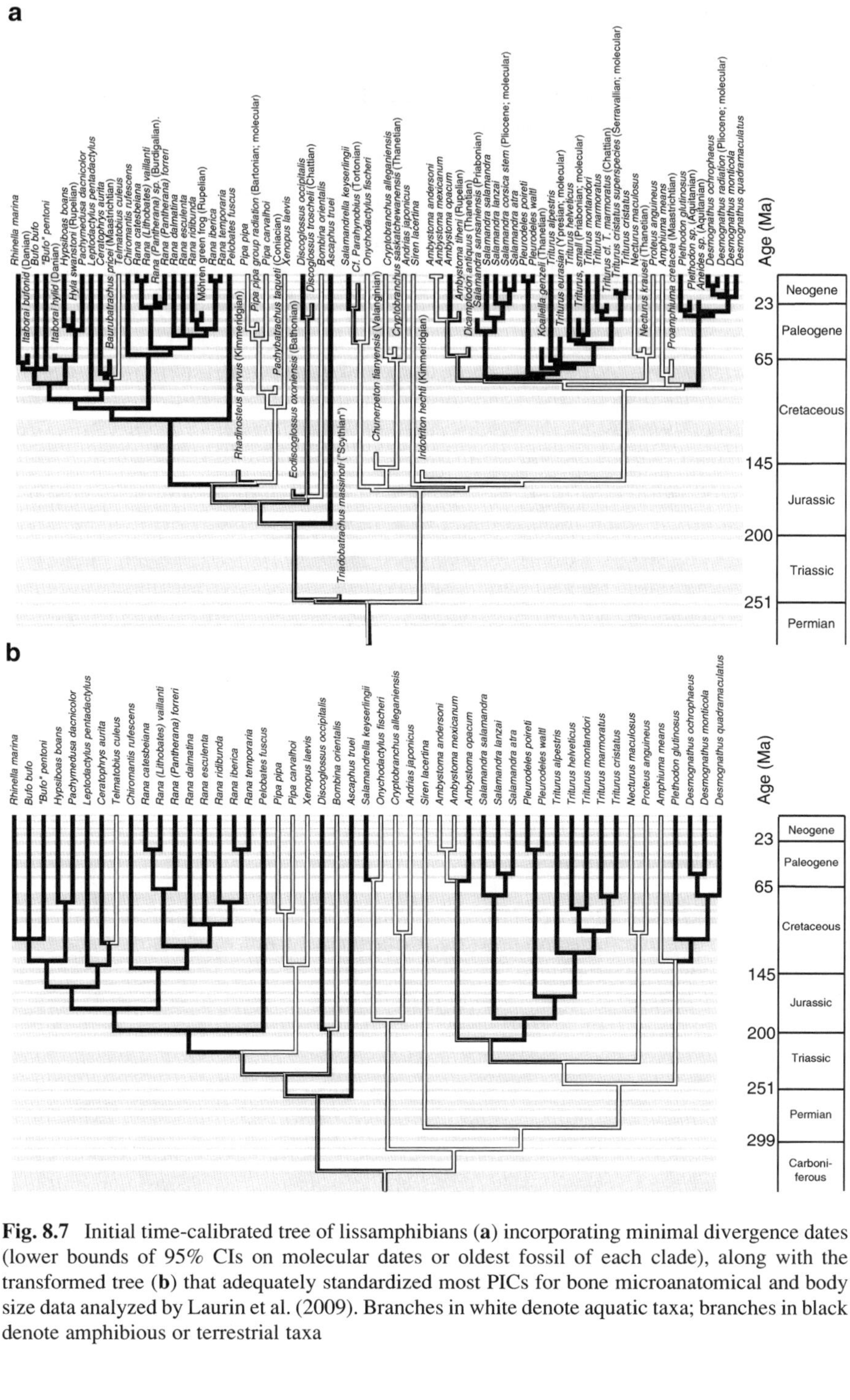

Fig. 8.7 Initial time-calibrated tree of lissamphibians (**a**) incorporating minimal divergence dates (lower bounds of 95% CIs on molecular dates or oldest fossil of each clade), along with the transformed tree (**b**) that adequately standardized most PICs for bone microanatomical and body size data analyzed by Laurin et al. (2009). Branches in white denote aquatic taxa; branches in black denote amphibious or terrestrial taxa

older than *Triadobatrachus,* the oldest known lissamphibian (Fig. 8.7a). This tree failed to adequately standardize most PICs, but a transformed tree (Fig. 8.7b) implying a divergence between anurans and urodeles in the Carboniferous (325 Ma) adequately standardized most PICs. This tree thus implies much greater ages of most nodes, but it remains biologically plausible (branch lengths may reflect evolutionary time) to the extent that most molecular dating studies propose a Carboniferous (or even Devonian) age for this divergence (e.g., San Mauro et al. 2005; Zhang et al. 2005).

The method outlined above eases somewhat the burden of compiling divergence time data because it allows paleontological and molecular data to be combined in a coherent way. Furthermore, given the branch manipulation methods implemented in the stratigraphic tools (Josse et al. 2006), if no data are available for some nodes, some minimal branch lengths can be inserted and lengthened (along with all other branches in the tree) to achieve adequate PIC standardization, thus minimizing the problem created by missing data (when neither molecular ages nor fossil data can be used to date a node, a common situation).

This method could clearly be pushed further. Developing an automated method to adjust minimal branch lengths using the algorithms implemented in the stratigraphic tools to obtain the shortest tree that adequately standardizes the PICs of a given dataset would be very useful, as the procedure of testing various settings and the resulting standardization on all characters of a dataset can be time-consuming. This has not been performed so far because of lack of time, but it could have widespread applications in comparative biology. Soon, most comparative biologists may be able to use time-calibrated trees in their analyses, rather than using trees with arbitrary branch lengths. This change will be facilitated by the growing number of molecular dating studies (e.g., San Mauro et al. 2005; Zhang et al. 2005; Hedges and Kumar 2009), although the meager level of funding of paleontological research, required for these fields to progress (especially, molecular dating), will be the limiting factor.

Acknowledgments I thank Pierre Pontarotti for inviting me to participate in this symposium and in this volume, and for his patience while waiting for this draft. Eli Amson provided comments that improved the draft.

References

Benton MJ (ed) (1993) Fossil record 2. Chapman & Hall, London

Benton MJ, Donoghue PCJ (2007) Paleontological evidence to date the tree of life. Mol Biol Evol 24:26–53

Bruggeman J, Heringa J, Brand BW (2009) PhyloPars: estimation of missing parameter values using phylogeny. Nucleic Acids Res 37:W179–W184

Butler RJ, Goswami A (2008) Body size evolution in Mesozoic birds: little evidence for Cope's rule. J Evol Biol 21:1673–1682

Canoville A, Laurin M (2010) Evolution of humeral microanatomy and lifestyle in amniotes, and some comments on paleobiological inferences. Biol J Linn Soc 100:384–406

Careau V, Morand-Ferron J, Thomas D (2007) Basal metabolic rate of canidae from hot deserts to cold arctic climates. J Mammal 88:394–400
Cubo J, Ponton F, Laurin M, de Margerie E, Castanet J (2005) Phylogenetic signal in bone microstructure of sauropsids. Syst Biol 54:562–574
Cubo J, Legendre P, de Ricqlès A, Montes L, de Margerie E, Castanet J, Desdevises Y (2008) Phylogenetic, functional, and structural components of variation in bone growth rate of amniotes. Evol Dev 10:217–227
Davis MC, Dahn RD, Shubin NH (2007) An autopodial-like pattern of Hox expression in the fins of a basal actinopterygian fish. Nature 444:473–477
de Buffrénil V, Rage J-C (1993) La "pachyostose" vertébrale de *Simoliophis* (Reptilia, Squamata): données comparatives et considérations fonctionnelles. Ann Paléontol 79(4):315–335
Desdevises Y, Legendre P, Azouzi L, Morand S (2003) Quantifying phylogenetically structured environmental variation. Evolution 57:2467–2652
Diniz-Filho JAF, de Sant'Ana CER, Bini LM (1998) An eigenvector method for estimating phylogenetic inertia. Evolution 52:1247–1262
Faith DP (1992) Conservation evaluation and phylogenetic diversity. Biol Conserv 61:1–10
Felsenstein J (1985) Phylogenies and the comparative method. Am Nat 125:1–15
Freckleton RP, Harvey PH, Pagel M (2002) Phylogenetics analysis and comparative data: a test and review of evidence. Am Nat 160:712–726
Germain D, Laurin M (2009) Evolution of ossification sequences in salamanders and urodele origins assessed through event-pairing and new methods. Evol Dev 11:170–190
Gittleman JL, Kot M (1990) Adaptation: statistics and a null model for estimating phylogenetic effects. Syst Zool 39:227–241
Grafen A (1989) The phylogenetic regression. Philos Trans R Soc Lond B 326:119–157
Hedges SB, Kumar S (eds) (2009) The timetree of life. Oxford University Press, New York
Hennig W (1965) Phylogenetic systematics. Annu Rev Entomol 10:97–116
Hone DWE, Dyke GJ, Haden M, Benton MJ (2008) Body size evolution in Mesozoic birds. J Evol Biol 21:618–624
Jeffery JE, Richardson MK, Coates MI, Bininda-Emonds ORP (2002) Analyzing developmental sequences within a phylogenetic framework. Syst Biol 51:478–491
Jeffery JE, Bininda-Emonds ORP, Coates MI, Richardson MK (2005) A new technique for identifying sequence heterochrony. Syst Biol 54:230–240
Josse S, Moreau T, Laurin M (2006) Stratigraphic tools for Mesquite. http://mesquiteproject.org/packages/stratigraphicTools/
Lamarck JB (1809) Philosophie zoologique. Flammarion, Paris
Laurin M (2004) The evolution of body size, Cope's rule and the origin of amniotes. Syst Biol 53:594–622
Laurin M (2010a) Assessment of the relative merits of a few methods to detect evolutionary trends. Syst Biol 59:689–704
Laurin M (2010b) How vertebrates left the water. University of California Press, Berkeley
Laurin M, Girondot M, de Ricqlès A (2000) Early tetrapod evolution. Trends Ecol Evol 15: 118–123
Laurin M, Canoville A, Quilhac A (2009) Use of paleontological and molecular data in supertrees for comparative studies: the example of lissamphibian femoral microanatomy. J Anat 215:110–123
Marjanović D, Laurin M (2007) Fossils, molecules, divergence times, and the origin of lissamphibians. Syst Biol 56:369–388
Martins EP, Hansen TF (1997) Phylogenies and the comparative method: a general approach to incorporating phylogenetic information into the analysis of interspecific data. Am Nat 149: 646–667
Martins EP, Diniz-Filho JAF, Housworth EA (2002) Adaptative constraints and the phylogenetic comparative method: a computer simulation test. Evolution 56:1–13
Mayr E (1982) The growth of biological thought—diversity, evolution, and inheritance. The Belknap Press of Harvard University Press, Cambridge

McShea DW (1996) Metazoan complexity and evolution: Is there a trend? Evolution 50:477–492
Midford P, Garland TJ, Maddison WP (2008) PDAP package for Mesquite. http://mesquiteproject.org/pdap_mesquite/index.html
Organ CL, Canoville A, Reisz RR, Laurin M (2011) Paleogenomic data suggest mammal-like genome size in the ancestral amniote and derived large genome size in amphibians. J Evol Biol 24:372–380
Pagel M (1997) Inferring evolutionary processes from phylogenies. Zoolog Scr 26:331–348
Pagel M, Meade A (2006) BayesTraits. http://www.evolution.rdg.ac.uk/BayesTraits.html
Pouydebat E, Laurin M, Gorce P, Bels V (2008) Evolution of grasping among anthropoids. J Evol Biol 21:1732–1743
Purvis A, Gittleman JL, Luh H-K (1994) Truth or consequences: effects of phylogenetic accuracy on two comparative methods. J Theor Biol 167:293–300
Rodríguez-Trelles F, Tarrío R, Ayala FJ (2002) A methodological bias toward overestimation of molecular evolutionary time scales. Proc Natl Acad Sci USA 99:8112–8115
Roelants K, Gower DJ, Wilkinson M, Loader SP, Biju SD, Guillaume K, Moriau L, Bossuyt F (2007) Global patterns of diversification in the history of modern amphibians. Proc Natl Acad Sci USA 104:887–892
Romer AS (1958) Tetrapod limbs and early tetrapod life. Evolution 12:365–369
San Mauro D, Vences M, Alcobendas M, Zardoya R, Meyer A (2005) Initial diversification of living amphibians predated the breakup of Pangaea. Am Nat 165:590–599
Sanderson MJ (1997) A nonparametric approach to estimating divergence times in the absence of rate constancy. Mol Biol Evol 14:1218–1231
Sanderson MJ (2002) Estimating absolute rates of molecular evolution and divergence times: a penalized likelihood approach. Mol Biol Evol 19:101–109
Schoch RR (2006) Skull ontogeny: developmental patterns of fishes conserved across major tetrapod clades. Evol Dev 8:524–536
Schoch RR, Carroll RL (2003) Ontogenetic evidence for the Paleozoic ancestry of salamanders. Evol Dev 5:314–324
Smith KK (1997) Comparative patterns of craniofacial development in eutherian and metatherian mammals. Evolution 51:1663–1678
Sordino P, van der Hoeven F, Duboule D (1995) Hox gene expression in teleost fins and the origin of vertebrate digits. Nature 375:678–681
Thorne JL, Kishino H (2002) Divergence time and evolutionary rate estimation with multilocus data. Syst Biol 51:689–702
Zhang P, Zhou H, Chen Y-Q, Liu Y-F, Qu L-H (2005) Mitogenomic perspectives on the origin and phylogeny of living amphibians. Syst Biol 54:391–400
Zuckerkandl E, Pauling L (1965) Molecules as documents of evolutionary history. J Theor Biol 8:357–366

Chapter 9
Seasonal Flowering and Evolution: Will Plant Species Be Under Stress from Global Warming?

Rod W. King

Abstract In regulating their seasonal flowering, plants have adapted to many environmental inputs including daylength, temperature, and light intensity. Adaptation to temperature alone explains flowering of *Pimelea ferruginea* (King et al. 1992), a Western Australian perennial shrub. In the laboratory, plants from the 650 km of its north–south coastal distribution show clinal adaptation for thermoregulated flowering: those from higher latitudes require cooler temperatures (15°C) than the more equatorial ones (21°C). This adaptation confers evolutionary advantage because, in reciprocal field transplant nurseries, all lines flowered after a cool field winter (13–15°C), but those originating from cooler sites failed to flower at the warmest (17–19°C) extreme of its latitudinal distribution. Thus, thermoregulated flowering of *P. ferruginea* provides an adaptive advantage. A less extreme response is evident in *Crowea exalata*, another Australian shrub. Increased temperature linearly reduces its flower numbers (16% loss per 3°C) and causes earlier flowering. Clearly, a 3–4°C global warming will restrict flowering and sometimes cause species extinction.

9.1 Introduction

For plants, the seasonal timing and extent of their flowering is critical for survival. They time their flowering and seed set to avoid extreme seasonal environments including frost, heat, and drought. Essentially, they use the seasonal changes in environment to predict and avoid imminent stress.

Studies in the early part of the last century established the importance for flowering time of winter cold (vernalization at temperatures below 10°C; see Gassner 1918; Lang 1965) and daylength (photoperiod; see Garner and Allard

R.W. King
CSIRO, Plant Industry, GPO Box 1600, Canberra, ACT 2601, Australia
e-mail: rod.king@csiro.au

P. Pontarotti (ed.), *Evolutionary Biology – Concepts, Biodiversity, Macroevolution and Genome Evolution*, DOI 10.1007/978-3-642-20763-1_9,

1920). In addition, in some species, rainfall, sunlight intensity, and mild temperatures can participate in this seasonal signaling (reviewed in King and Heide 2009). Thus, a species may not only adapt its flowering response to match latitudinal differences in daylength and vernalizing winter temperatures but also to a complex of other environmental inputs.

Recent molecular and genetic approaches have confirmed the heritability of flowering time response (see review in Koornneef et al. 2003), but it has been more difficult to establish evolutionary advantage and species fitness. For example, in a population of plants developed by intermating 19 accessions of the annual European herb, *Arabidopsis thaliana*, Scarcelli and Kover (2009) found that selection pressure in the laboratory for early flowering in the absence of vernalization (cold) led to a decrease in the frequency of alleles of *FRI*, a vernalization-requiring gene. However, *FRI* only accounted for 12% of the variation in flowering time. Furthermore, for another cold sensing gene, *FLC*, in field plantings, its action on flowering time could be overridden in ways yet to be explained (Wilczek et al. 2009).

In domesticated crop plants, the genetics of environmental regulation of flowering time is becoming increasingly well understood (see review in Trevaskis et al. 2007). As a consequence, cultivars can be deliberately adapted to particular growing zones. However, these studies do not directly test evolutionary fitness in natural populations: a limitation also evident in studies with derivative populations (see above for Arabidopsis).

With wild species, although adaptation can be seen from studies in a common field nursery or in controlled environments, a test for selective advantage may need reciprocal field nursery transplantation experiments (Mitchell-Olds et al. 2007). Even so, such field studies must meet a number of specific requirements:

1. To avoid site-to-site shifts in the mix of critical environmental determinants of flowering, its regulation should be as simple as possible and, preferably, involve a response to a single environmental input.
2. To avoid indirect effects including responses involving germination and vegetative growth, transplantation should involve plants, which are still vegetative but fully competent to flower.
3. Arbitrary site selection needs to be avoided. Sites selected for "convenience" may introduce complex and somewhat uncharacterized environmental differences. Even if selected for convenience, the sites should fall within the natural distribution of the species.

This review focuses on adaptation of flowering time to mild seasonal temperatures (10–20°C) coupled with possible effects of global warming on species survival. This scenario relates to a number of recent studies showing links between global temperatures and phenology (seasonal development including flowering time; see review in Forrest and Miller-Rushing 2010).

For one Australian perennial plant, *Pimelea ferruginea*, its past evolution and future survival of global warming is linked directly to its ability to thermoregulate its initiation of flowering. Both controlled environment and a reciprocal field transplant study show temperature regulation of its floral initiation (King et al. 1996). Difficulties with field transplantation studies (above) have been avoided, so the evidence points to evolutionary selection pressure/adaptive advantage. For a second Australian perennial species, *Crowea exalata*, similar increases in temperature (3°C) cause faster flowering but fewer flowers form (King et al. 2008).

9.2 Thermoregulation of Floral Initiation by Mild Temperatures

Plants of *P. ferruginea* grow vegetatively for at least 2 years if held in a controlled environment glasshouse at a high daily average temperature of 21°C. However, flowers initiate when 4- or 5-month-old plants are shifted to slightly cooler temperatures (12–18°C). The response shown in Fig. 9.1 is for one such line,

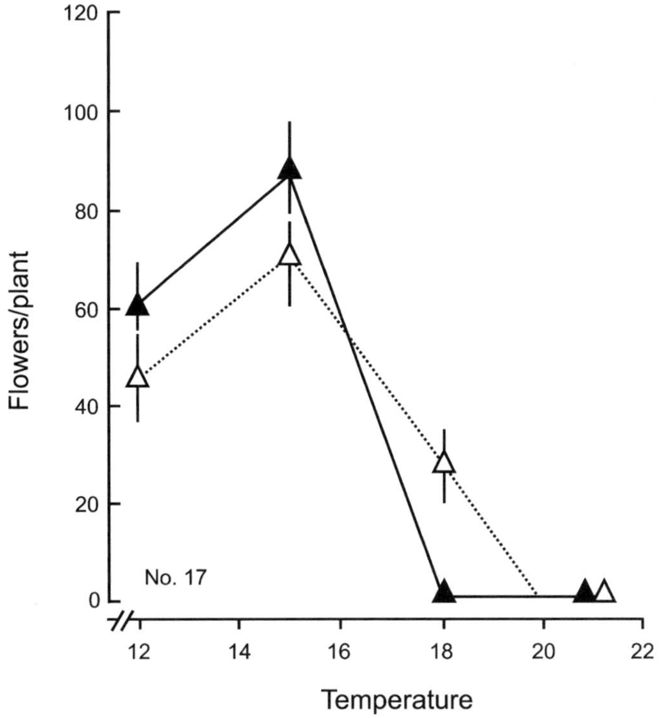

Fig. 9.1 Effect of temperature on flowering (number of open flowers plus buds) of *Pimelea ferruginea*. A batch of ca. 100 clonal plants was vegetatively propagated using cuttings taken from a single field plant (# 17) located at latitude 33° 34′S in Western Australia. There was no flowering over 2 years when the cuttings were growing at 21°C but they flowered rapidly when moved to temperatures of 12°C, 15°C, or 18°C in either short (8 h) or long days (16 h photoperiod). *Vertical bars* are 2× s.e. (Adapted from King et al. 1996)

17, which had been vegetatively propagated by taking cuttings from a single plant in the field.

When examined with a microscope, flower primordia have developed fully by 7 weeks (King et al. 1992). Clearly, mild temperatures with an optima of *ca* 15°C, directly and rapidly induce flowering of *P. ferruginea*. This temperature far exceeds the 4–10°C range accepted for promotion of flower by winter cold (vernalization: see Lang 1965).

Further lines of *P. ferruginea* covering the full latitudinal range of its natural distribution were vegetatively propagated at the same time in batches of approximately 100 plants. For all lines, mild temperatures regulated flowering and there was no photoperiodic effect when tested in daylengths, which span the natural seasonal differences across their latitudes of origin (cf. #17 Fig. 9.1). Thus, a single environmental factor, mild temperature, is the dominant and probably the only determinant of flowering of *P. ferruginea* (see requirement 1 above), There is none of the potential complexity found with Arabidopsis and other species (see King and Heide 2009), where multiple environmental inputs may regulate flowering either interchangeably or in concert.

Such thermoregulation of flower initiation by mild temperatures is not unique to *P. ferruginea* but has been observed for a number of other plant species across a range of families including the grasses (Heide 1994) and dicotyledonous species (see summaries in King et al. 1992 and King and Heide 2009).

All batches of cloned adult plants used in these studies were 4–5 months-old and vegetative when, at the onset of winter, they were either transferred to florally inductive temperatures in controlled environment conditions (above) or returned to field nurseries (see later). Thus, indirect environmental effects on plant establishment and juvenile development were avoided (requirement 2 above). The rapidity of flower initiation (7 weeks) also enhances the specificity of these studies.

9.3 Adaptive Differences in Controlled Environments

P. ferruginea is found over a north–south latitudinal distance of *ca* 650 km along the coast of Western Australia but in an extremely narrow strip of sand plain covering *ca* 200 m adjacent to the ocean (Rye 1988). Aside from occasional residential intrusions, the population is also undisturbed by human activity.

Lines vegetatively propagated from single plants sampled along the full range of their distribution, differed in their thermoregulation of flowering when tested in a controlled environment study (Fig. 9.2). There is clearly a latitudinal adaptation in the temperature permissive for flowering of *P. ferruginea*.

This latitudinal, permissive temperature gradient for flowering matches site temperatures (Fig. 9.2) and both show a 4°C range. Depending on the assumptions made, these two lines could be virtually identical as the upper permissive temperature limit for flowering might equally well be replaced by the 3°C lower optimal temperature for flowering (see Fig. 9.1). Further, site temperatures might

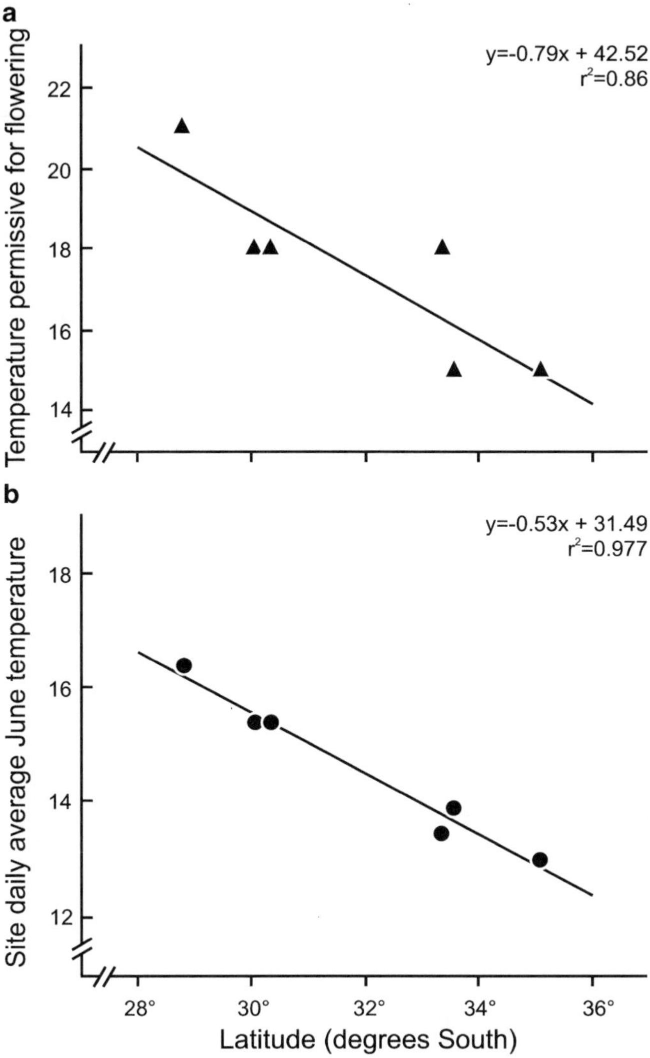

Fig. 9.2 Relationship between latitude of origin of selections of *P. ferruginea* and (**a**) the highest temperature which permitted flowering when tested in controlled environment conditions. (Values are derived from findings reported in full in King et al. 1996 and as shown in Fig. 9.1 for one location.) (**b**) Latitudinal differences in the average daily winter temperature for the month of June based on 100-year averages available from the Australian Bureau of Meteorology

be increased by a small amount as they are only for the first 4 weeks of June, the first month of winter. The estimate of average effective temperature could also have included metrological data on some of the warmer weeks prior to winter as flowering of *P. ferruginea* requires at least a 5-week exposure to mild temperatures (King et al. 1992).

Overall, the close match between site temperatures and adaptation of *P. ferruginea* for thermoregulated flowering implies that temperature has been a dominant evolutionary force in regulating its seasonal flowering and survival.

9.4 Mild Field Temperatures Regulate Flowering of *P. ferruginea*

A more definitive test of evolution and selective advantage was possible with reciprocal field nursery transfers performed across the latitudinal extremes of its distribution. Nurseries were at Denmark in Western Australia (Lat 34° 96′S), a cooler site, and Geraldton in Western Australia (Lat 28° 96′S), the warmest extreme of its natural distribution. The plants of *P. ferruginea* were growing vigorously and vegetatively when transferred at the start of winter from the controlled environment to the field. Thus, all three experimental requirements were satisfied (see Introduction)

At the cooler Denmark nursery, irrespective of their latitude of origin, all lines flowered after about 2 months (Fig. 9.3). Because winter temperatures at this site were always below the upper permissive limit for all lines, they all flowered rapidly. In other words, at this site there was no evidence of selective pressure on the known temperature adaptation between lines from cool or warm sites of origin. Further data (not shown) confirmed this finding for nurseries at this site and another cool site (King et al. 1996). In contrast, at the warmer nursery site (Geraldton), a selection from a warm site (Lat 31°01′) flowered as rapidly as it did at the cooler site but, now, flowering of lines from cooler sites (35° 07′S and 33° 34′S) was considerably delayed or the plants never flowered (33° 55′S). The 4–5°C higher temperature at Geraldton exceeds the temperature adaptation of these latter lines.

There are many reasons why earlier studies of flowering have shown adaptation without proving any evolutionary selective advantage. Here, an important and fortunate feature is the undisturbed and compact distribution *P. ferruginea* in a very narrow zone (*ca* 200 m) up and down the coast of Western Australia. In addition, there was a natural latitudinal temperature gradient of sufficient magnitude for the detection of differences between lines in their "thermostat" set point.

Overall, the action of temperature on flowering of *P. ferruginea* has been a significant selective force in its evolutionary adaptation, a conclusion based on three findings, namely:

1. *P. ferruginea* flowers in response to a single environmental input; mild winter temperatures. Flowering is blocked if the conditions are too warm (>18 to >21°C depending on the line).
2. The natural latitudinal temperatures at the time of its floral initiation match the temperatures tolerated for their thermoregulated flowering in controlled conditions.

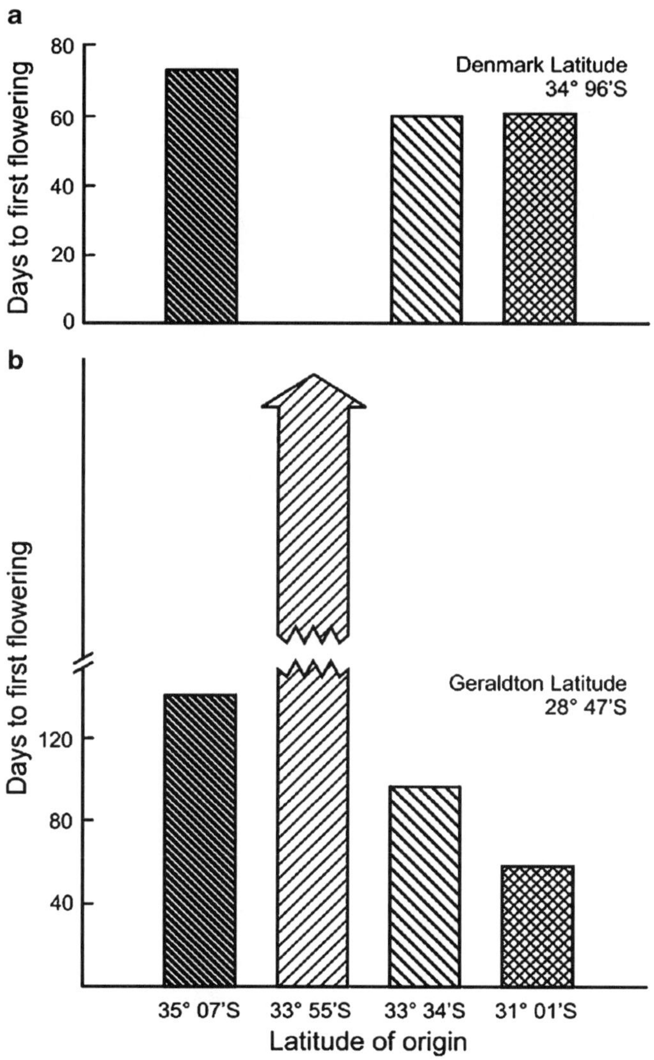

Fig. 9.3 Effect of latitude of origin of selections of *P. ferruginea* on their flowering in a nursery at Denmark (a cooler high latitude site with an average temperature of 13°C at transfer in June) or at Geraldton (a warmer lower latitude site with a high 17°C average temperature at transfer in June). The hatching of the bars is distinctive for the latitude of origin of each selection. (Adapted from King et al. 1996)

3. In reciprocal field nursery transplantation studies across the extremes of its distribution, all lines from warm or cool sites flowered at the cool site. In contrast, temperature at the 4°C warmer site exceeded the tolerance of lines from cooler sites.

9.5 Temperature Effects on Flower Number and Time to Flower

The role of temperature in flower development is well illustrated by findings with *Crowea exalata*, "Bindelong Compact." It initiates flowers when transferred from shade to a higher light intensity (full sunlight). Low temperature was not essential for flowering of *Crowea exalata*, and warmer conditions (up to 30°C) sped up the time to first flower but decreased total flower number at the peak of the display (Fig. 9.4).

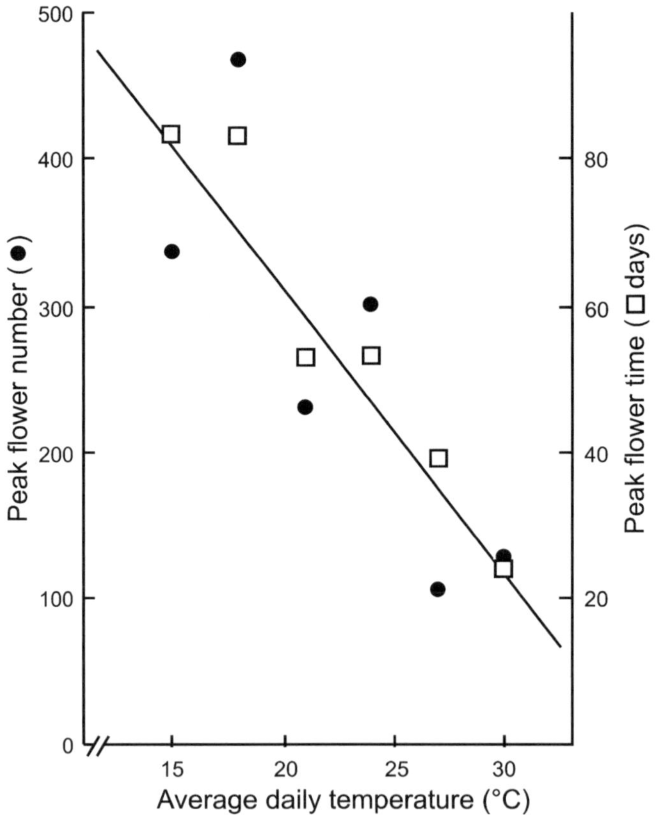

Fig. 9.4 Effect of temperature on flowering of *Crowea exalata*, "Bindelong Compact." The relationship between temperature and either days to peak flowering (*open square*) or, number of open flowers at peak flowering (*closed circle*). Initially, the plants were held vegetative in low irradiances in a green house operating at 21°C. Then the various temperatures were imposed at the time of transferring plants to full sunlight conditions for inducing flowering. The regression line between temperature and flower number and time showed a highly significant fit ($p < 0.001$) to the data. (Adapted from King et al. 2008)

There is no paradox between higher temperatures (>21°C) causing early flowering of *Crowea exalata* but blocking flowering of *P. ferruginea* (see above). These differences reflect independent effects of temperature on sequential developmental processes. Flower initiation may require mild temperatures but flowers develop faster the higher the temperature. These contrasting responses to temperature are evident with *P. ferruginea* where flower initiation is satisfied by a 7-week exposure to mild temperatures. Then, subsequently, higher temperatures are tolerated and speed up floral development (King et al. 1992). On the other hand, *Crowea exalata* only shows the effect of temperature to speed up floral development, its floral initiation is achieved by an increase in photosynthesis (King et al. 2008).

9.6 Thermoregulation and Seasonal Flowering of Other Plant Species

Unlike the findings with *P. ferruginea,* for many other species a combination of daylength and mild temperature is required either simultaneously (e.g., in the Australian perennial legume, *Hardenbergia*; King 1998) or over different seasons and stages of floral development as for some grasses (Heide 1994). Furthermore, the adaptations may be quite complex as with strawberry (*Fragaria* spp), which may tolerate short and long days at low growth temperatures but may only respond to long daylengths at higher temperatures (Heide and Sønsteby 2007).

To simply characterize the daylength and temperature environment, metrological data can be used to calculate seasonal photothermal envelopes for any particular latitude (Heide 1994). Such envelopes change from a pin-point size at the equator to a massive fan at very high latitudes. In theory, by superimposing a species temperature and daylength profile over its site photothermal envelope, inferences can be drawn about species survival. However, unlike the possible outcome of field transplantations, this latter approach provides only a limited test of selective advantage.

The analysis of seasonal regulation becomes even more complex when daylength and both positive and negative temperature responses are important. For example, with *Hardenbergia violaceae,* its flowers initiate in daylengths shorter than 12 h and for temperatures of 18°C and above (King 1998). Then, even as late as visible petal formation, warm conditions (21°C and above) cause all flowers to abort. The net result is that the plants only set seed in short days at 18°C. Thus, in nature, while the onset of flowering before winter allows rapid spring floral development, further rises in temperatures cause rapid loss of formed flowers as well as abortion of immature seed pods (King 1998).

Little is known of how flowering time "thermoregulators" might work in plants. For Arabidopsis, Blázquez et al. (2003) reported extensively on promotion of flowering at a higher temperature (23°C vs. 16°C). Their findings relate to promotion of flowering by warmer conditions imposed over the whole life cycle and were inconclusive in showing a role for known flowering time genes involved in floral initiation. Very likely, as with *Crowea exalata*, earlier flowering of Arabidopsis at higher temperature reflects a major effect of temperature on floral development not on floral initiation.

9.7 Global Warming, Phenology, and Species Survival

Increasingly, there are reports of effects of global warming on phenology (the timing of seasonal developmental events such as flowering of plants, and bird migration (see Forrest and Miller-Rushing 2010). Here, documentation of past selection pressure on thermoregulated flowering of *P. ferruginea* in its natural environment has provided a unique view of its future survival. Temperature alone controls the phenology of *P. ferruginea* and, based on the reciprocal transfer experiments, a 3–4°C global warming, at cooler sites will block or delay flowering. Furthermore, depending on upper "thermostat" set limits, permissible temperatures for flowering might also be exceeded for lines from warmer sites.

Heritability of temperature-regulated flowering of *P. ferruginea* is indicated by the match across latitudinal sites between a plants "thermostat" setting for flowering and site temperature (Fig. 9.2), a claim further supported by evidence that plants sampled from within one site, showed little or no distinction in their "thermostat" setting (line #17, Fig. 9.1; line # 18 and # 19: King et al. 1996). Nevertheless, the present site sampling is not sufficiently intensive to adequately address the question of natural genetic variation. Even one plant with a 3–4°C greater temperature tolerance could ensure species survival. It seems unlikely this species would survive by migration of warm-tolerant lines to cooler sites. The seed is not wind dispersed and the plant would need to migrate over 650 km in *ca* 50 years.

Compared with *P. ferruginea*, few plant species in their natural environment will provide such ideal material for studying responses to global warming. They may variously show responses to more complex sets of environmental inputs and if spread over wide geographical ranges, there is potential for divergent evolution. Nevertheless, as shown in a recent compilation (see Forrest and Miller-Rushing (2010) and associated reviews), there is growing evidence that global warming is affecting flowering. As one example, from 47 years of records of spring flowering times of 385 species, Fitter and Fitter (2002) reported that 16% had advanced their flowering date by 4.5 days on average but only in association with global temperature increases over the decade from 1990.

What is not yet clear from such phenological information is the impact of earlier flowering on species survival. As discussed above for *P. ferruginea* and *Crowea exalata*, misleading mechanistic assumptions can be introduced where warming will speed up, delay or block floral initiation but speed up floral development. Of even greater concern is the evidence for *Crowea exalata* that earlier flowering at higher temperatures results in fewer flowers forming (Fig. 9.4). Similar responses are also known for a number of species. Wheat, for example flowers earlier at higher temperatures but then forms fewer florets, grains, tillers, and leaves (see review in Evans 1993). Thus, heat sum models, while adequately predicting maturity and effects of global warming in speeding up plant development (Crauford and Wheeler 2009), could be of limited value in predicting seed production or yield of a crop or of species in the wild.

References

Blázquez M, Ahn JH, Weigel D (2003) A thermosensory pathway controlling flowering time in *Arabidopsis thaliana*. Nat Genet 33:168–171

Crauford PQ, Wheeler TR (2009) Climate change and the flowering time of crops. J Exp Bot 60:2529–2539

Evans LT (1993) Crop evolution, adaptation and yield. Cambridge University Press, Cambridge

Fitter AH, Fitter RS (2002) Rapid changes in flowering time in British plants. Science 296:1689–1691

Forrest J, Miller-Rushing AJ (2010) Toward a synthetic understanding of phenology in ecology and evolution. Phil Trans R Soc B 365:3101–3112

Garner WW, Allard HA (1920) Effect of the relative length of day and night and other factors of the environment on growth and reproduction in plants. J Agric Res 18:553–603

Gassner G (1918) Beiträge zur physiologischen characteristic sommerund winterannueller Gewächse, insbesondere der Getreidepflanzen. Z Botanik 10:417–480

Heide OM (1994) Control of flowering and reproduction in temperate grasses. New Phytol 128:347–362

Heide OM, Sønsteby A (2007) Interactions of temperature and photoperiod in the control of flowering of latitudinal and altitudinal populations of wild strawberry (*Fragaria vesca*). Physiol Plant 130:280–289

King RW (1998) Dual control of flower initiation and development by temperature and photoperiod in *Hardenbergia violacea*. Aust J Bot 46:65–74

King RW, Heide OM (2009) Seasonal flowering and evolution: the heritage from Charles Darwin. Funct Plant Biol 36:1027–1036

King RW, Dawson IA, Speer SS (1992) Control of growth and flowering in two Western Australian species of *Pimelea*. Aust J Bot 40:377–388

King RW, Pate JS, Johnston J (1996) Ecotypic differences in the flowering of *Pimelea ferruginea* (Thymelaeaceae) in response to cool temperatures. Aust J Bot 44:47–55

King RW, Worall R, Dawson IA (2008) Diversity in environmental controls of flowering in Australian plants. Sci Hort 118:161–167

Koornneef M, Alonso–Blanco C, Vreugdenhil D (2003) Naturally occurring genetic variation in Arabidopsis. Annu Rev Plant Biol 56:141–172

Lang A (1965) Physiology of flower initiation. In: Ruhland W (ed) Encyclopaedia of plant physiology, vol 15. Springer, Berlin, pp 1380–1536

Mitchell-Olds T, Willis JH, Goldstein DB (2007) Which evolutionary processes influence natural genetic variation for phenotypic traits? Nat Rev Genet 8:845–856

Rye BL (1988) Revision of Westen Australian Thymeleaceae. Nuytsia 6:129–278

Scarcelli N, Kover PX (2009) Standing genetic variation in FRIGIDA mediates experimental evolution of flowering time in Arabidopsis. Mol Ecol 18:2039–2049

Trevaskis B, Hemming MN, Dennis ES, Peacock WJ (2007) The molecular basis of vernalization-induced flowering in cereals. Trends Plant Sci 12:352–357

Wilczek AM, Roe JL, Knapp MC, Cooper MD, Lopez-Gallego C, Martin LJ, Muir CD, Sim S, Walker A, Anderson J, Egan JF, Moyers B, Petipas R, Giakountis A, Charbit E, Coupland G, Welch SM, Schmitt J (2009) Effects of genetic perturbation on seasonal life history plasticity. Science 323:930–934

Part III
Macroevolution

Chapter 10
The Emergence of Cellular Complexity at the Dawn of the Eukaryotes: Reconstructing the Endomembrane System with In Silico and Functional Analyses

Lila V. Koumandou and Mark C. Field

Abstract Eukaryotic cells depend on a complex network of intracellular organelles to perform endocytosis and exocytosis. These trafficking routes underlie many vital cellular processes, including nutrition, responses to environmental cues, defense from pathogens, and differentiation. Multiple disease mechanisms arise from defects in these pathways. How this complex system arose, especially when compared to the simpler trafficking systems of prokaryotes, remains largely unanswered. However, the availability of fully sequenced genomes from many diverse eukaryotic taxa and representing distinct lineages, increasingly facilitates the reconstruction of very early events in eukaryotic evolution. Studies based on comparative genomics and phylogenetics point to great complexity being already present in the last common ancestor of all eukaryotes and enriched with lineage-specific variability/flexibility. Here we describe the methodology and limitations behind such studies, how conclusions can be enhanced by functional analysis, as well as recent results relating to evolution of Rab small GTPases and the retromer complex.

10.1 Introduction

The endomembrane system of eukaryotic cells mediates uptake from the environment and transport of proteins, nutrients, and a variety of other molecules within the cell as well as release from the cell by secretion. The system comprises various

L.V. Koumandou
Department of Pathology, University of Cambridge, Tennis Court Road, Cambridge CB2 1QP, UK
Biomedical Research Foundation, Academy of Athens, Soranou Efesiou 4, 115 27 Athens, Greece
e-mail: koumandou@cantab.net

M.C. Field
Department of Pathology, University of Cambridge, Tennis Court Road, Cambridge CB2 1QP, UK

P. Pontarotti (ed.), *Evolutionary Biology – Concepts, Biodiversity, Macroevolution and Genome Evolution*, DOI 10.1007/978-3-642-20763-1_10,

organelles and membrane-bound transport intermediates, as well as recognition factors and effectors that direct traffic between the different compartments. Exocytosis can be thought of as starting at the endoplasmic reticulum where new proteins are synthesized and translocated across the ER membrane. From the ER, they are transported to the Golgi complex for post-translational modifications and packaged into vesicles that travel through the cell, and eventually fuse with the plasma membrane to release their contents to the outside of the cell. Conversely, endocytosis starts at the plasma membrane, where selected cargo is packaged into vesicles and trafficked to the endosomes, from where the cargo can either be rapidly recycled back to the plasma membrane, or proceed to the late endosome, the multivesicular body, and the lysosome for breakdown. Retrograde routes from the endosome to the Golgi complex, and from the Golgi complex to the ER also exist.

Common to all these routes are the processes of (a) cargo selection and vesicle formation, (b) vesicle transport along the cytoskeleton, and (c) recognition of the target compartment and vesicle fusion with the target membrane (Bonifacino and Glick 2004). Small GTPases of the Rab family mediate all these steps, and are crucial to orchestrating additional factors such as adaptins for cargo selection before vesicle budding, recruitment of the vesicle coat polymer, uncoating factors, effectors for interactions with the cytoskeleton, tethering factors for recognition of the target compartment, as well as SNARE proteins, which mediate membrane fusion for release of the vesicle's contents into the target compartment. Most of these proteins are members of large protein families, with paralogues restricted to specific cellular locations. For example, different members of the Rab GTPase family are restricted to specific cellular locations and responsible for specific trafficking routes (Stenmark and Olkkonen 2001). Members of the SNARE protein family and the adaptins are also restricted to specific compartments (Chen and Scheller 2001). Remarkably, the overall functions of orthologues appear to remain well conserved across the eukaryotes, so that, e.g., Rab11 is involved in recycling endocytic pathways in plants, mammals, chromalveolates, and excavates (Brighouse et al. 2010).

Distinct vesicle coats exist for endocytic vesicles (clathrin), for vesicles mediating ER to Golgi transport (COPII), and for vesicles traveling along the retrograde routes between the endosome and the Golgi (retromer) and the Golgi and ER (COPI). In a similar fashion to the Rabs, the vesicle coats (clathrin, COPI, and COPII) are related; the evolutionary history of these "protocoatomer" systems has yet to be fully elucidated and at present, some of the relationships are based on secondary structural conservation only (DeGrasse et al. 2009; Devos et al. 2004). Significantly, each coat system has a restricted cellular location (Devos et al. 2004). It is also possible that the protocoatomer has a direct prokaryotic origin as proteins with similar architectures have been reported in some bacterial lineages (Santarella-Mellwig et al. 2010). Introducing some variability on this theme, the tethering factors are protein complexes, each with a distinct cellular localization. Not all are members of the same protein family, although several may be structurally related to protocoatomer (Koumandou et al. 2007; Nickerson et al. 2009).

This pattern of evolution of large protein families with specific localization to distinct cellular compartments, poses the question of whether the families expanded as new compartments arose with increasing eukaryotic complexity. However, the general pattern that has emerged to date, from a variety of studies, points to the early emergence of a complex eukaryotic cell, the components of which are shared among all extant eukaryotic lineages. This argument has two correlates, one pointing to an ancient and possibly rapid diversification of eukaryotic lineages, and the other to the universal conservation of most proteins involved in endocellular trafficking among all eukaryotic lineages.

As most functional studies of the endomembrane trafficking system are carried out in yeast and mammals, these provide only a limited sampling of eukaryotic diversity. To examine the early evolution of the system's complexity across all eukaryotes, a much wider sampling of organisms is necessary. The evolution of eukaryotic lineages has been examined by phylogenetic, phylogenomic, and morphological methods, with increasingly available genomic data allowing fine-tuning of the overall picture. Molecular phylogenies group eukaryotes into five major lineages: (1) the Opisthokonta, including the animals and fungi, (2) the Amoebozoa, (3) the Archaeplastida, (4) the Excavata, and (5) the SAR clade, which includes the Rhizaria, the Alveolates (ciliates, dinoflagellates, Apicomplexa), and the Stramenopiles (brown algae and diatoms amongst others). It remains to be established whether the Haptophyta and Cryptophyceae possibly also belong to the SAR group (Adl et al. 2005; Burki et al. 2007; Hackett et al. 2007; Keeling et al. 2005; Simpson and Roger 2004). Regardless, all of these groups are uniformly deep-branching, meaning that the steps toward the formation of the last eukaryotic common ancestor (LECA) are largely inaccessible to us by phylogenetic methods. However, reconstructing the LECA is possible to a large extent (Field and Dacks 2009).

Previous studies on the evolution of various protein families, such as Rabs (Dacks and Field 2004; Pereira-Leal 2008), vesicle coats (Devos et al. 2004), tethers (Koumandou et al. 2007), ESCRTs (Leung et al. 2008), and SNAREs (Dacks and Doolittle 2004), reveal a highly complex LECA with evidence for all major organelles and trafficking routes having become established before the diversification of the eukaryotic lineages. Such analyses can also identify earlier and later diverging members within each protein family, as well as secondary losses in lineages where certain factors were nonessential. Although protein family expansion is common within the Metazoa, a surprising level of diversity is also shared across all eukaryotic lineages, so that innovation is a general phenomenon. This underscores the need for comparative cell biology to fully understand the functionality present in these diverse lineages. Here we describe our general strategy used in these studies, and extend the analysis to some new results for the Rab protein system and the retromer complex.

10.2 Bioinformatic Workflow

While most molecular cell biology studies tend to focus on a restricted group of select organisms, especially yeast, mammals, and invertebrate metazoan models, assessment of the origins of eukaryotic cellular functional diversity and capacity depends on sampling the full diversity of organisms. Plenty of fully sequenced genomes are now available for opisthokonts, representing both classical model systems (e.g., *Saccharomyces cerevisiae, Schizosaccahromyces pombe, Drosophila melanogaster, Ceanorhabditis elegans*), and species which diverged earlier in the lineage (e.g., *Rhizopus orhyzae* in the fungi; *Nematostella vectensis* and *Monosiga brevicollis* for the Metazoa). Plant and algal diversity can be encompassed by including both multicellualr and unicellular representatives, as well as both red and green algae. For the excavates, the amoebozoa, and the SAR group (stramenopiles, alveolates, rhizaria), fully sequenced species represent vastly differing lifestyles, and only recently does the set of organisms with fully sequenced genomes begin to cover the true group diversity. Inevitably, the focus has been on organisms of economic or public health priority, with an obvious bias toward highly derived species that are frequently pathogenic; this is, however, being overcome as the cost of sequencing even large eukaryotic genomes has fallen, and hence both better and denser sampling is now apparent. Genomes for the cryptophyte *Guillardia theta* and the haptophyte *Emiliana huxleyi* have become available only very recently, and the phylogenetic position of these organisms is still under debate. Where possible, two or more taxa must be included from any supergroup to facilitate detection of secondary losses versus absence from an entire group and to minimize detection failure because of species-specific divergence or incompleteness in the database.

For comparative genomic analysis, genomic databases can be retrieved from the NCBI BLAST interface (http://www.ncbi.nlm.nih.gov/BLAST/), the Joint Genome Institute (JGI) (http://genome.jgi-psf.org/euk_cur1.html), the Broad Institute (http://www.broadinstitute.org/), the Sanger Institute (http://www.genedb.org/), EuPathdb (http://eupathdb.org/eupathdb/), as well as organism-specific BLAST servers, e.g., for *C. merolae* (http://merolae.biol.s.utokyo.ac.jp/blast/blast.html), *T. gondii* (http://www.toxodb.org/), *C. parvum* (http://www.cryptodb.org/cryptodb/), *Giardia intestinalis* (http://www.giardiadb.org/giardiadb/). In addition, predicted proteomes for most species can be downloaded by ftp from the respective database for local analysis.

The strategy for finding orthologous genes has a number of steps to ensure robustness of the results, and at present is heuristic (Fig. 10.1). Searches are done using protein sequences, as they are overall more highly conserved than nucleotide sequences for distantly related species, and avoid any effects from codon bias. In our approach, BLASTp searches are largely performed manually, and checked by hand, as expect value (E-value) cutoff thresholds can vary considerably between different organisms, and for different proteins, especially for large vs. small proteins or those that retain more restricted structural features. The BLOSUM62

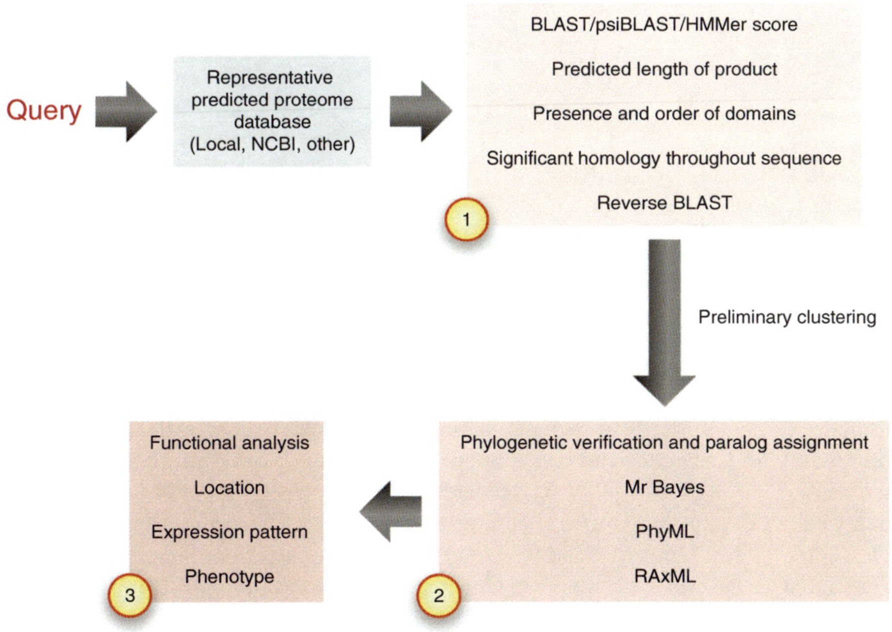

Fig. 10.1 An informatics workflow for comparative genomics. A query sequence is subjected to several tests to ensure that orthology is confidently predicted. These include a number of criteria designed to reduce miscalls due to regions of local similarity, and also frequently rely on the use of high-quality phylogenetic algorithms. See text for fuller discussion. Depending on the precise question being asked, progression from box 1 to box 2 or box 3 may not be required for fulfillment of an accurate call. The deeper intensity of the background color signifies the increased burden of moving from one box to the next

or BLOSUM45 substitution matrix is normally used, and, in general, E-values below e-3 are considered significant. Initial query sequences may be from yeast or human, and are used to search individually against each different organism; we find that this organism-by-organism approach leads to less overinterpretation when compared to a broader search. If the search identifies one clear hit, i.e., the top BLAST hit has an E-value much lower than all subsequent hits, then only the top hit is examined further. If the search identifies multiple top hits with similar E-values, then all top hits are examined further to determine if they represent paralogues. For example, small G proteins frequently generate multiple high-quality hits due to the conservation of the GTPase-binding site, necessitating further inspection for assignment (Fig. 10.1).

The candidate BLAST hits are subsequently tested by reverse BLAST, i.e., used to search the yeast or human proteome, or the nr database, and should return the original query or annotated orthologues from other species within the top five hits. In addition, the length of the putative orthologues should be similar to the original query, and any domains identified in the original query should also be conserved in the orthologue, which can easily be done by parsing through the NCBI conserved domain database (CDD) or similar. We find this to be simple to perform, but frequently is an excellent discriminator; many candidates have only moderate support from E-value alone, but taken with conserved size and domain architecture, a good case can often be made for an evolutionary relationship.

Finally, further support is provided if the alignment between orthologues spans the full length of the sequence, and is not only concentrated in the conserved domain regions. This is particularly important for common domains, as the presence of the domain alone does not guarantee that the hit corresponds to a true orthologue; many domains are very highly conserved, providing respectable BLAST scores, but only taking account of comparatively restricted regions of the protein; if in doubt, a region of high homology can be removed from the sequence and the BLAST analysis rerun to ascertain if the remaining portions of the candidate have any relationship to the initial query. For analyses of many proteins (e.g., for the retromer cargo proteins presented here), the BLAST and reverse BLAST searches can be automated, e.g., with a BioPerl script that retrieves the BLAST results, and only records homologues if the reverse BLAST to the original query has an E-value better than a set threshold (e.g., e-3). However, examination of at least some of the results by hand is strongly advised, to check for length and domain agreement, as described above, and any negative results (not found) should be treated with caution and may need to be reexamined (e.g., with HMMer, see below).

In cases where no hits are retrieved by the original query, or no correspondence is found by reverse BLAST, or where the length and domain information are dubious, more detailed searches can be performed. One strategy is loosely termed "genome walking," i.e., using as an original query an orthologue from a closely related organism (e.g., using an *Arabidopsis* protein as query to search in *Chlamydomonas*). If this also fails, HMMer or PSI-BLAST can be used; these use the entire set of sequences for each protein family to generate a profile or consensus sequence and search based on that against any proteome in which BLAST did not recover a homologue. These are more sensitive than BLAST, and usually guaranteed to identify even highly divergent hits. The important downside here is that the sequences retrieved may have very weak sequence similarity and in fact lack a true evolutionary relationship – this is a particular issue when sequences contain coiled-coil regions, which are rather frequent in trafficking factors, and underscores the importance of manual curation of datasets.

In cases where all these attempts fail, or the results are rejected based on phylogeny (see below), the conclusion is that an orthologous protein is not found in this organism. This may be due to a true loss, due to the limits of detection of similarity search algorithms for highly divergent sequences, or due to sampling/misannotation errors in the available genome sequence. Examination of the genomic context of the gene may provide further clues here. For example, if the gene is within a syntenic region, where the order of genes is conserved between different species, one can examine whether the neighboring genes are conserved, although this is only of use when examining closely related taxa. If that is indeed the case, and the protein-coding sequence is absent, this is a powerful argument for secondary loss, which usually means that the gene's function was redundant or nonessential. Presence or absence in closely related species can also be examined before secondary loss is invoked for a whole lineage. Conversely, if a protein is retained only in certain species or lineages where it might not be expected by parsimonious evolution (i.e., suggesting multiple independent acquisitions or losses), related factors can be examined to add robustness to the results; in some cases, examination for genes that contribute toward a given pathway can be helpful. For example, in the analysis of retromer cargo, the cation-independent mannose 6-phosphate receptor (CIMPR) is classically responsible for lysosomal delivery of proteins bearing the mannose-6-phosphate modification. As only some species outside the Metazoa were found to posses CIMPR, it was necessary to look for the presence of genes encoding the two enzymes required for mannose-6-phosphate modification, N-acetylglucosamine-1-phosphotransferase subunits α/β precursor (GNPTAB) and N-acetylglucosamine-1-phosphodiester α (NAGPA).

Finally, all candidate orthologues are examined by phylogeny to confirm orthology. Bayesian as well as maximum likelihood methods are used, including repeated sampling rounds to obtain posterior probability, and bootstrap support values, respectively. In cases where multiple orthologues are found in certain species, phylogeny can distinguish whether these are from ancient or recent gene duplications. Bayesian phylogeny can be run locally using Mr Bayes (http://mrbayes.csit.fsu.edu/), but for large datasets, a processor cluster is essential. Finally, maximum likelihood methods can be performed using remote web servers (http://www.hiv.lanl.gov/content/sequence/PHYML/interface.html and http://phylobench.vital-it.ch/raxml-bb/index.php provide good options), or can be run locally.

10.3 Results

One of the most important families of proteins involved in membrane trafficking are the Rab GTPases, and these proteins have received considerable attention (Stenmark 2009). They function to coordinate the actions of vesicle budding, targeting and fusion, and interact with a large number of proteins (Lee et al. 2009). As Ras-like GTPases, their intrinsic enzymatic activity is poor and hence hydrolysis of GTP requires the intervention of a GTPase activating protein (GAP). Rabs constitute a large family, with over 70 in *H. sapiens* and over 300 in *T. vaginalis*. As this topic has been reviewed extensively recently (Brighouse et al. 2010; Elias 2010), we will focus on a few specific issues here and the reader is referred elsewhere for a broader perspective.

A major goal has been the derivation of a Rab phylogeny (Pereira-Leal and Seabra 2001). As Rab orthologues are almost always associated with the same organelle, even across large evolutionary distances, these proteins conceptually provide an atlas of the compartments present, and critically this makes such information accessible for organisms that are hard to analyze experimentally for technical reasons. Hence, being able to determine the Rab complement in any lineage would provide an extremely valuable insight into the structure of the endomembrane system. Further, understanding how Rab proteins have evolved would also solve the problem of when specific compartments arose, how they were expanded, and how they were lost. Additionally, such an analysis avoids the asymmetry problem, whereby it is easy to find conservation or secondary losses in divergent taxa, but due to significantly less direct experimental data, identification of true novel features in divergent lineages is rather more challenging.

Deriving a robust Rab phylogeny is more complex than might appear, as the database is misannotated, Rabs are small proteins (~250 amino acids) with a combination of highly variable C-termini and extremely well-conserved GTP-binding regions, plus the dataset is huge. This combination of factors has confounded many attempts to provide high-resolution phylogenies as there are too few informative character states for accurate resolution. However, a new method, "ScrollSaw," that essentially examines subsets of sequence data, and then combines these has been derived, which facilitates great improvement over traditional methods (Elias et al. manuscript in preparation). What is surprising is that the reconstruction predicts a considerable Rab complement in LECA, but which is consistent with the emerging view of great complexity in this organism (Fig. 10.2). This also indicates that secondary loss has played a significant role in evolution of the Rab protein family.

As a means to validate this conclusion, we have also performed detailed analysis of the TBC (Tre-2, Bub2, Cdc16) domain Rab GAP family (Gaberet-Castello et al. manuscript in preparation). This family accounts for most known Rab GAPs (Pan et al. 2006), and was selected as it is paralagous and the TBC domain facilitates reliable identification. The number of TBC proteins encoded in the genomes of most organisms is similar to the number of Rabs. The specificity of most TBC GAPs is poorly defined (Barr and Lambright 2010; Will and Gallwitz 2001), raising the issue of how activity is regulated. Applying the ScrollSaw approach to TBC evolution, it is clear that innovation of TBC GAPs is complex. Again, a large cohort is predicted to be present in the LECA, but it is also clear that lineage-specific innovations postdate the LECA; it is remarkable that the TBC GAPs and Rab phylogenies achieve such an overall degree of congruence, which gives confidence that the conclusion is probably correct.

As an example of a smaller and more restricted system, we have also analyzed the retromer complex, which is involved in retrograde traffic from the endosome to the Golgi. In *S. cerevisiae* and mammalian cells, retromer comprises five subunits: a sorting nexin dimer (Vps5 and Vps17 in yeast, SNX1 and SNX2 in mammals) which mediates membrane binding via PX domains and senses membrane curvature via BAR domains, and a trimeric subcomplex formed of Vps26, Vps29, and Vps35, which is responsible for cargo selection.

Retromer mediates recycling of vacuolar hydrolase receptors in yeast and mammals (Seaman et al. 1998; Arighi et al. 2004; Mari et al. 2008), as well as trafficking of the polymeric immunoglobulin receptor (Verges et al. 2004), plasma membrane iron transporters (Strochlic et al. 2007), Wntless (Eaton 2008), and processing of the amyloid precursor protein (He et al. 2005). Recently, retromer was also implicated in clearance of apoptotic bodies (Chen et al. 2010) and trafficking from the mitochondria to the peroxisome (Braschi et al. 2010). Using the comparative

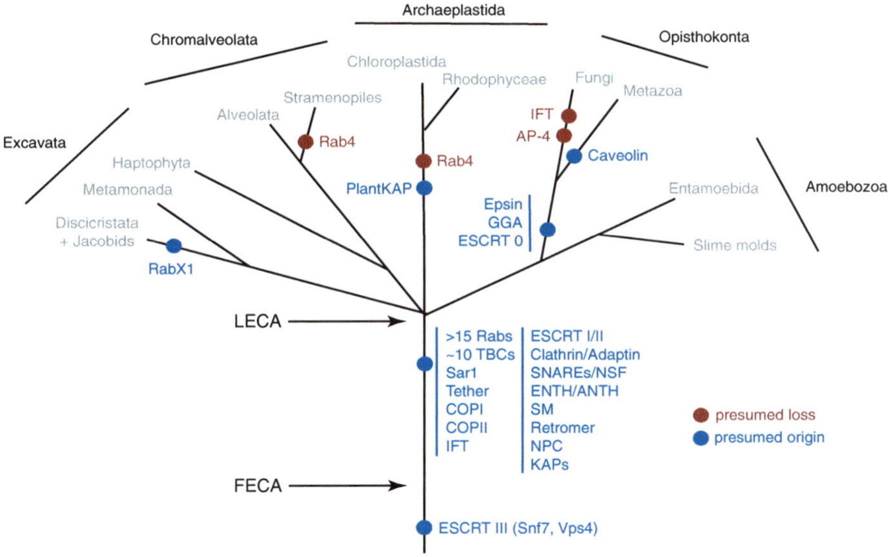

Fig. 10.2 Schematic eukaryotic phylogeny (sensu Adl) highlighting origins of trafficking components. Much of the basic bauplan for the eukaryotic cell predates the radiation of the eukaryotes, so that major coats, the factors required for vesicle specificity and the various control elements were all in place by the last eukaryotic common ancestor (LECA). The tree omits much detail, especially the origins and losses of a great many members of paralagous families from specific taxa, and also ignores any potential lateral gene transfer. *ANTH* AP180 N-Terminal Homology (ANTH) domain, *COP* coatomer, *ENTH* Epsin N-terminal homology (ENTH) domain, *ESCRT* endosomal sorting complex required for transport (a late endosomal membrane bending system also involved in cytokinesis), *FECA* first eukaryotic common ancestor (broadly equivalent to the eukaryogenesis event itself), *IFT* intraflagellar transport, *KAP* karyopherin (nucleocytoplasmic transport receptors), *NSF* NEM-sensitive factor (an ATPase that mediates SNARE protein complex disassembly), *NPC* nuclear pore complex or nucleoporins, *SM* Sec1/Munc18-like proteins (involved in SNARE-mediated vesicle fusion), *SNARE* SNAP (Soluble NSF attachment protein) receptors (coiled coil proteins required for vesicle fusion), *TBC* Tre-2, Bub2, Cdc16 domain (Rab GTPase activating proteins)

genomics workflow described above, we find that the Vps26/Vps29/Vps35 subcomplex is extremely well conserved. Interestingly, all cargo recognition subunits show expansions, with Vps26 the most widely expanded. Phylogenetics indicates that most expansions are species-specific (Koumandou et al. 2011).

The Vps5/Vps17 membrane-attachment subcomplex is also well conserved but less well than the cargo recognition complex. Vps17 is specific to the fungi, but SNX5/6 are probable functional analogues in Metazoa (Wassmer et al. 2007). Vps5 has been duplicated into SNX1/2 in Metazoa. Humans possess a total of 33 sorting nexins, fungi about eight, and non-Opisthokonta about four, indicating a huge and specific expansion in Metazoa. While speculative, this may reflect the huge complexity of endosomal systems in metazoan organisms where these trafficking systems perform important roles in cell–cell adhesion, signaling pathways, immune defense, and development, which may require specific adaptors to traffic distinct cohorts of molecules through the late endosomal pathway.

Given the extremely good conservation of retromer throughout the eukaryotes, we also examined conservation of the retromer cargo. Vps10, the best characterized of this group, is a transmembrane lysosomal hydrolase receptor orthologous to mammalian sortilins (Mari et al. 2008). We find that Vps10 is broadly conserved with expansions in *Homo sapiens, Danio rerio, Nematostella vectensis, Monosiga brevicollis, Saccharomyces cerevisiae, Rhizopus oryzae,* and *Tetrahymena thermophila.* The *S. cerevisiae, R. oryzae*, and *T. thermophila* expansions are species-specific while metazoan-specific expansions have generated several distinct metazoan sortilin/Vps10 families. However, Vps10 is absent from several lineages, suggesting multiple secondary losses and, importantly, a role for retromer in sorting distinct sets of cargo in different organisms.

We therefore performed comparative genomics for the 14 previously reported retromer cargo proteins, as well as the retromer-interacting protein EHD1 (Gokool et al. 2007). Most studies of retromer and its cargo are from opisthokonts and indeed many of the reported cargo proteins are specific to the Metazoa, e.g., PIGR, EGFR, and Wntless, as they are involved in metazoan-specific signaling or immune defense. The vacuolar sorting receptor VSR1, originally identified in *A. thaliana* (Yamazaki et al. 2008), is restricted to the Archaeplastida. However, we also found widely distributed cargo, namely, EHD1, STE13, KEX2, and the FET3/FTR1 iron transporter. Importantly, all species lacking Vps10 contain at least one putative alternative cargo (Koumandou et al. 2011).

Sequence conservation alone does not immediately signify conservation of function. As part of our workplan, we use *Trypanosoma brucei* as a both accessible and highly divergent organism to facilitate comparisons with mammalian or other model systems. *In S. cerevisiae*, retromer mutants exhibit variable deficiencies with highly fragmented vacuoles in Vps5 and Vps17 mutants, moderate fragmentation in Vps26 mutants and no observable morphological defects for Vps29 and Vps35 mutants (Raymond et al. 1992). Mammalian SNX1/2 colocalize with endosomal markers EEA1 and Rab5, and with the mammalian Vps26-Vps29-Vps35 trimer (Haft et al. 2000; Kurten et al. 2001; Teasdale et al. 2001). Mouse SNX1 and SNX2 double knockouts arrest embryonic development, as do mutations in mammalian Vps26, and transcriptome analysis implicates Vps35 in Alzheimer's disease, underlining the importance of retromer to mammalian systems (Schwarz et al. 2002; Radice et al. 1991; Small et al. 2005).

In trypanosomes Vps5/Vps26/Vps29 and Vps35 mRNA expression is strongly upregulated in the mammalian form (Koumandou et al. 2008, 2011) where endocytosis is also more active (Natesan et al. 2007) and consistent with a role for retromer in endocytic activity. Retromer is also clearly essential as knockdowns of several subunits arrest cell proliferation (Koumandou et al. 2011). Trypanosome Vps26 and Vps5 exhibit both diffuse cytoplasmic localization plus distinct puncta located between the nucleus and kinetoplast, likely corresponding to endosomes (Field and Carrington 2009). TbVps26 partially colocalized with the clathrin heavy chain, and markers of the early and recycling endosomes; it was also proximal to

Vps28, which marks the multivesicular body (MVB), and to the lysosomal marker p67. Overall, these data indicate an endosomal location (Koumandou et al. 2011). Knockdowns also confirm a role in endosomal trafficking and result in a modest increase in p67 expression, representing a possible increase to lysosomal traffic via a block of retrograde retromer traffic from the endosome to the Golgi. Further, there is a decrease in intracellular levels of ISG75, a transmembrane protein that undergoes ubiquitin-dependent degradation (Chung et al. 2008; Leung et al. 2008), which is likely due to accelerated turnover. Finally, silencing Vps26 in mammalian cells results in fragmentation of the Golgi complex (Seaman 2004); a similar effect was found in Vps26-silenced trypanosomes (Koumandou et al. 2011). Together, the locations and effects of retromer subunit knockdowns suggest functional conservation between trypanosome, mammalian and yeast retromer.

10.4 Conclusions and Challenges

Understanding how the modern eukaryotic cell architecture arose, and was subsequently modified by differential selective pressures, has been a major goal in evolutionary cell biology. This is not solely of academic interest as many eukaryotic pathogens invest considerably in their cell surface as a host–pathogen interface and a site for immune evasion. An understanding of what such lineages have on board in terms of molecular components and function is a potentially potent weapon for combating infection and agricultural pathogens.

What is now very clear is the great complexity of LECA, and that this complexity encompasses many different families of proteins, adding confidence that this view is correct. We are also beginning to suspect that many extant organisms are in fact simpler than LECA with respect to their trafficking systems. This suggests that secondary losses, as well as paralogous expansions, are a major evolutionary driver responsible for the diversification of different lineages. A few factors appear to have been carried over from prokaryotic origins, but the majority are probably *de novo* innovations restricted to eukaryotes (Fig. 10.3).

Combined obstacles have made the elucidation of the evolutionary history of cellular functions a challenging task, and it remains incompletely addressed. In part, our view of some of the earliest events in eukaryogenesis is still very uncertain, but with the increase in genome sequencing, it is now possible to utilize molecular sequence data to address such problems. While still capable of generating equivocal or poorly supported models, with the inevitable controversies, such approaches have significant advantages. For the unwary, however, there remain some major pitfalls, which become even more pressing with increased size and complexity of datasets, and which necessitate the use of automated search algorithms.

We have developed a workflow that attempts to address at least some of these issues, and which is predicated on a reliance for heuristic analysis and the application of some biological principles. Such approaches are labor intensive and slower than fully automated approaches, but we consider them to be ultimately more accurate. The full workflow can incorporate functional studies, as we describe

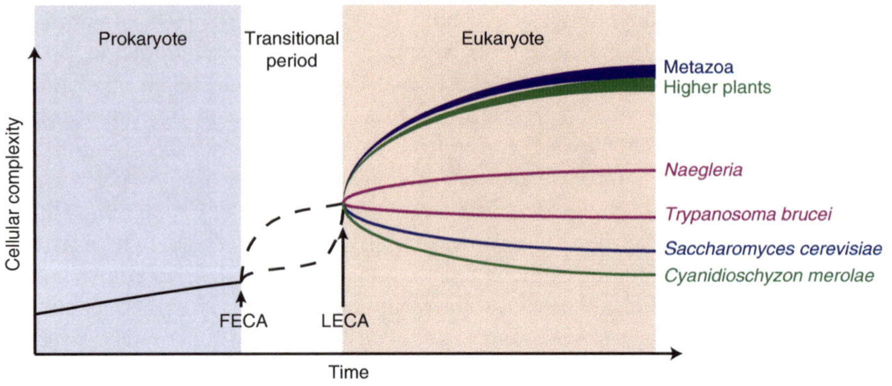

Fig. 10.3 Schematic for evolution of complexity in eukaryotic cells. Prokaryotic evolution (*blue*) proceeds in this model to increase complexity over time, but at the point of eukaryogenesis, equivalent to the first eukaryotic common ancestor (FECA, *arrow*), the acquisition of a nucleus propels the ability to increase complexity. Two extreme potential evolutionary trajectories are shown (*dotted lines*), where the initial event was followed by a period of rapid innovation (*top*), or a period of little innovation followed by rapidly increased complexity (*lower*). All trajectories between these extremes are possible, with the eventual arrival at the last eukaryotic common ancestor (LECA, *arrow*), which likely also represents an extreme bottleneck as all eukaryotes appear to radiate from a single lineage. Following eukaryotic radiation, many taxa evolved to increased complexity, of which the most potent examples are the Metazoa and higher plants. Some lineages appear to resemble the LECA in complexity, e.g., *Naegleria gruberi*, while many other taxa, including *Trypanosoma*, yeasts, and the extremophile red algae *C. merolae* have become less complex due to secondary losses of many components. Opisthokonts are in *blue*, Archaeplastida in *green*, and Excavata in *purple*. The diagram is heavily schematic and seeks to make general points only

here for retromer and several Rab GTPases. Such studies are complex, expensive, and prone to interpretive error as a great deal of interpretive expertise needs to be used, and one is frequently operating in organisms where the level of knowledge is sparse. However, such analysis can provide substantial support for *in silico* calls. For example, Tsg101/Vps23, an ESCRT complex subunit, in trypanosomes has very low similarity to the mammalian orthologue; knockdown and localization studies, however, confirm that the gene product plays a role in late endosomal transport, providing a strong argument that the *in silico* assignment is correct (Leung et al. 2008). Additionally, localization of SNARE proteins in trypanosomes has helped increase confidence in *in silico* assignments, which for these proteins can be difficult (Besteiro et al. 2006).

Several issues remain as challenges, of which at least three are paramount. First is the ongoing issue of asymmetry, whereby most *ab initio* identification of proteins involved in trafficking pathways is performed in a very small number of opisthokont taxa (Dacks and Field 2007). While using these organisms as a basis for comparative genomics will identify conserved elements and potential secondary losses or opisthokont lineage-specific innovations, by definition it fails to capture innovations in other supergroups. Analyses that focus on broader paralogous families can go some way to solving this problem, and as more genome data becomes available, this may fade as a major issue. Second, search algorithms themselves are problematic. BLAST itself is rather insensitive, but using pattern recognition as implemented by HMMER or PSI-BLAST greatly increases the potential for false positives. Regardless, even these algorithms can fail to capture candidates, as demonstrated recently with a HMMER-based reconstruction of nuclear pore complex evolution – this is significantly better than BLAST alone, but still failed to identify many gene products (DeGrasse et al. 2009; Neumann et al. 2010). Third, sequence relationships are not the same as functional equivalence, which necessitates the expense and expertise required to gain direct functional insight. However, without such evidence, much valuable insight can be simply overlooked.

In summary, it is clear that the LECA was a complex organism. While some processes were likely inherited directly from prokaryotic predecessors, a spectacular level of innovation seems to have accompanied progression from the eukaryogenesis event itself to LECA.

Acknowledgments Many of these studies have been supported in part by the Wellcome Trust. We are also most grateful to collaborators and colleagues for discussions and access to unpublished data, and especially Carme Gabernet-Castello, Joel B. Dacks, Michael P. Rout, and Marek Elias.

References

Adl SM, Simpson AG, Farmer MA, Andersen RA, Anderson OR, Barta JR, Bowser SS, Brugerolle G, Fensome RA, Fredericq S, James TY, Karpov S, Kugrens P, Krug J, Lane CE, Lewis LA, Lodge J, Lynn DH, Mann DG, McCourt RM, Mendoza L, Moestrup O, Mozley-Standridge SE, Nerad TA, Shearer CA, Smirnov AV, Spiegel FW, Taylor MF (2005) The new higher level classification of eukaryotes with emphasis on the taxonomy of protists. J Eukaryot Microbiol 52:399–451

Arighi CN, Hartnell LM, Aguilar RC, Haft CR, Bonifacino JS (2004) Role of the mammalian retromer in sorting of the cation-independent mannose 6-phosphate receptor. J Cell Biol 165:123–133

Barr F, Lambright DG (2010) Rab GEFs and GAPs. Curr Opin Cell Biol 22:461–470

Besteiro S, Coombs GH, Mottram JC (2006) The SNARE protein family of *Leishmania major*. BMC Genomics 7:250

Bonifacino JS, Glick BS (2004) The mechanisms of vesicle budding and fusion. Cell 116:153–166

Braschi E, Goyon V, Zunino R, Mohanty A, Xu L, McBride HM (2010) Vps35 mediates vesicle transport between the mitochondria and peroxisomes. Curr Biol 20:1310–1315

Brighouse A, Dacks JB, Field MC (2010) Rab protein evolution and the history of the eukaryotic endomembrane system. Cell Mol Life Sci 67:3449–3465

Burki F, Shalchian-Tabrizi K, Minge M, Skjaeveland A, Nikolaev SI, Jakobsen KS, Pawlowski J (2007) Phylogenomics reshuffles the eukaryotic supergroups. PLoS ONE 2:e790

Chen YA, Scheller RH (2001) SNARE-mediated membrane fusion. Nat Rev Mol Cell Biol 2:98–106

Chen D, Xiao H, Zhang K, Wang B, Gao Z, Jian Y, Qi X, Sun J, Miao L, Yang C (2010) Retromer is required for apoptotic cell clearance by phagocytic receptor recycling. Science 327:1261–1264

Chung WL, Leung KF, Carrington M, Field MC (2008) Ubiquitylation is required for degradation of transmembrane surface proteins in trypanosomes. Traffic 9:1681–1697

Dacks JB, Doolittle WF (2004) Molecular and phylogenetic characterization of syntaxin genes from parasitic protozoa. Mol Biochem Parasitol 136:123–136

Dacks JB, Field MC (2004) Eukaryotic cell evolution from a comparative genomic perspective: the endomembrane system. In: Hirt R, Horner D (eds) Organelles, genomes and eukaryote phylogeny: an evolutionary synthesis in the age of genomics. GRC Press, Boca Raton, pp 309–334

Dacks JB, Field MC (2007) Evolution of the eukaryotic membrane-trafficking system: origin, tempo and mode. J Cell Sci 120:2977–2985

DeGrasse JA, DuBois KN, Devos D, Siegel TN, Sali A, Field MC, Rout MP, Chait BT (2009) Evidence for a shared nuclear pore complex architecture that is conserved from the last common eukaryotic ancestor. Mol Cell Proteomics 8:2119–2130

Devos D, Dokudovskaya S, Alber F, Williams R, Chait BT, Sali A, Rout MP (2004) Components of coated vesicles and nuclear pore complexes share a common molecular architecture. PLoS Biol 2:e380

Eaton S (2008) Retromer retrieves wntless. Dev Cell 14:4–6

Elias M (2010) Patterns and processes in the evolution of the eukaryotic endomembrane system. Mol Membr Biol 27:469–489

Field MC, Carrington M (2009) The trypanosome flagellar pocket. Nat Rev Microbiol 7:775–786

Field MC, Dacks JB (2009) First and last ancestors: reconstructing evolution of the endomembrane system with ESCRTs, vesicle coat proteins, and nuclear pore complexes. Curr Opin Cell Biol 21:4–13

Gokool S, Tattersall D, Seaman MN (2007) EHD1 interacts with retromer to stabilize SNX1 tubules and facilitate endosome-to-Golgi retrieval. Traffic 8:1873–1886

Hackett JD, Yoon HS, Li S, Reyes-Prieto A, Rummele SE, Bhattacharya D (2007) Phylogenomic analysis supports the monophyly of cryptophytes and haptophytes and the association of rhizaria with chromalveolates. Mol Biol Evol 24:1702–1713

Haft CR, de la Luz SM, Bafford R, Lesniak MA, Barr VA, Taylor SI (2000) Human orthologs of yeast vacuolar protein sorting proteins Vps26, 29, and 35: assembly into multimeric complexes. Mol Biol Cell 11:4105–4116

He X, Li F, Chang WP, Tang J (2005) GGA proteins mediate the recycling pathway of memapsin 2 (BACE). J Biol Chem 280:11696–11703

Keeling PJ, Burger G, Durnford DG, Lang BF, Lee RW, Pearlman RE, Roger AJ, Gray MW (2005) The tree of eukaryotes. Trends Ecol Evol 20:670–676

Koumandou VL, Dacks JB, Coulson RM, Field MC (2007) Control systems for membrane fusion in the ancestral eukaryote; evolution of tethering complexes and SM proteins. BMC Evol Biol 7:29

Koumandou VL, Natesan SK, Sergeenko T, Field MC (2008) The trypanosome transcriptome is remodelled during differentiation but displays limited responsiveness within life stages. BMC Genomics 9:298

Koumandou VL, Klute MJ, Herman EK, Nunez-Miguel R, Dacks JB, Field MC (2011) Evolutionary reconstruction of the retromer complex and its function in *Trypanosoma brucei*. J Cell Sci 124:1496–1509

Kurten RC, Eddington AD, Chowdhury P, Smith RD, Davidson AD, Shank BB (2001) Self-assembly and binding of a sorting nexin to sorting endosomes. J Cell Sci 114:1743–1756

Lee MT, Mishra A, Lambright DG (2009) Structural mechanisms for regulation of membrane traffic by rab GTPases. Traffic 10:1377–1389

Leung KF, Dacks JB, Field MC (2008) Evolution of the multivesicular body ESCRT machinery; retention across the eukaryotic lineage. Traffic 9:1698–1716

Mari M, Bujny MV, Zeuschner D, Geerts WJ, Griffith J, Petersen CM, Cullen PJ, Klumperman J, Geuze HJ (2008) SNX1 defines an early endosomal recycling exit for sortilin and mannose 6-phosphate receptors. Traffic 9:380–393

Natesan SK, Peacock L, Matthews K, Gibson W, Field MC (2007) Activation of endocytosis as an adaptation to the mammalian host by trypanosomes. Eukaryot Cell 6:2029–2037

Neumann N, Lundin D, Poole AM (2010) Comparative genomic evidence for a complete nuclear pore complex in the last eukaryotic common ancestor. PLoS ONE 5:e13241

Nickerson DP, Brett CL, Merz AJ (2009) Vps-C complexes: gatekeepers of endolysosomal traffic. Curr Opin Cell Biol 21:543–551

Pan X, Eathiraj S, Munson M, Lambright DG (2006) TBC-domain GAPs for Rab GTPases accelerate GTP hydrolysis by a dual-finger mechanism. Nature 442:303–306

Pereira-Leal JB (2008) The Ypt/Rab family and the evolution of trafficking in fungi. Traffic 9:27–38

Pereira-Leal JB, Seabra MC (2001) Evolution of the Rab family of small GTP-binding proteins. J Mol Biol 313:889–901

Radice G, Lee JJ, Costantini F (1991) H beta 58, an insertional mutation affecting early postimplantation development of the mouse embryo. Development 111:801–811

Raymond CK, Howald-Stevenson I, Vater CA, Stevens TH (1992) Morphological classification of the yeast vacuolar protein sorting mutants: evidence for a prevacuolar compartment in class E vps mutants. Mol Biol Cell 3:1389–1402

Santarella-Mellwig R, Franke J, Jaedicke A, Gorjanacz M, Bauer U, Budd A, Mattaj IW, Devos DP (2010) The compartmentalized bacteria of the planctomycetes-verrucomicrobia-chlamydiae superphylum have membrane coat-like proteins. PLoS Biol 8:e1000281

Schwarz DG, Griffin CT, Schneider EA, Yee D, Magnuson T (2002) Genetic analysis of sorting nexins 1 and 2 reveals a redundant and essential function in mice. Mol Biol Cell 13:3588–3600

Seaman MN (2004) Cargo-selective endosomal sorting for retrieval to the Golgi requires retromer. J Cell Biol 165:111–122

Seaman MN, McCaffery JM, Emr SD (1998) A membrane coat complex essential for endosome-to-Golgi retrograde transport in yeast. J Cell Biol 142:665–681

Simpson AG, Roger AJ (2004) The real 'kingdoms' of eukaryotes. Curr Biol 14:R693–R696

Small SA, Kent K, Pierce A, Leung C, Kang MS, Okada H, Honig L, Vonsattel JP, Kim TW (2005) Model-guided microarray implicates the retromer complex in Alzheimer's disease. Ann Neurol 58:909–919

Stenmark H (2009) Rab GTPases as coordinators of vesicle traffic. Nat Rev Mol Cell Biol 10:513–525

Stenmark H, Olkkonen VM (2001) The Rab GTPase family. Genome Biol 2:REVIEWS3007

Strochlic TI, Setty TG, Sitaram A, Burd CG (2007) Grd19/Snx3p functions as a cargo-specific adapter for retromer-dependent endocytic recycling. J Cell Biol 177:115–125

Teasdale RD, Loci D, Houghton F, Karlsson L, Gleeson PA (2001) A large family of endosome-localized proteins related to sorting nexin 1. Biochem J 358:7–16

Verges M, Luton F, Gruber C, Tiemann F, Reinders LG, Huang L, Burlingame AL, Haft CR, Mostov KE (2004) The mammalian retromer regulates transcytosis of the polymeric immunoglobulin receptor. Nat Cell Biol 6:763–769

Wassmer T, Attar N, Bujny MV, Oakley J, Traer CJ, Cullen PJ (2007) A loss-of-function screen reveals SNX5 and SNX6 as potential components of the mammalian retromer. J Cell Sci 120:45–54

Will E, Gallwitz D (2001) Biochemical characterization of Gyp6p, a Ypt/Rab-specific GTPase-activating protein from yeast. J Biol Chem 276:12135–12139

Yamazaki M, Shimada T, Takahashi H, Tamura K, Kondo M, Nishimura M, Hara-Nishimura I (2008) *Arabidopsis* VPS35, a retromer component, is required for vacuolar protein sorting and involved in plant growth and leaf senescence. Plant Cell Physiol 49:142–156

Chapter 11
Neurophylogeny: Retracing Early Metazoan Brain Evolution

Rudi Loesel

Abstract The current view of early metazoan phylogeny suggests that the bilaterian body plan arose only once during evolution. This first urbilaterian animal was most likely equipped with an anterior condensation of nerve cells – a brain – from which all brains of modern animals have diverged. Until recently, the ancestor of all bilaterian phyla was viewed as a very simple animal with an accordingly simple brain. Molecular studies, however, demonstrate a multitude of homologous genes that are expressed in similar patterns in the developing brains of vertebrates, insects, and annelids. Taken together, these findings imply that the anatomy of the urbilaterian cerebrum might have been more elaborate than previously assumed. If true, ancient architectural features might have been conserved during evolution and should be identifiable in distantly related modern animal phyla. Comparative studies on representatives of arthropods, onychophorans, and annelids suggest that this is indeed the case. This chapter summarizes recent neuroanatomical surveys that aim to retrace the early evolution of the metazoan brain and to use neuroanatomical data to test conflicting hypothesis on phylogenetic relationships between major animal phyla.

11.1 Introduction

> As long as the brain is a mystery, the universe will also remain a mystery.
>
> Santiago Ramon y Cajal (Nobel Laureate 1906)

This chapter is based on the habilitation thesis of R.Loesel

R. Loesel
Unit of Developmental Biology and Morphology of Animals, Institute for Biology II (Zoology), RWTH Aachen University, Lukasstrasse 1, 52070 Aachen, Germany
e-mail: loesel@bio2.rwth-aachen.de

P. Pontarotti (ed.), *Evolutionary Biology – Concepts, Biodiversity, Macroevolution and Genome Evolution*, DOI 10.1007/978-3-642-20763-1_11,

The brain is the most complicated structure that has evolved in the 640 million years since the emergence of multicellular animals. Yet our knowledge of early evolutionary events that have brought about this fascinating organ is marginal. Has the brain as an anterior aggregation of neurons that governs behavior evolved only once, implying that the brains of all animals are a variation of a common ancient scheme? Do common neuroarchitectural features shared by distantly related taxa reflect the evolutionary history of the brain or have these features evolved in parallel due to computational needs? Can brain characters help to resolve phylogenetic relationships between animal taxa? These are some of the important questions raised in the field of research that is now termed "neurophylogeny." This chapter summarizes recent findings on the brain architecture of some of the animal groups richest in species.

11.1.1 Neurophylogeny: History, Concepts, and Methods

While the term "neurophylogeny" that links neuroanatomy and phylogeny together is rather new (introduced by Harzsch 2002), the method of inferring evolutionary events by comparing cerebral characters is not. An early pioneer in the field of comparative brain anatomy was Santiago Ramon y Cajal. Together with Camillo Golgi, he was awarded with the Nobel Prize in 1906 for his superb neurohistological stainings and his precise reconstructions of neurons in a variety of animal groups, such as insects, birds, and mammals (Cajal 1911). Later, two Swedish neuroanatomists, Nils Holmgren (1916) and Bertil Hanström (1928), published elaborate studies that described the gross morphology of the brains in a wide variety of invertebrate taxa. Their descriptions, though, were sometimes rather superficial and in many cases, they did not present original data. However, they clearly demonstrated that the basic neuroarchitecture of the brain is quite conserved, at least at the taxonomic level of orders, i.e., that the brain is a slowly evolving organ, which renders comparative neuroanatomy a suitable tool for deep time evolutionary studies.

Classical neurohistological methods like the ones developed by Golgi (1873) or Bodian (1937) are still used today and have remained powerful tools to analyze the fine structure of brain tissue (Strausfeld 2005; Strausfeld et al. 2006b to name just a few recent studies). However, with the advent of modern staining techniques, the field of comparative neuroanatomy has gained new momentum. These modern tools include immunohistological stainings that allow specific labeling of neurons that express a certain transmitter or gene product. In combination with recent developments in imaging techniques like confocal laser scanning microscopy and 3D-reconstruction software, we now have the tools to analyze the architecture of the brain on different levels of complexity.

Once the data are acquired, strict standards are needed for comparing the neuroanatomy of different species to identify similarities that might be due to the emergence from a common ancestor. Such criteria that cover the entire range of structural resolution – from the gross morphology of the brain down to molecular details of individual neurons – have been postulated by Kutsch and Breidbach in 1994:

- Number and arrangement of neuropils in the brain
- Connections of a given neuropil to other main areas of the brain
- Number and arrangement of subunits the neuropil comprises
- Basic three-dimensional neuroarchitectural design
- Distribution of identifiable biochemical markers (presence of neurotransmitters, expression of genes) in neurons of a given brain center
- Role of a brain center in perception of stimuli and/or control of behavior
- Presence of the neuropil in basal representatives of the taxa under investigation

These are just a few examples for features of the nervous system that can be analyzed for phylogenetic considerations. Others extend to physiological properties of neurons or developmental events during embryogenesis. For a comparison of neuroanatomical characters between animal groups that have diverged over half a billion years ago, one needs an established ground pattern of the brain structure in one of the major metazoan taxa as a point of reference. Since many of the criteria listed above have emerged from the descriptions of hexapod neuroanatomy, this taxon represents an ideal reference to start a comparison with other invertebrate groups.

11.1.2 The Insect Brain: The Best Described Invertebrate Brain as a Point to Start From

The literature on brain architecture in insects is vast. This is in part due to the fact that many insect species are established model organisms for investigating various neurobiological issues like neuronal development during embryogenesis (Urbach and Technau 2003), learning and memory (Heisenberg 2003), brain plasticity (Okada et al. 2007), locomotor control and navigation (Ilius et al. 2007), mechanisms of the internal clock (Homberg et al. 2003), odor perception (Schmucker and Schneider 2007), and the role of neuropeptides in brain function (Nässel and Homberg 2006) to name just a few. Thus, detailed descriptions of the brains of a variety of insect species are available. In many cases, the morphology, physiological properties, and transmitter content of individually identifiable cerebral neurons have been described (e.g., Kanzaki et al. 1989, 1991; Loesel and Homberg 1998, 1999, 2001).

Figure 11.1 depicts the neuroarchitecture of a generalized insect brain. Like all arthropod brains, it is clearly divided into an outer cortex that contains the neuronal cell bodies and into central neuropils that exclusively comprise dendritic and axonal arborizations of neurons. Neuropils are usually surrounded by a glial sheath, which

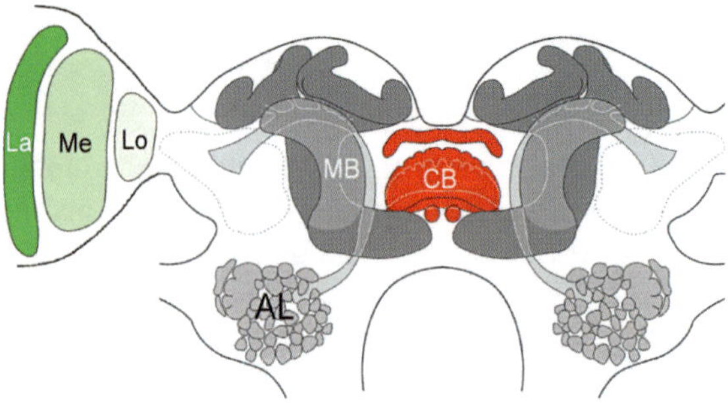

Fig. 11.1 Internal anatomy of the insect brain (supraoesophageal ganglion): *ACT* antennocerebral tract, *La* alpha-lobe, *AL* antennal lobe, *bL* beta-lobe, *Ca* calyx, *CB* central body, *LA* lamina, *Lo* lobula, *Me* Medulla, *MB* mushroom body, *P* peduncle (modified from Strausfeld 1998)

makes their demarcation from neighboring brain areas an easy task. Connections between neuropils are established by fiber tracts.

The optic lobes contain a set of at least three visual neuropils (shown in different shades of green). These are the lamina, the medulla, and the lobula. The outermost of these neuropils, the lamina, receives direct inputs from photoreceptor axons of the compound eye. Interneurons link the lamina to the medulla and the medulla to the lobula, respectively. The lobula as a third-order optic neuropil is pivotal for higher computational tasks such as object discrimination and movement detection (Egelhaaf and Borst 1993). In several orders of the Pterygota (winged insects), the medulla as well as the lobula are split into two separate neuropils (outer and inner medulla, lobula proper, and lobula plate). Apart from that, deviations from this scheme are extremely sparse in insects (Loesel 2006).

The antennal lobes (light gray) are primary olfactory brain centers that receive direct input from olfactory receptor neurons of the antennae. Several authors have emphasized the common architecture of primary olfactory brain areas across animal phyla, most conspicuously their compartmentation into anatomical subunits termed glomeruli (Hildebrand and Shepherd 1997; Strausfeld and Hildebrand 1999; Eisthen 2002). The question, whether these similarities could be the result of a common selective pressure to perform the same computational task or whether they are derived from a common deep time ancestor will be discussed later. In insects, glomeruli are usually arranged in one or two layers around a central coarse neuropil. The number of glomeruli is species-specific and ranges from about 40 in Diptera and Ensifera to approximately 250 in ants. In some groups, a glomerular organization is completely absent, while in other groups (Caelifera), individually identifiable glomeruli have been replaced by several thousand isomorphic so-called microglomeruli (numbers from Schachtner et al. 2005). There is no correlation between the number of glomeruli in the antennal lobe and the taxonomic position of a given species. Thus, the fine structure of the antennal lobe is clearly an inadequate character for phylogenetic studies (Loesel 2006).

The mushroom bodies (dark gray) are prominent protocerebral neuropils that act as centers for sensory integration (Gronenberg 2001) and memory formation (Heisenberg 2003). They are the neuronal basis for associative and flexible behaviors (Farris and Roberts 2005). With the exception of the archaeognathans, where mushroom bodies have probably been secondarily reduced (Farris 2005), the remaining insect taxa share a common ground plan in terms of mushroom body cellular architecture and connectivity. Mushroom bodies consist of several thousand parallel fibers of intrinsic neurons, called Kenyon cells. Dendritic arborizations of these neurons form the calyces, the major synaptic input region to the mushroom bodies. The most prominent inputs to the calyces originate in the antennal lobes through collaterals of olfactory interneurons that connect the antennal lobe with the protocerebrum via an antennocerebral tract. Mushroom bodies, however, are not merely higher order olfactory neuropils, but are present even in anosmic insects (Strausfeld et al. 1998). In a variety of social hymenopterans and in the cockroach *Periplaneta americana*, additional inputs originate in the optic lobes. The axons of Kenyon cells project from the calyx into the peduncle. They then bifurcate and form the lobes (usually an α- and a β-lobe), the major output regions of the mushroom bodies.

The central complex (red) is a set of protocerebral midline neuropils that plays a role in limb coordination (Strausfeld 1999), locomotion control (Strauss 2003), and navigation (Homberg 2004). In insects, the main components of the central complex are the central body and the protocerebral bridge. The neuroarchitecture of the central body is characterized by several layers, the most prominent of which are the ellipsoid body (lower division) and the fan-shaped body (upper division). Williams (1975) described the fan-shaped and ellipsoid bodies of the locust as comprising reiterative columns. These columns are spread out like the staves of a fan (hence, the name fan-shaped body). Columnar neurons provide connections with the protocerebral bridge via a complicated arrangement of chiasmata. These features are highly conserved in neopteran insects and have been described to be principally identical in the locust *Schistocerca gregaria* (Williams 1975), the flies *Musca domestica*, and *Drosophila melanogaster* (Strausfeld 1976; Hanesch et al. 1989; Renn et al. 1999), the beetle *Tenebrio molitor* (Wegerhoff et al. 1996), the cockroach *Periplaneta americana* (Loesel et al. 2002), the bee *Apis mellifera* (Homberg 1985, 1987), and the wasp *Polistes canadensis* (Strausfeld 1999).

11.1.3 Comparative Neuroanatomy as a Tool to Address Two Open Questions: Phylogenetic Relationships Between Major Invertebrate Groups and the Evolution of Prominent Brain Centers

The analysis of the insect brain has demonstrated the presence of each of the major neuropils (visual neuropils, glomeruli of the antennal lobe, mushroom body, central body) in all hexapod orders investigated. This raises the possibility that these architectural characters of the brain are symplesiomorphic, i.e., that they might be more ancient than the taxon hexapoda itself. When during animal evolution did these brain centers arise? Are certain neuropils older than others? This question will be addressed throughout the remainder of this thesis by comparisons with the

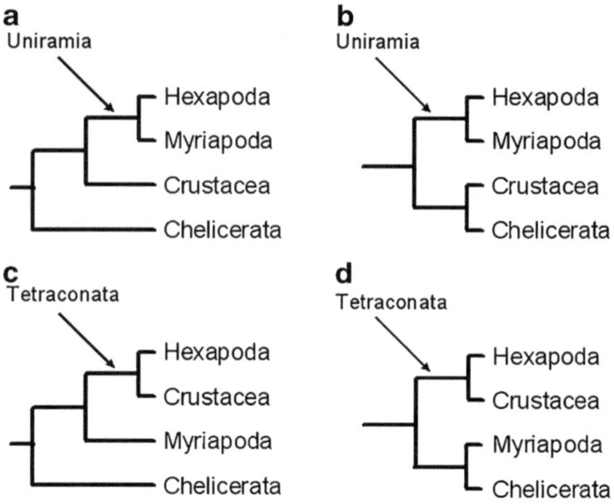

Fig. 11.2 Four conflicting hypotheses of the relationships of arthropod groups (summarized by Mallat et al. 2004)

neuroanatomy of arthropods other than insects and then by comparing arthropod brains with the brains of other major invertebrate taxa, namely, onychophorans, annelids, and molluscs.

The second question that will be addressed is whether cerebral features can be utilized as characters to analyze phylogenetic relationships at higher taxonomic levels. The brain offers a multitude of independent morphological characters, which have scarcely been used in the construction of phylogenetic trees. The need for supplying additional character sets is clearly illustrated by the fact that many phylogenetic trees are not satisfyingly resolved or generally accepted. This is exemplified by published trees of the arthropods, in which various contradicting sister-group relationships of the four major taxa (hexapods, millipedes, crustaceans, chelicerates) have been postulated. Figure 11.2 exemplifies four trees that are currently being discussed.

Can the field of comparative neuroanatomy add new data that might help to resolve these conflicting hypotheses? The next two chapters provide examples of neurophylogenetic studies that support the Tetraconata and the sister-group relationship of this taxon to the Myriapoda by retracing the evolution of major brain centers in these groups.

11.2 Evolution of the Arthropod Brain

11.2.1 Neuropils of the Optic Lobe

One anatomical character that has been proposed to unify the hexapods and the crustaceans in the taxon Tetraconata is the presence of four cone cells in each ommatidium of the compound eye (Dohle 2001). We investigated (Sinakevitch

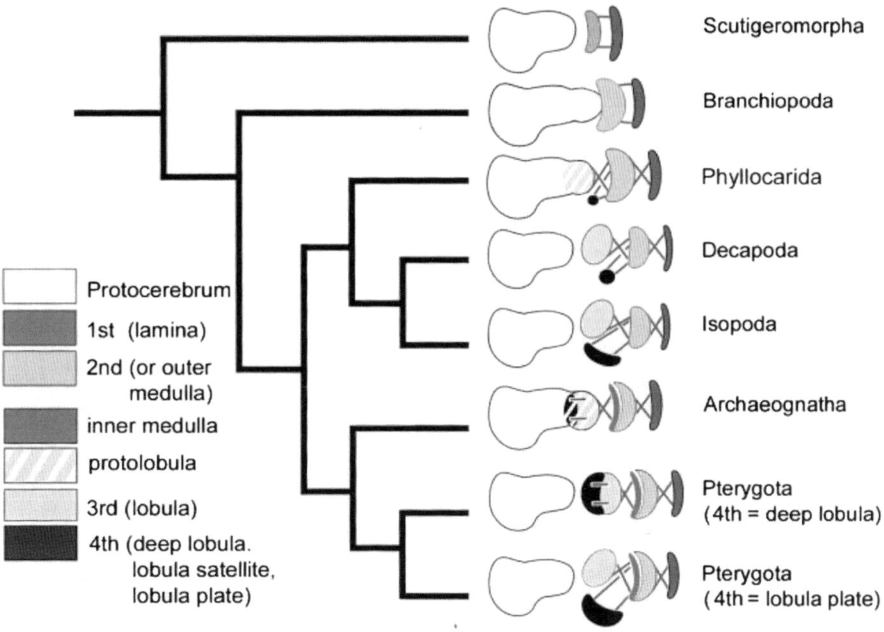

Fig. 11.3 Proposed relationships of hexapods, crustaceans, and chilopods with reference to optic lobe organization (from Sinakevitch et al. 2003)

et al. 2003) whether this concordance in receptor morphology is mirrored by similarities in optic lobe neuropils that process the visual information. The individual optic neuropils (see Chap. 11.1.2) are identifiable and distinguishable by their unique neuroarchitectural features, i.e., their arrangement and connectivity of neurons (for details, see attached reprint of Sinakevitch et al. 2003). The comparison of visual neuropils in a variety of representatives of hexapods and crustaceans and the inclusion of data on scutigeromorphs (Chilopoda, Myriapoda) has resulted in a phylogenetic tree that proposes a scenario of optic lobe evolution in these groups (Fig. 11.3).

In this phylogenetic study, insects and malacostracan crustaceans are united by the presence of at least four optic neuropils and two chiasmata that link the lamina to the medulla and the medulla to the lobula. This architectural setup contrasts with the simpler optic lobe organization in branchiopods and chilopods. The latter perhaps reflecting the ancestral condition.

Our analysis of optic lobe architecture supports the Tetraconata and implies that the hexapods are not a sister-group to but rather a group within the crustaceans. The Tetraconata in turn are viewed as the sister-taxon to the Myriapoda, which were represented by the Scutigeromorpha, the only myriapod group with compound eyes (Müller et al. 2003).

11.2.2 The Central Body

The phylogenetic analysis of optic lobe neuropils remained restricted to those arthropod groups with compound eyes, and therefore putatively homologous visual systems, thereby excluding the chelicerates whose visual neuropils bears no resemblance to the Tetraconata–Myriapoda clade (Strausfeld and Barth 1993; Strausfeld et al. 1993). Higher-order sensory integration centers in the brain like the central body or the mushroom bodies, however, might be of a more ancient evolutionary origin.

An unpaired midline neuropil reminiscent of the central body in insects is present in all major arthropod taxa. The available data on the neuroarchitecture in insects (see Chap. 11.1.2) and in one representative of the decapoda (Utting et al. 2000) supported a common origin of the central body in these animals. However, there are differences in the architectural complexity and in the number of subunits in the central bodies of those species investigated. It has been common text book knowledge that the midline neuropil of chelicerates (for historical reasons called arcuate body) is not homologous to the central body of the hexapods (Roth and Wullimann 1996). This assessment was based on differences in size and relative position of the arcuate body as compared to the central body. Concerning central brain architecture in myriapods, no detailed accounts were available from the literature.

In a broad taxonomic comparison (Loesel et al. 2002, Loesel 2004), we utilized a spectrum of different immunocytological and neuroanatomical staining techniques to analyze the neuroarchitecture of unpaired midline neuropils throughout the arthropods, including representatives of the hexapods, crustaceans, chilopods, diplopods, and chelicerates. The data were used to reconstruct a scenario for central body evolution based on a parsimony assumption. A polychaete species was selected for the out-group comparison, because at that time no central body had been described in annelids.

Our analysis demonstrated that all unpaired midline neuropils investigated (including the chelicerate arcuate body) share structural similarities:

1. The neuropil is subdivided into discrete horizontal layers that are innervated by tangential neurons.
2. Columnar fibers that are oriented perpendicular to the horizontal layers have been identified in all arthropod midline neuropils.
3. A subset of columnar fibers crosses the midline of the brain to form a chiasm, which presumably serves to facilitate interhemispherical information exchange within the central body.
4. The relative size and complexity of the neuropil reflects the mobility of the animal. A comparison between representatives of insects and chelicerates suggest that the elaboration of the central body's neuroarchitecture correlates to the complexity of the locomotive repertoire the animal is equipped with (predatory and highly mobile species have the comparatively largest central bodies).

Again, our results support the Tetraconata hypothesis and suggest a common evolutionary origin of the central body and the chelicerate arcuate body, with the chilopod central body representing the intermediate condition. Interestingly, several staining techniques carried out on the brains of diplopods, new world as well as old world species, failed to identify an unpaired midline neuropil in this taxon, resulting in a basal position of the diplopods within the arthropod tree (Fig. 11.4). This last finding, however, is not in accordance with the majority of recent phylogenetic studies (Mallat et al. 2004; Wheeler et al. 2004; Sierwald and Bond 2007) that place the diplopods together with the chilopods unequivocally within the group Myriapoda. Even though other accounts doubt the monophyly of the Myriapoda (e.g., Shear 1998), the diplopods have never been viewed as the most basal arthropod taxon.

One likely explanation for this contradiction is that the complexity of the diplopod brain has been secondarily reduced. Like in all phylogenetic trees based on morphological data, the least complex organisms tend to appear in a basal position. Neuroanatomical characters might be especially susceptible to secondary reductions in complexity since nervous tissue is particularly costly in development and maintenance due to its high metabolic requirements (Laughlin 2001). At just 2% of body mass, the human brain consumes 20% of resting metabolic energy (Clarke and Sokoloff 1999), and the brain of an electric fish may consume 60% (Nilsson 1996). Blowfly photoreceptors take 8% of resting oxygen consumption (Howard et al. 1987), and their specific metabolic rate (rate per gram) exceeds that of most striated muscles (Laughlin et al. 1998). Diplopods do not execute complicated and variable motion sequences with their legs (Hopkin and Read 1992). It seems likely that the central body as a higher brain center that putatively controls complex locomotory maneuvers in other arthropods has been reduced in diplopods. We therefore assume that the diplopods should be regarded as an outlier group in our study, i.e., due to a secondary reduction of neuroanatomical characters, the position of the diplopods is not correctly resolved.

11.3 Phylogenetic Origin of Arthropod Brain Centers: A Comparison with the Sister Taxon, the Onychophorans

Our studies on the neuroarchitecture of brains across all major arthropod groups suggest that neuropils computing optical information share a common origin in those taxa with compound eyes but differ from visual centers in chelicerates. On the other hand, the available data imply that higher-order sensory integration centers are homologous within the arthropods. Our findings demonstrated the presence of an unpaired midline neuropil that shares common architectural features in the Tetraconata, the Chilopoda, and the Chelicerata. Since this neuropil is present in basal representatives of these groups, it is the parsimonious assumption that the tetraconate and chilopod central body is homologous to the arcuate body of

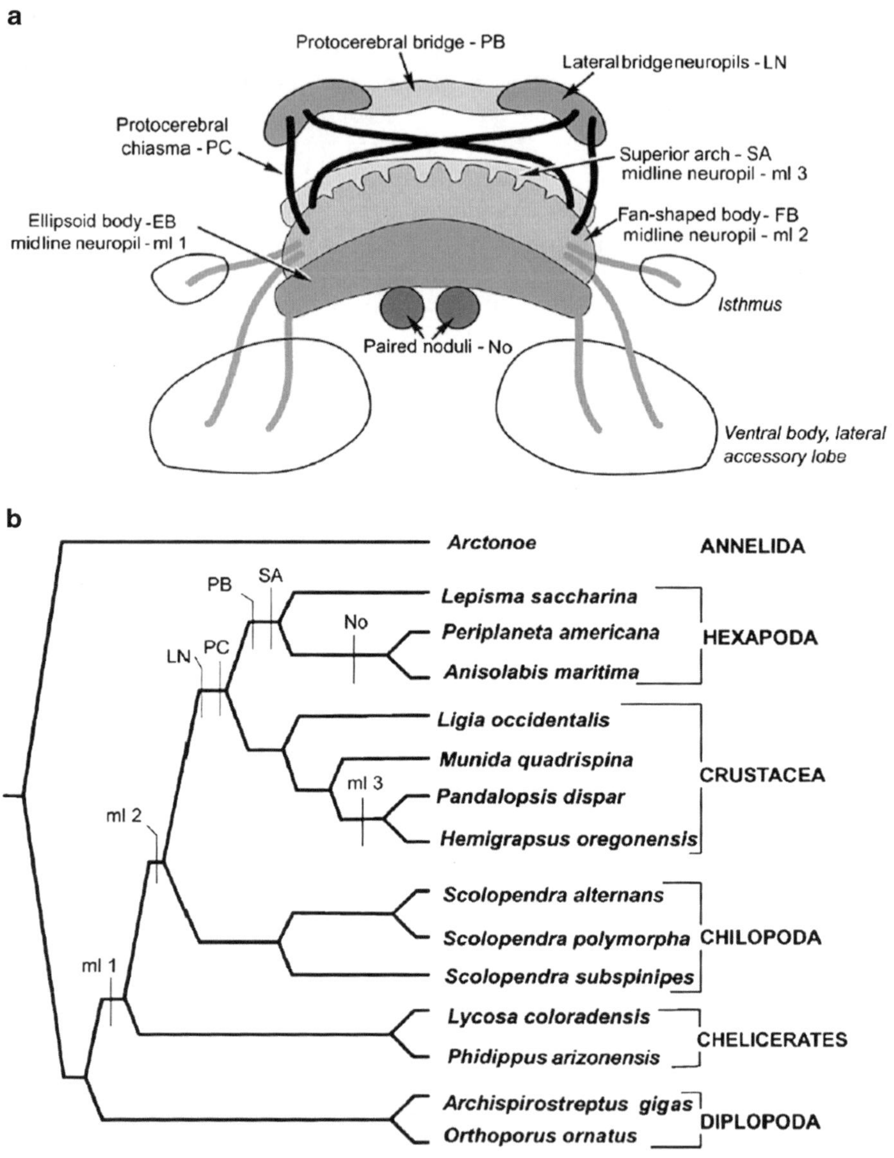

Fig. 11.4 (**a**) Schematic representation of the principal components of the central complex of insects, including the central body, which comprises distinct layers (ml 1–3). (**b**) Hypothetical scenario of central body evolution based on parsimony (from Loesel et al. 2002)

chelicerates. Further prominent neuropils of the arthropod central brain are the paired mushroom bodies. While the principal neuroarchitecture of mushroom bodies (consisting of numerous parallel fibers of intrinsic neurons) is similar in chelicerates, diplopods, chilopods, and hexapods, their homology to mushroom bodies (also named Corpora pedunculata or hemiellipsoid bodies, see Hanström 1928; Farris 2005) of crustaceans, which are functionally equivalent to their namesakes in other arthropods but lack parallel fiber-containing lobes, are still under debate (Strausfeld 1998; McKinzie et al. 2003; Farris 2005).

Are these central brain neuropils synapomorphies of arthropods or of a more ancient evolutionary origin? To answer this question, we conducted an in-depth analysis of the brain architecture of the putative sister taxon of the arthropods, the onychophorans. The following two chapters summarize the findings that have been published in Loesel (2004); Strausfeld et al. (2006a, b).

11.3.1 The Onychophoran Brain Resembles the Brain of Chelicerates, the Basal Arthropod Taxon

The onychophoran representative *Euperipatoides rowelli* possesses a tripartite brain. The paucity of discrete axon tracts distinguishes the onychophoran brain from the brains of tetraconates and chilopods but not from chelicerate brains. Like in the brains of chelicerates, neuropils of the onychophoran brain lack obvious glial demarcations, other than those denoting the anterior borders of the trito- and deutocerebrum. The protocerebrum contains the majority of the cerebral mass along with several neuropils (Fig. 11.5) that are reminiscent of neuropils in the arthropod brain, namely, the mushroom bodies and the central body. The neuroarchitecture of these brain centers in onychophorans share characteristics specifically with the situation found in chelicerates.

As in all arthropods (with exception of the Crustacea), the principal sensory input region of the onychophoran mushroom body, the calyx, is embedded in a dense cluster of small-diameter perikarya of intrinsic neurons, whose parallel axons form a peduncle and several output lobes. One characteristic the mushroom bodies of *E. rowelli* share with chelicerates but with no other arthropod group is the presence of a commissure that renders the mushroom bodies into one confluent structure. Another feature of the onychophoran mushroom body is its direct connection to a second-order visual neuropil. This character is again found exclusively in chelicerates but not in mandibulate arthropods.

The central body (arcuate body) of *E. rowelli* is a crescent-shaped neuropil that spans almost the entire width of the protocerebrum, its outward shape and position within the brain being identical to the situation found in chelicerates. Even more striking similarities are observed when analyzing its internal neuroarchitecture: In both, the onychophoran as well as in chelicerates, the central body is composed of discrete layers of tangential neurons that extend into the neuropil from its lateral margin. These layers are provided by successive strata of collaterals of columnar fibers that originate from thousands of cell bodies dorsal of the neuropil and which enter the central body from its anterior surface. A detailed comparison of the central

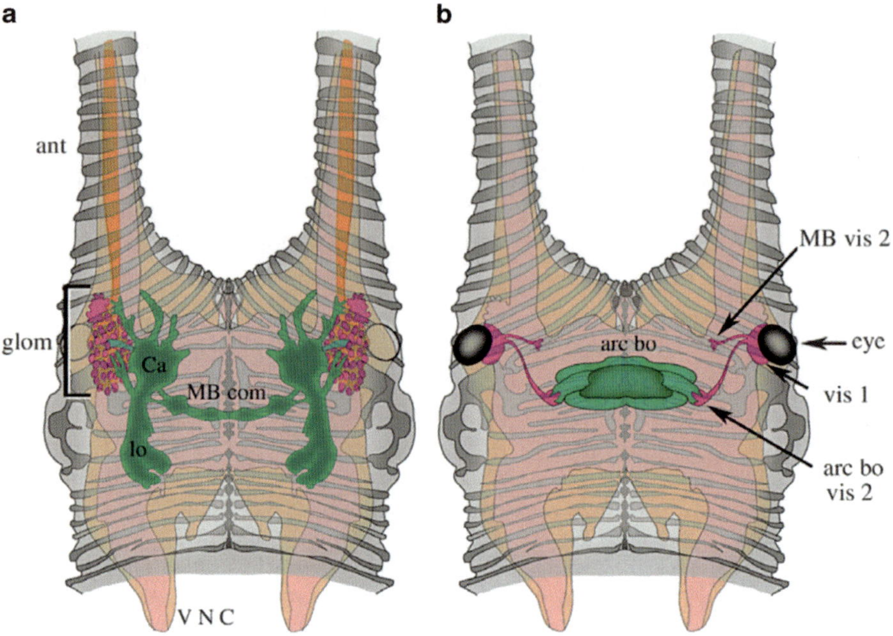

Fig. 11.5 General brain organization of the onychophoran *Euperipatoides rowelli* (from Strausfeld et al. 2006a). (**a**) Shows receptor fibers (*yellow*) that project through the antennae (ant) and innervate the olfactory glomeruli (glom, *purple*). The mushroom body (*green*) consists of a calical input region (Ca) and output lobes (lo). As in chelicerates, the mushroom bodies are connected via a commissure (MB com). *VCN* ventral nerve cord. (**b**) The central body (= arcuate body, arc bo, *green*) is connected to a second-order visual neuropil (arc bo vis 2, *purple*). Another second-order visual neuropil (MB vis 2, *purple*) provides optic information to the mushroom bodies. A first-order visual neuropil (vis 1, *purple*) is located proximal to the eye

bodies of *E. rowelli* and the spider *Cupiennius salei* demonstrated an almost identical arrangement of horizontal layers and columnar fibers in these two species.

One prominent architectural feature of the brain of arthropods and onychophorans that we have not discussed yet is the presence of olfactory glomeruli, first-order integration centers for odor information. Glomeruli are spherical neuropils that receive inputs from axons of odor receptor neurons. Olfactory glomeruli usually appear in clusters of a few dozens to several hundreds, their number corresponding to the number of odorant receptors that are expressed in receptor neurons (Mori et al. 1999). Olfactory glomeruli are common not only in the brain of arthropods and onychophorans but are also present in annelids (see Chap. 4), in molluscs (Chase and Tolloczko 1993), and in the telencephalon of vertebrates (Cajal 1911). Due to the widespread occurrence of these neuroarchitectural units across animal phyla, one might be tempted to infer a common evolutionary origin of olfactory glomeruli dating back to an early bilaterian predecessor. Our analysis, however, suggests that olfactory glomeruli are not even homologous within the onychophoran–arthropod clade. This conclusion is based on the observation that in onychophorans, olfactory glomeruli are situated in the protocerebrum, while these structures occur in the deutocerebrum in insects and crustaceans. In chelicerates, the olfactory glomeruli are located in the neuromere of whichever segment provides an appendage equipped with odor receptors. Olfactory appendages can be the pedipalps (in solfugids), the first leg pair (amblypygids, uropygids), or every leg pair as in pycnogonids.

Thus, in a classical sense, olfactory glomeruli cannot be considered homologous structures (sensu Hennig 1950) because the criterion of relative position is not fulfilled. Rather, the occurrence of these structures in different neuromers seems to be "driven" by olfactory inputs. Future experiments will focus on the question whether a common genetic mechanism is responsible for the appearance of olfactory glomeruli at different positions in the brain. This molecular machinery might be homologous within animals that possess olfactory glomeruli, while the structures themselves are obviously not.

11.3.2 Constructing the Phylogenetic Tree of the Arthropod–Onychophoran Clade Utilizing Neuroarchitectural Characters

The phylogenetic trees presented in Chap. 2 are based on existing hypotheses of arthropod evolution that were chosen because our data comply with them in the most parsimonious way. To test whether neuroanatomical characters alone contain sufficient phylogenetic information to construct a tree, we examined the relationships of 27 arthropod–onychophoran and one annelid representative based on 118 independent neuroarchitectural characters. Complex brain structures such as the central body or the mushroom bodies were not treated as single characters but were partitioned, e.g., yielding 22 separate characters for the central body and 26 characters for the mushroom bodies.

The resulting tree (Fig. 11.6) supports the Tetraconata. Chilopods are viewed as sister taxon to the Tetraconata, Chelicerates as sister taxon to the mandibulata. Out of the four conflicting hypothesis on arthropod relationships presented in Chap. 11.1.3, our results are in accordance with the phylogram in Fig. 11.2c.

This study also revealed the limitations of phylogenetic research based exclusively on brain anatomy. One problem is that neuroanatomical data results in an unsatisfactory resolution at low taxonomic levels. For example, in our tree *Tarantula* spec. is not grouped together with the other araneans but with an uropygid representative. This low resolution at the level of orders is due to the fact that the principal neuroarchitecture of the brain within the four major arthropod groups is quite conserved. Minor variations of the common scheme that may occur are likely to be linked to the behavioral ecology of the species under consideration and not to its phylogenetic position (Loesel 2006). Besides the lack of morphological differences at low taxonomic levels, a secondary reduction in brain complexity

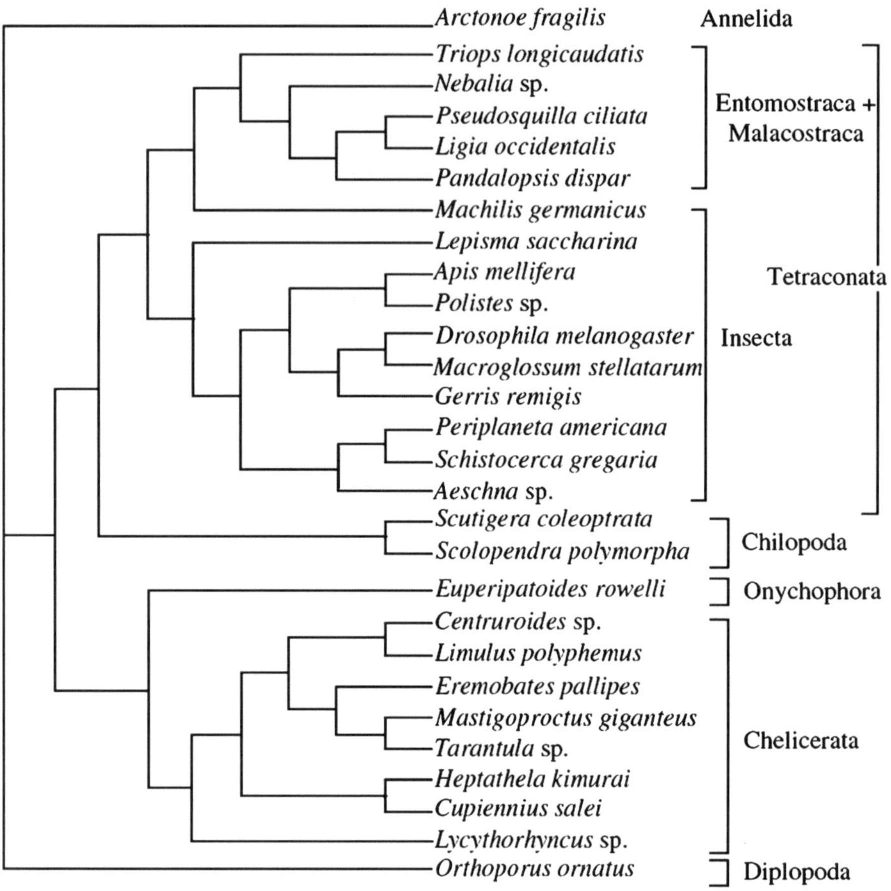

Fig. 11.6 Phylogenetic tree of arthropods and onychophorans based on 118 neuroanatomical characters (from Strausfeld et al. 2006a)

(see Chap. 11.2.2) may also be the source of errors, as is clearly the case with the position of the diplopod *Orthoporus ornatus* that came out basal in our tree. The similarity between protocerebral brain centers in onychophorans and chelicerates resulted in a sister-group relation of these taxa and not in the generally accepted sister-group relation of onychophorans and arthropods.

Apart from the possible sources of error discussed above, neuroanatomical data provide an appropriate amount of characters for phylogenetic reconstructions at higher taxonomic levels. Moreover, our analysis implies the parsimonious assumption that the two most prominent protocerebral brain centers, the central body and the mushroom bodies are part of the arthropod–onychophoran ground plan. Are these neuropils derived characters pertaining only to the (pan-)arthropods or of a more ancient origin? The problem in answering this question is that there is

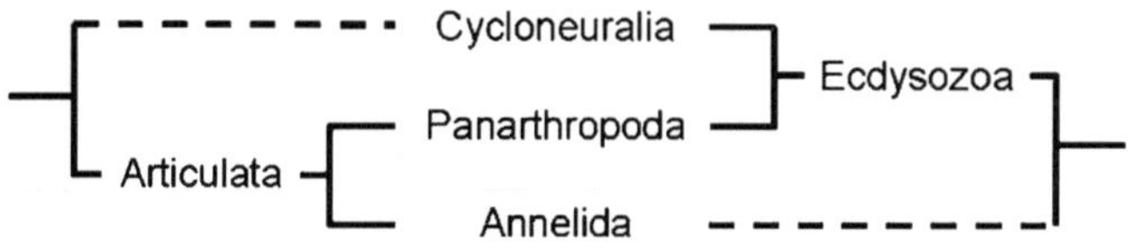

Fig. 11.7 Simplified tree illustrating the two competing hypothesis of arthropod–annelid relationship currently under discussion

no consent among phylogenists what the sister-taxon of the arthopods is. Traditionally, the annelids were grouped together with the arthopods in the taxon Articulata (Cuvier 1817; for a recent discussion of the Articulata concept, see Scholtz 2002). This view has been dramatically challenged by two simultaneously published molecular analyses using 18S rDNA (Aguinaldo et al. 1997; Eernisse 1997). In these studies, the arthropods appeared as close relatives of several Nemathelminthes groups, now unified under the taxon Cycloneuralia (Fig. 11.7). Because the members of the resulting clade share the character of molting a cuticle, the group has been named Ecdysozoa (Aguinaldo et al. 1997). The annelids were united with molluscs and other spiralian groups widely separated from the Ecdysozoa.

Despite mounting support for the Ecdysozoa from molecular studies, many specialists are still in favor of the Articulata (Wägele et al. 1999; Wägele and Misof 2001; Nielsen 2001; Scholtz 2003). This demonstrates that the uncertain relationship between arthropods and annelids is at the core of an ongoing debate on early metazoan radiation and phylogeny. Therefore, we were interested whether neuroanatomical findings can contribute to resolve this question.

11.4 A Further Step Back in Time: Similarities Between the Arthropod–Onychophoran Clade and Annelids: Homology or Homoplasy?

While the internal phylogeny of the annelids is still largely unresolved, vagile polychaetes most likely resemble the ancestral condition (Bartolomaeus et al. 2005). In order to compare the brain architecture of annelids to that of arthropods, we described the neuroanatomy of two predatory polychaete representatives in detail (Heuer and Loesel 2008, 2009). Our analyses mainly focused on the internal architecture and connectivity of the mushroom bodies (termed corpora pedunculata in the older annelid literature) that have been found to be present in all vagile polychaetes investigated so far (Hanström 1928; Åkesson 1963; Bullock and Horridge 1965; Strausfeld et al. 1995).

In *Nereis diversicolor*, the neuroarchitecture of the mushroom bodies matches that found in euarthropods/onychophorans in several aspects (Fig. 11.8). Afferents carrying sensory information invade the anterior part of the mushroom bodies. Among arthropods, the morphological appearance of this region varies in different clades.

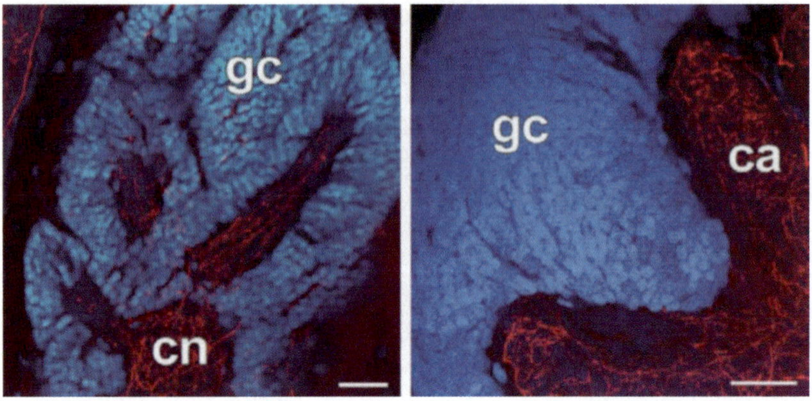

Fig. 11.8 High magnification images of the mushroom body's input region of *Nereis diversicolor* (*left*) and the cockroach *Leucophae maderae* (*right*). The input region (termed calyx *ca* in insects) is embedded in a dense cluster of intrinsic globuli cells (*gc*, shown in *blue*). Olfactory input is provided through serotonergic fibers (*red*) in both animals. *cn* core neuropil of the annelid mushroom body, scale bars: 20 μm (from Heuer and Loesel 2008)

In many insects, for instance, it is of a cup-like shape, whereas in onychophorans, it forms finger-like protrusions; the latter resembles the state observed in *N. diversicolor*. Independent of its diverse shape, the anterior region is always surrounded by perikarya of thousands of small-diameter globuli cells (termed Kenyon cells in insects). The axonal outgrowths of these cells form a bundle of parallel fibers, called the peduncle, which extends posteriorly and medially. In insects, diplopods, and onychophorans, the peduncle splits up into several lobes that represent the main output regions of the mushroom bodies. In *N. diversicolor*, the peduncle breaks up into three lobes. The occurrence of fine extrinsic fibers that are possibly dendritic, raises the possibility that the lobes also serve as output structures in this species. One important difference between the mushroom bodies of euarthropods/onychophorans and of *N. diversicolor* is the origin of fibers that provide sensory information to the mushroom bodies. In insects, chelicerates, millipedes, centipedes, and onychophorans, neuroanatomical evidence suggests that mushroom bodies are second-order neuropils of the olfactory pathway. In these taxa, primary sensory input is provided to the olfactory glomeruli from where the information is passed on to the mushroom bodies (Strausfeld et al. 1995). This contrasts with our findings for *N. diversicolor* in which fibers of the palpal nerve directly innervate the finger-like protrusions of the mushroom bodies. These fibers presumably convey chemosensory information (Dorsett and Hyde 1969). Olfactory glomeruli have not been identified at any location in the brain. Our results are in accordance with findings in *Nereis virens* in which the palpal nerve also directly innervates the mushroom bodies (Hanström 1928).

An interesting single finding in this context has been provided by Rössler et al. (1999). In their study, the normal development of olfactory glomeruli in the hawk moth *Manduca sexta* was artificially inhibited during ontogenesis. Ingrowing axons of olfactory receptor cells that therefore did not find their original targets kept on growing into the protocerebrum and directly innervated the calyces of the mushroom bodies. This demonstrates that, in an insect, a targeting mechanism that directs olfactory receptors to the mushroom bodies is existent but normally not used. This targeting mechanism might be interpreted as an evolutionary relict dating back to deep time ancestors of arthropods that did not have glomeruli as part of the olfactory pathway.

The cerebral architecture of another vagile polychaete, *Harmothoe areolata*, differs from *N. diversicolor* mainly with regard to the relative size and elaboration of the mushroom bodies (Fig. 11.9). In *H. areolata*, the mushroom bodies are of enormous size, occupying approximately two-thirds of the total brain volume. The core neuropil is divided into two lobes with the anterior lobe receiving direct sensory input, while the posterior lobe is connected to a cluster of spherical neuropils. In this respect, the anterior lobe of the mushroom body of *H. areolata* is reminiscent of the state found in *N. diversicolor*, whereas the posterior lobe exhibits features of the olfactory pathway found in arthropods. Besides the mushroom bodies, the brain also contains an unpaired crescent-shaped central neuropil, its contour and relative position being reminiscent of the central body of hexapods. Such a crescent-shaped neuropil has also been identified in *N. diversicolor*, albeit less conspicuous in this species.

Although relationships within the polychaete annelids – as well as in the Annelida as a whole – are still poorly understood, current phylogenetic reconstructions indicate a rather close relationship between *N. diversicolor* (Nereidiformia) and *H. areolata* (Aphroditiformia), grouping them both into the Phyllodocida

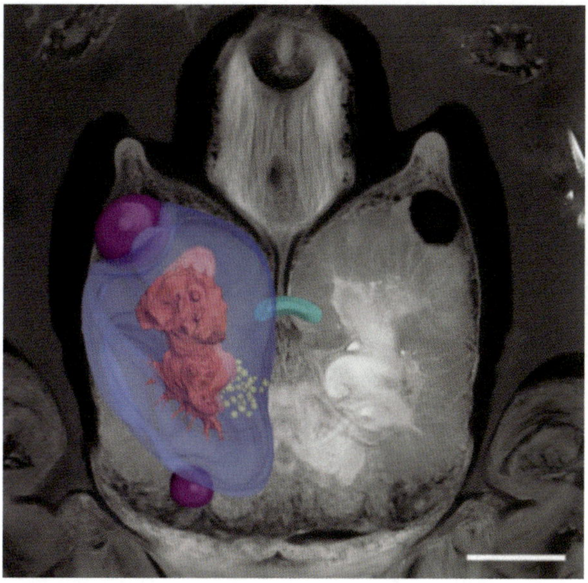

Fig. 11.9 A three-dimensional reconstruction of major brain compartments superimposed onto a section through the head of *H. areolata* reveals similarities to the cerebral architecture of arthropods (compare to Figs. 11.1 and 11.5). *blue* globuli cells mass, *red* mushroom body core neuropil, *yellow* glomeruli, *green* putative central body, *purple* eyes, scale bar: 200 μm (from Heuer and Loesel 2009)

(Struck et al. 2007). The differences in cerebral anatomy we encountered in two closely related species demonstrate that we cannot draw a conclusion about the ancestral neuroarchitecture of annelids yet. Studies to analyze the brain of additional polychaete representatives, errant, sedentary, and meiobenthic ones, as well as representatives of the Clitellata, are in progress. So far, our investigations suggest that at least the presence of mushroom bodies pertains to the ground plan of annelid neuroarchitecture. On the basis of this alone, our data are a strong support for the Articulata, i.e., for a close relationship between annelids and arthropods. In the light of mounting molecular evidence supporting the Ecdysozoa-Lophotrochozoa, however, other explanations for the commonalities in brain morphology between annelids and arthropods should also be considered. We will explore these alternative scenarios in the next chapter.

11.5 Summary and Perspectives: The Search for the Urbilaterian Brain

The preceding chapters summarized our studies on the cerebral architecture of a variety of invertebrate clades with the aim to retrace early brain evolution and resolve disputed phylogenetic relationships. We were able to present scenarios for the evolution of major brain centers such as optic neuropils and the central body in panarthropods. Our data strongly support the Tetraconata, i.e., a close relationship between hexapods and crustaceans. The studies also revealed that mushroom bodies and the central body pertain to the ancestral condition of the panarthropod brain. Similarities in brain architecture between arthropods and basal annelids might reflect a close relationship of these two taxa. Yet, in the light of molecular studies that place annelids and arthropods at widely separated positions in the bilaterian tree, two alternative interpretations will be tested in future studies.

One obvious alternative explanation for commonalities found in annelids and arthropods is that shared features such as the mushroom bodies and the central body have evolved independently. It could be argued that once a cerebral ganglion evolves to a certain level of neuronal complexity, similar architectural principles have to evolve accordingly due to specific computational needs and restrictions. This view, however, is not supported by our present state of knowledge. Other animal taxa with elaborate brain architectures do not possess identifiable mushroom bodies or central bodies. This has long been known for the well-investigated vertebrate brain (Cajal 1911; Butler and Hodos 2005). Next to vertebrates, cephalopods possess the most complex cerebral ganglia in the animal kingdom. We therefore investigated the neuroanatomy of the central brain of the pygmy squid *Idiosepius notoides*, but found no structural entities reminiscent of neuropils in arthropods or annelids (Wollesen et al. 2008).

This leads to a third scenario to explain commonalities in arthropod/annelid brain architecture: Cerebral centers such as the mushroom body or the central body might have been present already in the last common ancestor of the protostomes or even the entire Bilateria and might have been retained in some taxa and reduced in others. Recent molecular investigations on gene expression patterns in the developing cerebrum indeed suggest that the brains of all animals are derived from a common ancestor, i.e., the urbilaterian brain. Taken together, these studies not only provide data to support a common origin of the brain of protostomes and deuterostomes but also imply that this urbilaterian brain might have been a rather complex, perhaps tripartite organ (reviewed by Lichtneckert and Reichert 2005). Others argue that genetic patterning alone does not provide conclusive evidence for a homology of the resulting anatomical structures. This is exemplified by a basal deuterostome, the hemichordate *Saccoglossus kowalevskii*, which expresses genes in a pattern resembling that in the developing brain of *Drosophila melanogaster* and of vertebrates, although *S. kowalevskii* does not have a localized CNS (Holland 2003; Lowe et al. 2003).

Currently we cannot resolve the question whether the presence of shared brain centers in arthropods and annelids should be interpreted as support for the Articulata or whether they represent ancient architectural features of the protostome or even urbilaterian brain. The first interpretation would imply that molecular studies on animal phylogeny are unreliable while the second scenario would require a repeated loss of brain complexity during animal evolution. Much needs to be done for neurophylogenists in the future, for those working anatomically as well as genetically. Especially, studies on poorly examined taxa like basal molluscs, priapulids, chaetognaths, or even flatworms (for which the presence of mushroom body-like structures has been occasionally claimed but never investigated in depth) will help to close gaps in our knowledge so that we finally might be able to retrace early metazoan brain evolution and bring anatomical and molecular data into agreement.

References

Aguinaldo AMA, Turbeville JM, Linford LS, Rivera MC, Garey JR, Raff RA, Lake JA (1997) Evidence for a clade of nematodes, arthropods and other moulting animals. Nature 387:489–493

Åkesson B (1963) The comparative morphology and embryology of the head in scale worms (Aphroditidae, Polychaeta). Ark Zool 16:125–163

Bartolomaeus T, Purschke G, Hausen H (2005) Polychaete phylogeny based on morphological data – a comparison of current attempts. Hydrobiologia 535(536):341–356

Bodian D (1937) A new method for staining nerve fibers and nerve endings in mounted paraffin sections. Anat Rec 69:153–162

Bullock TH, Horridge GA (1965) Structure and function in the nervous system of invertebrates. Freeman, San Francisco

Butler AB, Hodos W (2005) Comparative vertebrate neuroanatomy. Evolution and adaption. Wiley, New York

Cajal SR (1911) Histologie du système nerveux de l'Homme et des vertébrés. Maloine, Paris. This is the French translation of Cajal's original textura del sistema nervioso del Hombre y los vertebrados from 1894

Chase R, Tolloczko B (1993) Tracing neural pathways in snail olfaction: from the tip of the tentacles to the brain and beyond. Micrsc Res Tech 24:214–230

Clarke DD, Sokoloff L (1999) Circulation and energy metabolism of the brain. In: Siegel GJ, Agranoff BW, Albers RW, Fisher SK, Uhler MD (eds) Basic neurochemistry: molecular. Cellular and medical aspects. Lippincott-Raven, Philadelphia, pp 637–669

Cuvier G (1817) Le règne animal, vol II. Déterville, Paris

Dohle W (2001) Are the insects terrestrial crustaceans? A discussion of some new facts and arguments and the proposal of the proper name 'Tetraconata' for the monophyletic unit Crustacea + Hexapoda. In: Deuve T (ed) Origin of the hexapoda. Ann Soc Entomol Fr 37:85–103

Dorsett DA, Hyde R (1969) The fine structure of the compound sense organs on the cirri of *Nereis diversicolor*. Z Zellforsch 97:512–527

Eernisse DJ (1997) Arthropod and annelid relationships reexamined. In: Fortey RA, Thomas RH (eds) Arthropod relationships. Chapman and Hall, London, pp 43–56

Egelhaaf M, Borst A (1993) Motion computation and visual orientation in flies. Comp Biochem Physiol 104A:659–673

Eisthen HL (2002) Why are olfactory systems of different animals so similar? Brain Behav Evol 59:273–293

Farris SM (2005) Evolution of insect mushroom bodies: old clues, new insights. Arthropod Struct Dev 34:211–234

Farris SM, Roberts NS (2005) Coevolution of generalist feeding ecologies and gyrencephalic mushroom bodies in insects. Proc Natl Acad Sci USA 102:17394–17399

Golgi C (1873) Sulla struttura della sostanza grigia del cervello. Gazz Med Ital Lomb 33:244–246

Gronenberg W (2001) Subdivisions of hymenopteran mushroom body calyces by their afferent supply. J Comp Neurol 436:474–489

Hanesch U, Fischbach KF, Heisenberg M (1989) Neuronal architecture of the central complex in *Drosophila melanogaster*. Cell Tissue Res 257:343–366

Hanström B (1928) Vergleichende anatomie des nervensystems der wirbellosen tiere unter berücksichtigung seiner funktion. Springer, Berlin

Harzsch S (2002) Neurobiologie und evolutionsforschung: "Neurophylogenie" und die stammesgeschichte der euarthropoda. Neuroforum 4(02):267–273

Heisenberg M (2003) Mushroom body memoir: from maps to models. Nat Rev Neurosci 4(4):266–275

Hennig W (1950) Grundzüge einer theorie der phylogenetischen systematik. Deutscher Zentralverlag, Berlin

Heuer CM, Loesel R (2008) Immunofluorescence analysis of the internal brain anatomy of *Nereis diversicolor* (Polychaeta, Annelida). Cell Tissue Res 331:713–724

Heuer CM, Loesel R (2009) Three-dimensional reconstruction of mushroom body neuropils in the polychaete species *Nereis diversicolor* and *Harmothoe areolata* (Phyllodocida, Annelida). Zoomorphology 128:219–226

Hildebrand JG, Shepherd GM (1997) Mechanisms of olfactory discrimination: convergent evidence for common principles across phyla. Annu Rev Neurosci 20:595–611

Holland ND (2003) Insights into the urbilaterian brain: conserved genetic patterning mechanisms in insect and vertebrate brain development. Nat Rev Neurosci 4:617–627

Holmgren N (1916) Zur vergleichenden anatomie des gehirns von Polychaeten, Onychophoren, Xiphosuren, Arachniden, Crustaceen, Myriapoden und Insekten. K Sven Vetensk Akad Handl 56:1–303

Homberg U (1985) Interneurons of the central complex in the bee brain (*Apis mellifera*, L.). J Insect Physiol 31:251–264

Homberg U (1987) Structure and functions of the central complex in insects. In: Gupta AP (ed) Arthropod brain: its evolution, development, structure, and functions. Wiley, New York, pp 347–367
Homberg U (2004) In search of the sky compass in the insect brain. Naturwissenschaften 91:199–208
Homberg U, Reischig T, Stengl M (2003) Neural organization of the circadian system of the cockroach *Leucophaea maderae*. Chronobiol Int 20(4):577–591
Hopkin SP, Read HJ (1992) The biology of millipedes. Oxford University Press, New York
Howard J, Blakeslee B, Laughlin SB (1987) The intracellular pupil mechanism and photoreceptor signal – noise ratios in the fly *Lucilia-cuprina*. Proc R Soc Lond B 231:415–435
Ilius M, Wolf R, Heisenberg M (2007) The central complex of *Drosophila melanogaster* is involved in flight control: studies on mutants and mosaics of the gene ellipsoid body open. J Neurogenet 21(4):321–338
Kanzaki R, Arbas EA, Strausfeld NJ, Hildebrand JG (1989) Physiology and morphology of projection neurons in the antennal lobe of the male moth *Manduca sexta*. J Comp Physiol A 165:427–453
Kanzaki R, Arbas EA, Hildebrand JG (1991) Physiology and morphology of protocerebral olfactory neurons in the male moth *Manduca sexta*. J Comp Physiol A 168:281–298
Kutsch W, Breidbach O (1994) Homologous structures in the nervous system of Arthropoda. Adv Insect Physiol 24:1–113
Laughlin SB (2001) Energy as a constraint on the coding and processing of sensory information. Curr Opin Neurobiol 11(4):475–480
Laughlin SB, de van Ruyter Steveninck RR, Anderson JC (1998) The metabolic cost of neural information. Nat Neurosci 1:36–41
Lichtneckert R, Reichert H (2005) Insights into the urbilaterian brain: conserved genetic patterning mechanisms in insect and vertebrate brain development. Heredity 94:465–477
Loesel R (2004) Comparative morphology of central neuropils in the brain of arthropods and its evolutionary and functional implications. Acta Biol Hung 55:39–51
Loesel R (2006) Can brain structures help to resolve interordinal relationships in insects? Arthropod Syst Phylogeny 64(2):101–106
Loesel R, Homberg U (1998) Sustained oscillations in an insect visual system. Naturwissenschaften 85:238–240
Loesel R, Homberg U (1999) Histamine-immunoreactive neurons in the brain of the cockroach *Leucophaea maderae*. Brain Res 842:408–418
Loesel R, Homberg U (2001) Anatomy and physiology of neurons with processes in the accessory medulla of the cockroach, *Leucophaea maderae*. J Comp Neurol 439(2):193–207
Loesel R, Nässel DR, Strausfeld NJ (2002) Common design in a unique midline neuropil in the brains of arthropods. Arthropod Struct Dev 31:77–91
Lowe CJ, Wu M, Salic A, Evans L, Lander E, Stange-Thomann N, Gruber CE, Gerhart J, Kirschner M (2003) Anteroposterior patterning in hemichordates and the origins of the chordate nervous system. Cell 113:853–865
Mallat JM, Garey JR, Schultz JW (2004) Ecdysozoan phylogeny and Bayesian inference: first use of nearly complete 28S and 18S rRNA gene sequences to classify the arthropods and their kin. Mol Phylogenet Evol 31:178–191
McKinzie ME, Benton JL, Beltz BS, Mellon D (2003) Parasol cells of the hemiellipsoid body in the crayfish *Procambarus clarkii*: dendritic branching patterns and functional implications. J Comp Neurol 462:168–179
Mori K, Nagao H, Yoshihara Y (1999) The olfactory bulb: coding and processing of odor molecule information. Science 286:711–715
Müller CHG, Rosenberg J, Richter S, Meyer-Rochow VB (2003) The compound eye of *Scutigera coleoptrata* (Linnaeus, 1758) (Chilopoda: Notostigmophora): an ultrastructural reinvestigation that adds support to the Mandibulata concept. Zoomorphology 122:191–209

Nässel DR, Homberg U (2006) Neuropeptides in interneurons of the insect brain. Cell Tissue Res 326(1):1–24
Nielsen C (2001) Animal evolution, 2nd edn. Oxford University Press, Oxford
Nilsson GE (1996) Brain and body oxygen requirements of *Gnathonemus petersii*, a fish with an exceptionally large brain. J Exp Biol 199:603–607
Okada R, Rybak J, Manz G, Menzel R (2007) Learning-related plasticity in PE1 and other mushroom body-extrinsic neurons in the honeybee brain. J Neurosci 27(43):11736–11747
Renn SC, Armstrong JD, Yang M, Wang Z, An K, Kaiser K, Taghert PH (1999) Genetic analysis of the *Drosophila* ellipsoid body neuropil: organization and development of the central complex. J Neurobiol 41:189–207
Rössler W, Oland LA, Higgins MR, Hildebrand JG, Tolbert LP (1999) Development of a glia-rich axon-sorting zone in the olfactory pathway of the moth *Manduca sexta*. J Neurosci 22:9865–9877
Roth G, Wullimann MF (1996) Evolution der nervensysteme und sinnesorgane. In: Dudel J, Menzel R, Schmidt RF (eds) Neurowissenschaften. Springer, Berlin, pp 1–31
Schachtner J, Schmidt M, Homberg U (2005) Organization and evolutionary trends of primary olfactory centers in Tetraconata (Crustacea + Hexapoda). Arthropod Struct Dev 34(3):257–299
Schmucker M, Schneider G (2007) Processing and classification of chemical data inspired by insect olfaction. Proc Natl Acad Sci USA 104(51):20285–20289
Scholtz G (2002) The Articulata hypothesis – or what is a segment? Org Divers Evol 2:197–215
Scholtz G (2003) Is the taxon articulata obsolete? Arguments in favour of a close relationship between annelids and arthropods. In: Legakis A, Sfenthourakis S, Polymeni R, Thessalou-Legaki M (eds) The new panorama of animal evolution. Proceedings of the 18th international congress of zoology, Pensoft, Sofia, pp 489–501
Shear WA (1998) The fossil record and evolution of the Myriapoda. In: Fortey RA, Thomas RH (eds) Arthropod relationships. Systematics association, special vol series 55. Chapman and Hall, London, pp 211–219
Sierwald P, Bond JE (2007) Current status of the Myriapod class Diplopoda (Millipedes): taxonomic diversity and phylogeny. Annu Rev Entomol 52:401–410
Sinakevitch I, Douglass JK, Scholtz G, Loesel R, Strausfeld NJ (2003) Conserved and convergent organization in the optic lobes of insects and isopods, with reference to other crustacean taxa. J Comp Neurol 467:150–172
Strausfeld NJ (1976) Atlas of an insect brain. Springer, Heidelberg
Strausfeld NJ (1998) Crustacean-insect relationships: the use of brain characters to derive phylogeny amongst segmented invertebrates. Brain Behav Evol 52:186–202
Strausfeld NJ (1999) A brain region in insects that supervises walking. Prog Brain Res 123:273–284
Strausfeld NJ (2005) The evolution of crustacean and insect optic lobes and the origins of chiasmata. Arthropod Struct Dev 34(3):235–256
Strausfeld NJ, Barth FG (1993) Two visual systems in one brain: neuropils serving the secondary eyes of the spider *Cupiennius salei*. J Comp Neurol 328:43–62
Strausfeld NJ, Hildebrand JG (1999) Olfactory systems: common design, uncommon origins? Curr Opin Neurobiol 9:634–639
Strausfeld NJ, Weltzien P, Barth FG (1993) Two visual systems in one brain: neuropils serving the principal eyes of the spider *Cupiennius salei*. J Comp Neurol 328:63–75
Strausfeld NJ, Buschbeck EK, Gomez RS (1995) The arthropod mushroom body: its functional roles, evolutionary enigmas and mistaken identities. In: Breidbach O, Kutsch W (eds) The nervous systems of invertebrates: an evolutionary and comparative approach. Birkhäuser, Basel, pp 349–381
Strausfeld NJ, Hansen L, Li Y, Gomez RS, Ito K (1998) Evolution, discovery, and interpretation of arthropod mushroom bodies. Learn Mem 5:11–37

Strausfeld NJ, Strausfeld CM, Loesel R, Rowell D, Stowe S (2006a) Arthropod phylogeny: onychophoran brain organization suggests an archaic relationship with a chelicerate stem lineage. Proc R Soc B 273:1857–1866

Strausfeld NJ, Strausfeld CM, Stowe S, Rowell D, Loesel R (2006b) The organization and evolutionary implications of neuropils and their neurons in the brain of the onychophoran *Euperipatoides rowelli*. Arthropod Struct Dev 35(3):169–196

Strauss R (2003) Control of *Drosophila* walking and orientation behavior by functional subunits localized in different neuropils in the central brain. In: N Elsner, H Zimmermann (eds) Proceedings of the 29th göttingen neurobiol conference, Thieme, 2003, p 206

Struck TH, Schult N, Kusen T, Hickman E, Bleidorn C, McHugh D, Halanych KM (2007) Annelid phylogeny and the status of Sipuncula and Echiura. BMC Evol Biol 7:57

Urbach R, Technau GM (2003) Early steps in building the insect brain: neuroblast formation and segmental patterning in the developing brain of different insect species. Arthropod Struct Dev 32(1):103–123

Utting M, Agricola HJ, Sandeman R, Sandeman D (2000) Central complex in the brain of crayfish and its possible homology with that of insects. J Comp Neurol 416:245–261

Wägele JW, Misof B (2001) On quality of evidence in phylogeny reconstruction: a reply to Zrzavy's defence of the 'Ecdysozoa' hypothesis. J Zool Syst Evol Res 39:165–176

Wägele JW, Erikson T, Lockhart P, Misof B (1999) The Ecdysozoa: artifact or monophylum? J Zool Syst Evol Res 37:211–223

Wegerhoff R, Breidbach O, Lobemeier M (1996) Development of locustatachykinin immunopositive neurons in the central complex of the beetle *Tenebrio molitor*. J Comp Neurol 375:157–166

Wheeler WC, Giribet G, Edgecombe GD (2004) Arthropod systematics: the comparative study of genomic, anatomical, and paleontological information. In: Cracraft J, Donoghue MJ (eds) Assembling the tree of life. Oxford University Press, New York, pp 281–318

Williams JLD (1975) Anatomical studies of the insect central nervous system: a ground-plan of the midbrain and an introduction to the central complex in the locust, *Schistocerca gregaria* (Orthoptera). J Zool 176:67–86

Wollesen T, Loesel R, Wanninger A (2008) Distribution of FMRFamidergic neurons in the central nervous system of the cephalopod mollusc *Idiosepius notoides*. Acta Biol Hung 59:111–116

Chapter 12
A New Early Cambrian Lobopod-Bearing Animal (Murero, Spain) and the Problem of the Ecdysozoan Early Diversification

José Antonio Gámez Vintaned, Eladio Liñán, and Andrey Yu. Zhuravlev

Abstract A new xenusian, *Mureropodia apae* gen. and sp. nov., is found in the lower Cambrian of the Murero Lagerstätte in the Cadenas Ibéricas, NE Spain. In *Mureropodia*, the lobopod length/body width ratio reveals that this animal hardly was able to walk on the bottom surface. Possibly, it could use the limbs for anchoring the body to the substrate. A well-developed dermomuscular sac of circular and longitudinal muscular systems as well as probably retractile proboscis fit such an interpretation. The ground plan of the Xenusia includes a vermiform body; a proboscis or mouth cone; paired lobopods with claws; a cuticle displaying a repeated anatomical patterning; a straight digestive tract with terminal mouth and anus. Morphologically heterogeneous xenusians, which crawled with their lobopods along the bottom, might give rise to four morphofunctional lineages – to cephalorhynch worms by adaptation for burrowing with retractable proboscis; to tardigrades by adaptation for interstitial habitat; to euarthropods by adaptation to walking on joint appendages; and to anomalocaridids by adaptation to swimming with lateral flaps.

J.A. Gámez Vintaned
Área de Paleontología, Depto. de Geología, Fac. de Biológicas, Universitat de València, c/Dr. Moliner, 50, Burjassot E-46100, Spain
e-mail: j.antonio.gamez@uv.es; gamez@unizar.es

E. Liñán
Área y Museo de Paleontología, Depto. de Ciencias de la Tierra, Fac. de Ciencias, Universidad de Zaragoza, c/Pedro Cerbuna, 12, Zaragoza E-50009, Spain
e-mail: linan@unizar.es

A. Yu. Zhuravlev
Área y Museo de Paleontología, Depto. de Ciencias de la Tierra, Fac. de Ciencias, Universidad de Zaragoza, c/Pedro Cerbuna, 12, Zaragoza E-50009, Spain

Geological Institute, Russian Academy of Sciences, Pyzhevskiy pereulok, 7, Moscow 119017, Russia
e-mail: ayzhur@mail.ru; andrey@unizar.es

P. Pontarotti (ed.), *Evolutionary Biology – Concepts, Biodiversity, Macroevolution and Genome Evolution*, DOI 10.1007/978-3-642-20763-1_12,

12.1 Introduction

In 1859, Charles Darwin published his famous now "The Origin of Species...", where he wrote that "by the theory of natural selection all living species have been connected with the parent species of each genus, by a difference not greater than we see between the natural and domestic varieties of the same species at the present day; and these parent species, now generally extinct, have in their turn been similarly connected with more ancient forms; and so on backward, always converging to the common ancestor of each great class" but "what geological research has not revealed, is the former existence of infinitely numerous gradations, as fine as existing varieties, connecting together nearly all existing and extinct species" (Darwin 1859: 289, 303).

Looking on flooding creationist booklets, it seems that evolutionary biology and palaeontology are unable still to find any gradations (intermediate forms) supporting Darwin's ideas. Until the very end of the last century, among the reasons of such an apparent lack of intermediates connecting principal animal phyla were pure neontological speculations on general phylogeny (e.g., an arthropod ancestor was thought to be somewhat polychaetan-like). Thus, any unusual fossils were shoehorned into phyla established by neontological methods.

The history of xenusians is a typical example of such an approach to the fossil record. The first of them – *Aysheaia pedunculata* and *Hallucigenia sparsa* – were discovered as soft-bodied imprints in the middle Cambrian of Western Canada as early as 100 years ago and interpreted as polychaetans (Walcott 1911). These were worm-like animals bearing numerous paired telescopic limbs (lobopods) alike those of onychophorans and tardigrades. Another one – lower Cambrian *Xenusion auerswaldae* – was found in ice age erratics originated from the Cambrian of Scandinavia and described as an onychophoran (Pompeckj 1927). Introduction of soft-bodied Precambrian vendobionts to the scientific society by the late 1960s brought out new thoughts about the nature of such fossils and *Xenusion* was affiliated with Ediacaran frond-like vendobionts (Tarlo 1967) while *Hallucigenia* was turned into a legendary conundrum evoked during the Cambrian explosion and extinct soon after this event (Conway Morris 1977). Until the late 1980s, this triad was placed among ancestors of either onychophorans or tardigrades due to an absence of antennas and jaws and a terminal position of the mouth (Hutchinson 1969; Delle Cave and Simonetta 1975; Whittington 1978).

Discoveries of lower Cambrian Chengjian biota in southern China and Sirius Passet biota in northern Greenland, both of which teemed with diverse lobopod-bearing fossils, happened right in time of molecular revolution in animal phylogeny given the birth to the Ecdysozoa and the Lophotrochozoa clades (Adoutte et al. 2000). These clades separated annelids and arthropods but united instead arthropods and some former aschelminthes (cephalorhynchs or cycloneuralians). Cambrian lobopod-bearing creatures (class Xenusia Dzik and Krumbiegel 1989; phylum Tardipolypoda Chen and Zhou 1997) occurred right in place to be variously interpreted as a stem group of the Arthropoda within the Ecdysozoa. The Xenusia are though to be descendents of cephalorhynch-like worms (a recidivism of annelid-like arthropod ancestor hypothesis) and ancestors of the Onychophora and the Anomalocaridida, which in turn were stem groups of the Euarthropoda (chelicerates, trilobites, crustaceans, insects, and various myriapods) (Dzik and Krumbiegel 1989; Budd 2001a; Liu et al. 2008b; Budd and Telford 2009; Daley et al. 2009; Ma et al. 2009). However, pedestrian analyses of Cambrian fossil cephalorynchs revealed that these vermiform ecdysozoans were too derived to be a stem group for any appendage-bearing ecdysozoans (Harvey et al. 2010; Zhuravlev et al. 2011 in press). Thus, xenusians turned to be the key group for understanding of the ecdysozoan early evolution and, actually, for proving of the unity of this "molecular" clade. Nowadays, over 20 soft-bodied xenusian taxa of a generic level are known in the lower Cambrian – middle Silurian interval. These fossils are ubiquitous elements of almost each lower Palaeozoic Lagerstätte and their microscopic phosphatized sclerites are widespread in lower–middle Cambrian strata (Fig. 12.1). However, even the anterior–posterior orientation of these animals is still disputable and, thus, each new find of such a body fossil provides us with a crucial information on the ecdysozoan origins and relationships.

12.2 A New Xenusian from the Murero Lagerstätte of Spain

12.2.1 Geological Setting

The Murero Lagerstätte, where a new xenusian is found, occurs in the Cadenas Ibéricas (Aragón, northeastern Spain), in the Badules Unit, which forms part of the West Asturian-Leonese tectonostratigraphic zone (Gozalo and Liñán 1988) (Fig. 12.2). Stratigraphically, the sites of exceptionally preserved fossils are restricted to the Mesones Group (comprising the Valdemiedes, Mansilla, and Murero formations) and the Acón Group, of upper Bilbilian (Cambrian Series 2, Stage 4) through lower Languedocian (Cambrian Series 3, Drumian Stage) age (Gozalo 1995; Gozalo et al. 2011 in press). The succession represents a significant portion of the Cambrian taphonomic window interval, from its middle interim until the very closure (Fig. 12.1). A number of exceptionally preserved fossils include various seaweeds, sponges, palaeoscolecidan and eucephalorhynch worms (stem group cephalorhynchs), complete trilobite carapaces preserving tiny details, lingulate brachiopods having peduncles, chancelloriid scleritomes, articulated echinoderms (cinctans, eocrinoids, edrioasteroids), and others (Conway Morris and Robison 1986; Gámez Vintaned 1995; Liñán 2003; Gozalo et al. 2003; García-Bellido et al. 2007; Zamora et al. 2009).

The only specimen of *Mureropodia apae* gen. et sp. nov. (Fig. 12.3) is derived from the base of the level RV1/5 (*Protolenus jilocanus* Zone, upper Bilbilian Stage, uppermost lower Cambrian) of the upper Valdemiedes Formation, which crops out along the Rambla de Valdemiedes 1 (RV1) section of Liñán and Gozalo

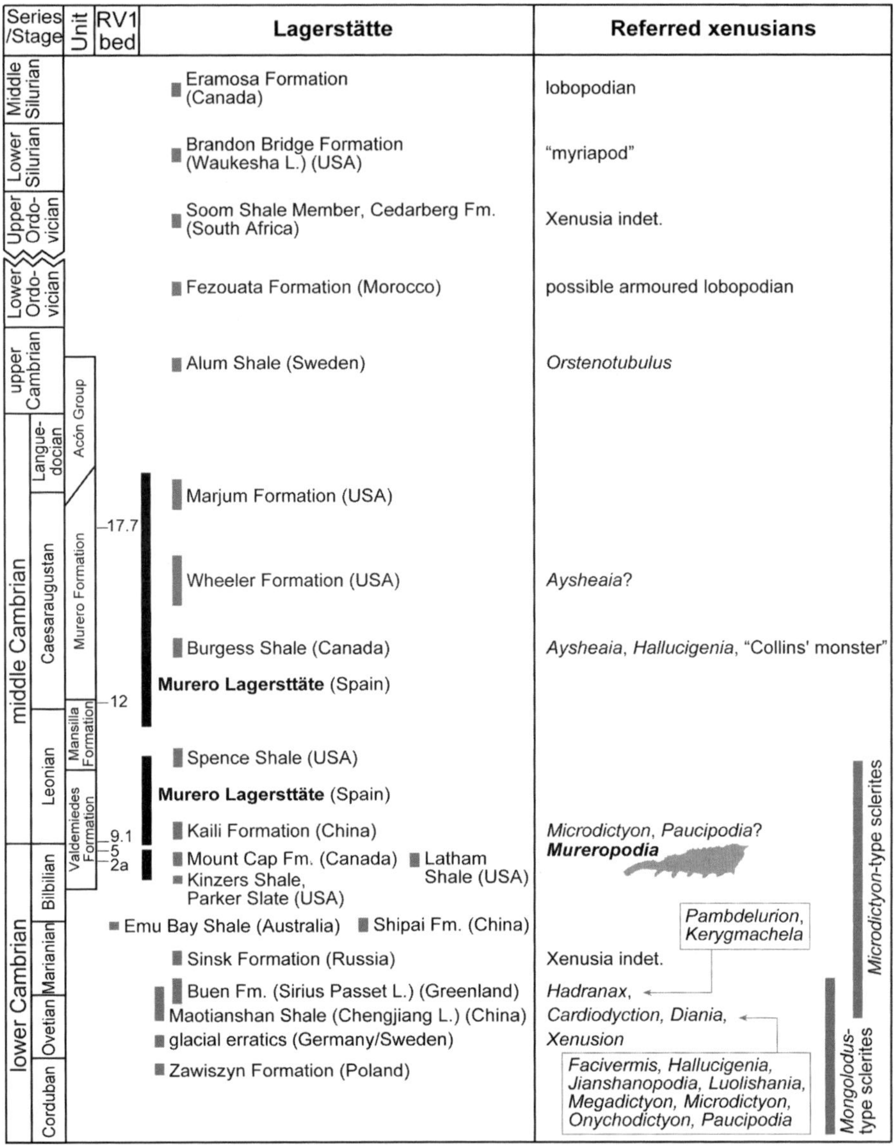

Fig. 12.1 Generalized stratigraphic sketch of the Rambla de Valdemiedes 1 section at Murero according to Spanish chronostratigraphy and relative position of major Cambrian – Silurian Lagerstätten with indication of xenusian taxa (Bengtson et al. 1986; Robison 1991; Budd 1997; Wilson et al. 2004; Zhao et al. 2005; Zhuravlev 2005; Maas et al. 2007; von Bitter et al. 2007; Liu et al. 2008b; Whittle et al. 2009; Van Roy et al. 2010; Liu et al. 2011)

(1986), in the outskirts of the village of Murero (80 km to the southwest of Zaragoza; Fig. 12.2). At this locality, the partially exposed Mesones Group (195 m in thickness) represents a transgressive-highstand-transgressive systems

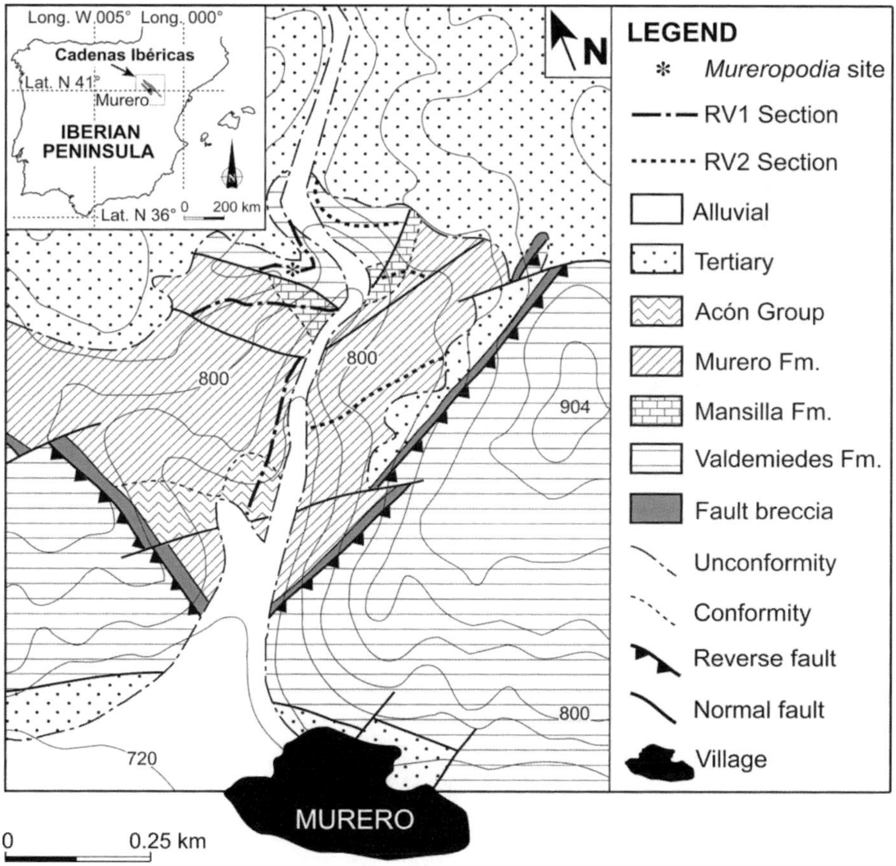

Fig. 12.2 Location of the Murero Lagerstätte, Cadenas Ibéricas, NE Spain (index map and geologic map, updated from Liñán and Gozalo (1986))

tract succession (sequences S5 and S6 of Gámez Vintaned et al. 2009b). This succession is represented mainly by fine-grained siliciclastic rocks – silty mudstone and shale – with carbonate nodules and occasional thin dolostone interbeds; dolostone is more abundant in the Mansilla Formation. Mineralogical data (textural rock features, mineral association, illite crystallinity, and illite crystal size determined by transmission electron microscopy) indicates the shales of the Mesones Group experienced an incipient anchimetamorphism (Bauluz et al. 1998, 2000).

The level RV1/5 itself comprises mainly tabular strata of homogenous, greenish gray silty mudstone and shale – with occasional parallel lamination – composed of silt-sized quartz and feldspar grains and clay flakes (illite and chlorite) oriented randomly. Thin (1–3 mm) discontinuous interbeds of very fine-grained micaceous sandstone are also present. Some of the bedding planes are covered with moderately sorted, current aligned, skeletal fragments. Polymeroid trilobites (mostly the Ellipsocephalidae) dominate, while phosphatic and calcitic inarticulate

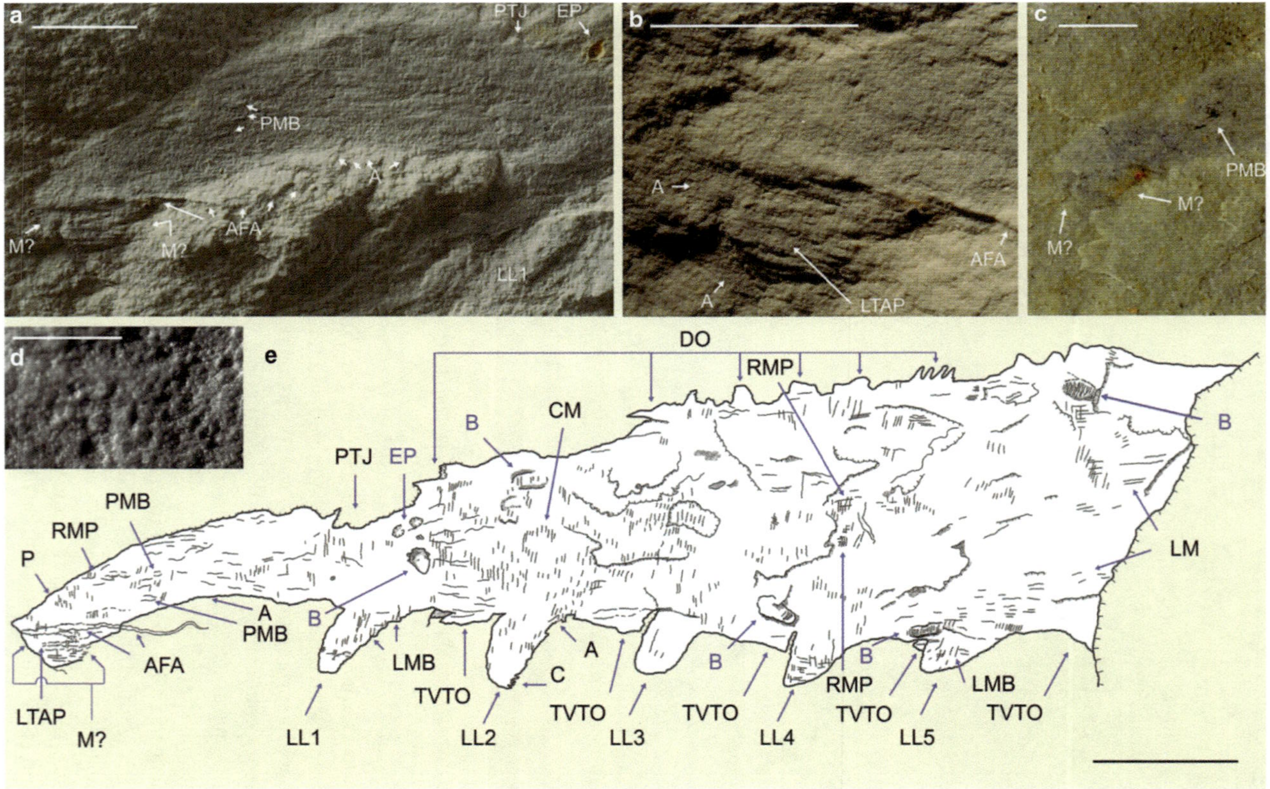

Fig. 12.3 *Mureropodia apae* gen. and sp. nov., MPZ 2009/1241. (**a**) proboscis and left lobopod 1. Inset: enlargement in "d". (**b**) Detail of proboscis tip. (**c**) proboscis tip; water cover (note carbonaceous remains). (**d**) Dermal papillae at the upper margin of the mid proboscis (inset in "a"). (**e**) Camera lucida drawing of Fig. 12.4a, b. (Lighting from NW is simulated.) Lighting in others: from N (**a**, **b**), W (**d**), and vertical (**c**). Scale bar = 5 mm (**a–c**), 1 mm (**d**), 1 cm (**e**). *A* appendicules, *AFA* antenniform frontal appendage, *B* microburrows, *C* claws, *CM* circular muscle bundles, *DO* dorsal outgrowths, *EP* edrioasteroid plates, *LL1–LL5* left lobopods, *LM* longitudinal muscle bundles, *LMB* lobopod muscle bundles, *LTAP* lower triangular area of the proboscis, *M?* mouth?, *P* proboscis, *PMB* longitudinal proboscis muscle bundles, *PTJ* proboscis/trunk junction, *RMP* reticulated muscular pattern, *TVTO* triangular ventral trunk outgrowths

brachiopods (*Trematobolus simplex*) and edrioasteroids are also present indicating the polymeroid trilobite biofacies of Dies Álvarez (2004). Trace fossils of the *Sericichnus* ichnoassociation are not uncommon (including *Cylindrichnus concentricus*, *Helminthopsis hieroglyphica*, *Planolites terraenovae*, and *Sericichnus mureroensis*; Gámez Vintaned and Mayoral 1995), but they do not occur on bedding planes bearing exceptionally preserved fossils (only meiobioturbation occurs here). The depositional environment of the level RV1/5 is interpreted as a low-to-moderate energy, relatively warm water open sublittoral shelf (Gozalo 1995).

Mineralogically, the lower level RV1/5 consists of quartz (55%), phyllosilicate (43%), and feldspar (2%); carbonates are absent. Mica is the main component (66%) of the clay fraction, while chlorite composes 30% and chlorite–smectite interstratified mineral composes 4% (Liñán et al. 1993).

Soft-bodied fossils, also including keratinous demosponges and palaeoscolecidans, intact complete skeletons, and algal thalli are preserved on the bedding planes and within beds as either slightly deformed skeletons or flattened compressions that easily split into part and counterpart. These fossils are mostly restricted to the argillaceous lithologies and do not exhibit evidences of being transported.

Mureropodia itself occurs 1 cm above the base of a massive, non-laminated greenish gray siltstone being preserved as a horizontally spread compression, and replicated with chlorite (Fig. 12.4a). Although the chlorite cover is composed solely of euhedral chlorite blades, it is visually distinct from the rock, because the crystals are much larger than those of matrix and because blade facets are preferentially oriented along the body surface imparting it a dark greenish glitter (Fig. 12.4b). An original composition of the *Mureropodia* integument could be either carbonaceous or phosphatic because both primarily carbonaceous algal thalli and phosphatic palaeoscolecidan cuticles are preserved here as chlorite replicas. Energy-dispersive X-ray spectroscopy (Figs. 12.5d, e, 12.6, and 12.7) reveals a presence of both carbon and phosphorus in the *Mureropodia* integument but the carbon content is higher (Figs. 12.5j and 12.8). Because these elements are absent from the surrounding matrix, originally carbonaceous-phosphate composition of the cuticle is implied.

It is suggested that soft tissues have been preserved in the Murero Lagerstätte through rapid postmortem mineralization by authigenic clay minerals (Gámez Vintaned et al. 2009a).

12.2.2 The Systematics, Morphology, and Preservation of the Murero xenusian

Class Xenusia Dzik and Krumbiegel 1989
Order Xenusiida Dzik and Krumbiegel 1989

Fig. 12.4 *Mureropodia apae* gen. and sp. nov., MPZ 2009/1241, upper Bilbilian Stage, uppermost lower Cambrian, Murero Lagerstätte, Aragón, NE Spain. Anterior body compression, replaced

Genus *Mureropodia* Gámez Vintaned, Liñán and Zhuravlev, gen. nov.

Derivation of name. From the village of Murero, Zaragoza Province, Aragón, NE Spain, and from the Greek πούς, δός (leg). The gender is feminine.

Type species. *Mureropodia apae* Gámez Vintaned, Liñán and Zhuravlev, sp. nov.

Diagnosis. Large fusiform xenusiid bearing several pairs of stubby lobopods, possessing an anterior proboscis and lacking any morphological limb differentiation and trunk ornamentation.

Comparison. *Mureropodia* differs from the closest morphologically *Paucipodia* Hou et al. (2004) by a distinct proboscis, a rather smooth body surface lacking annulation, a reverse limb length/trunk width ratio (from 0.92 to 0.26, with a mean of 0.47, against 1.3–2.5), and by a fusiform rather than cylindrical body shape.

Mureropodia apae Gámez Vintaned, Liñán and Zhuravlev, sp. nov.

1996 Onicóforo indet. – Liñán et al., p 77.

2008 Lobopodan – Liñán et al., pp 5, 38.

2011 Xenusian gen. and sp. indet. – Zhuravlev et al., Fig. 12.4a.

Derivation of name. In honor of the APA (Asociación Paleontológica Aragonesa), the doyen of Spanish amateur palaeontologists' associations. One of its members, Mr. Javier Andreu Comín (Zaragoza) found the fossil in 1996 during sampling campaign led by EL in cooperation with APA and SAMPUZ members.

Holotype. Specimen MPZ 2009/1241 deposited in the collection of Museo Paleontológico de la Universidad de Zaragoza (Spain); an incomplete specimen showing a proboscis and a trunk with several limbs from the Valdemiedes Formation, uppermost Bilbilian Stage, lower Cambrian, level RV1/5 of the Rambla de Valdemiedes 1 section, Murero Lagerstätte, Aragón, NE Spain (Figs. 12.3–12.5, and 12.9).

Diagnosis. As for genus.

Description. *Mureropodia apae* is a sizable xenusiid: the specimen length as preserved is 86 mm, the trunk width is up to 18.5 mm.

The head bears a long, slightly arcuate, tubular proboscis 22 mm in length. Its width varies from 4.0 mm at the tip to 4.7 mm at the midpoint, and to 6.5 mm abutting the trunk/proboscis junction. A slight constriction (up to 6.0 mm) is observed at the area of the proboscis/trunk junction. The anteriormost part of the proboscis bears a long slender, slightly curling, outgrowth running rearward

Fig. 12.4 (continued) with chlorite, showing proboscis and stubby lobopods. (**a**) Ammonium chloride coating (very low-angle lighting from NE). (**b**) Water cover (high-angle lighting from N). (**c**) Detail of "a" showing circular and longitudinal muscular systems in the trunk (inset: area pictured in "d"). (**d**) Detail of inset in "c" where the outermost layer of the integument is flaked off, showing the reticulated muscular pattern (very low-angle lighting from NW). (**e**) Microburrow at the upper, rear part of the trunk (*arrow* in "a"). Reticulated muscular pattern and dermal papillae are indicated. (Lighting from W.) (**f**) Partial view of left lobopod 5 showing telescopic structure and muscle bundles associated to it (lighting from NW). Scale bar = 1 cm (**a**–**c**), 1 mm (**d**–**f**). *B* microburrow, *DO* dorsal outgrowths, *DP* dermal papillae, *LL1*–*LL5* left lobopods, *LMB* lobopod muscle bundles, *P* proboscis, *PTJ* proboscis/trunk junction, *RMP* reticulated muscular pattern, *T* trunk, *TVTO* triangular ventral trunk outgrowths

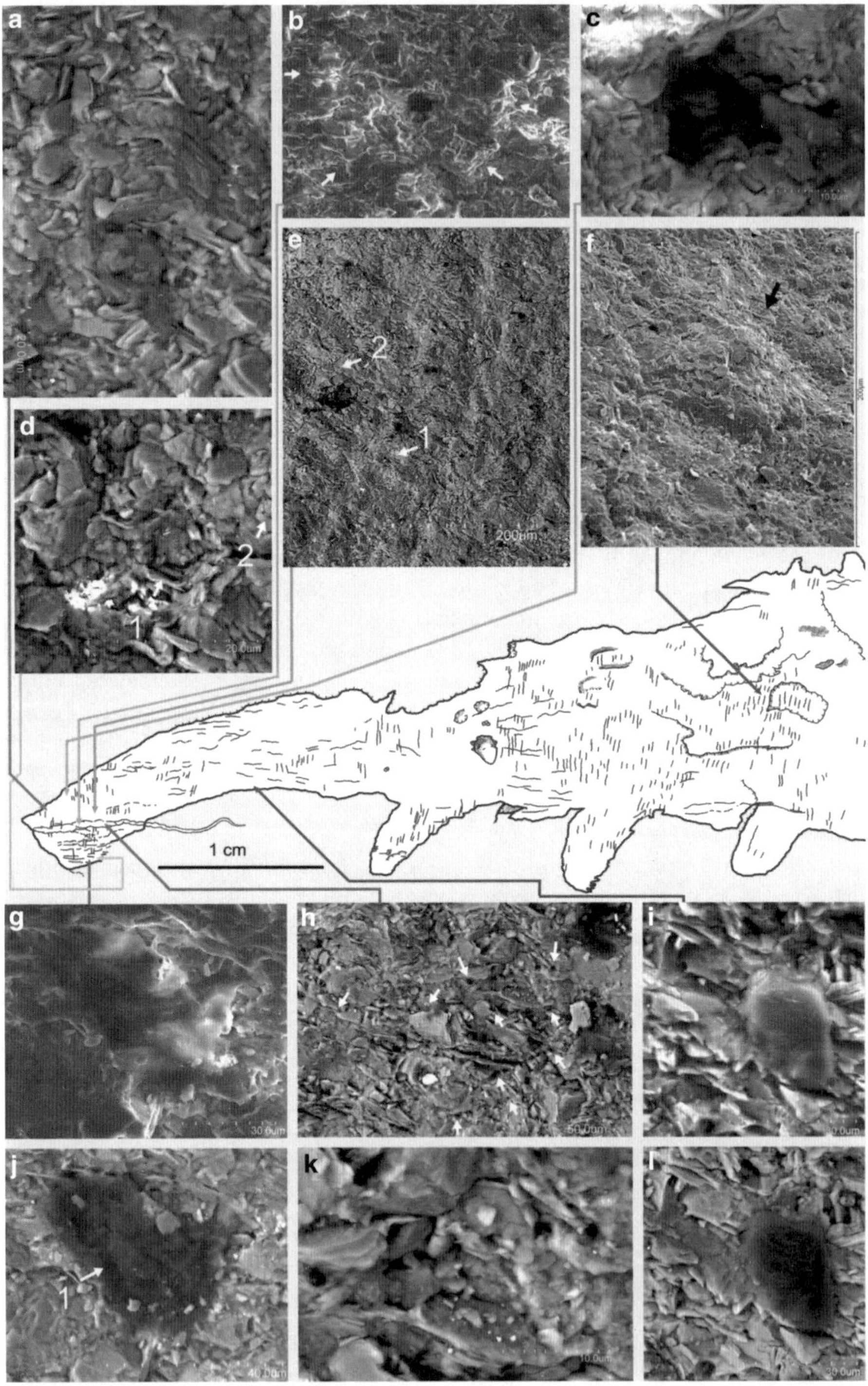

Fig. 12.5 SEM micrographs of *Mureropodia apae* gen. and sp. nov. (MPZ 2009/1241) with precise location of the takes (environmental SEM, except for "f") (Secondary electrons imaging:

(Figs. 12.3 a, b, e and 12.4a, b). The lower terminal area of the proboscis has a different preservation possessing a clear microrelief and showing diverse types of microstructures, including various pores (Fig. 12.5b, c), facet-like polygonal microplates (Fig. 12.5d), and pectinate structures (Fig. 12.5g, j). Some minute pores appear on the slender long outgrowth projecting rearward and are organized in rows (Fig. 12.5h, k) running obliquely to it and parallel to the proboscis margin. A curved pointed claw-like sclerite occurs in the lower area of the proboscis about its midpoint (Fig. 12.5i, l).

The trunk is fusiform and tapers anteriorly. Both the proboscis and the trunk bear a system of fine transverse and longitudinal, closely spaced, ridges imparting the fossil a regular reticulated surface pattern (Figs. 12.3, 12.4a, c, d, e, and 12.5e). Transverse ridges are 0.075–0.125 mm across; in some areas, they are closely packed, without intervening spaces. Two types of longitudinal ridges are present differing by their width; the thicker are sparse, 0.25–0.50 mm across and are more common on the anterior part of the specimen and near the limbs (Fig. 12.9b, e). The finer ridges are 0.075–0.150 mm across. Longitudinal and transverse ridges are visible on the surface of the fossil but also within the trunk, underneath the outermost layer (0.085 mm in thickness) of the integument, in areas where the integument is flaked off (Fig. 12.4d). A finer reticulated surface pattern is formed by thin ridges at the proboscis tip (Fig. 12.5e); under SEM, such ridges reveal fine transverse striation (Fig. 12.5a).

The trunk bears five stubby, triangular to elliptical in outline, limbs of similar lobopod type (LL1–LL5). They are slightly directed forward. The length of limbs (5.50–4.50 mm) and the distance between them (11.8–9.2 mm) decrease progressively to the rear. The tips of the limbs are moderately pointed. Crossing of thick longitudinal and transversal ridges are visible on all limbs, imparting them a telescopic structure (Figs. 12.4f and 12.9b, c, e, f).

LL2 is terminated with four curved claws. Claws are triangular in outline, flattened laterally, posteriorly directed, and pointed upward (Fig. 12.9b,c). Claw length is 0.30 mm; width at their base, 0.10–0.15 mm; spacing of 0.20–0.21 mm.

The proximal and middle posterior side of LL2 also bears threadlike slim outgrowths – appendicules of 0.09–0.15 mm in length, 0.05 mm in width at the base; spacing of about 0.12–0.16 mm (Fig. 12.9d); they also appear to present on the proximal and middle posterior side of LL4 and LL5. Somewhat

Fig. 12.5 (continued) (**a**, **c**–**e**, **h**, **j**–**l**). Backscattered electrons imaging: (**b**, **f**, **g**, **i**)). (**a**) Fine transverse striation (running from *left* to *right*) on thinner muscular ridges at the proboscis tip. (**b**) Tetraradial pore surrounded by a circular ridge on the antenniform frontal appendage. (**c**) Star-shaped pore. (**d**) Polygonal cuticular areoles (*arrows* indicate EDX analyses of Fig. 12.7; whitish areas have a phosphatic composition). (**e**) Thinner muscular reticulated pattern at the proboscis tip (*arrows* indicate EDX analyses of Fig. 12.6). (**f**) Oblique view of dermal papilla from the trunk. (**g**, **j**) Carbonaceous, pectinate sclerite (*arrow* indicate EDX spectrum analysis 1 of Fig. 12.8). (**h**, **k**) Minute pores (*arrows*) on the antenniform frontal appendage, organized in subparallel rows (general and detail views). (**i**, **l**) Carbonaceous curved, pointed claw-like sclerite. Inset orientations with respect to the general drawing are maintained. Scale bars, as indicated on the pictures

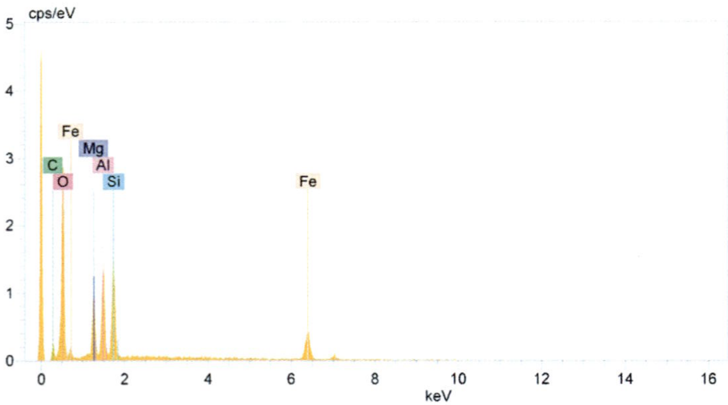

HV: 15.0 kV Pulsd.: 1.27 kcps

Spectrum: 1

El	AN	Series	unn. C [wt.-%]	norm. C [wt.-%]	Atom. C [at.-%]	Error [%]
O	8	K-series	34.54	40.85	56.71	6.4
Fe	26	K-series	18.20	21.53	8.56	4.0
Si	14	K-series	11.09	13.11	10.37	1.4
Al	13	K-series	10.38	12.27	10.10	1.3
Mg	12	K-series	7.56	8.94	8.17	1.1
C	6	K-series	2.78	3.29	6.09	0.8

Total: 84.6 %

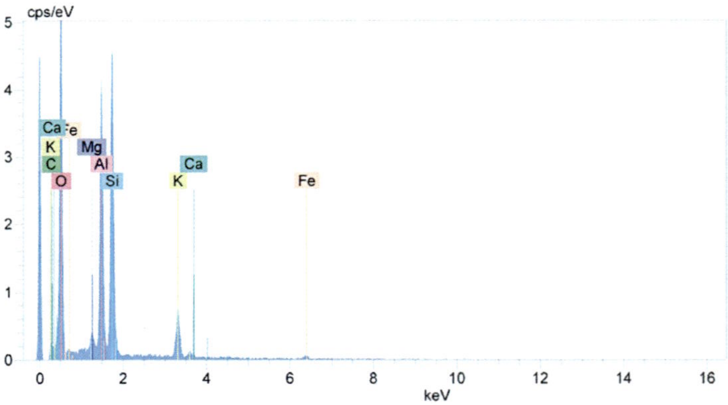

HV: 15.0 kV Pulsd.: 2.14 kcps

Spectrum: 2

El	AN	Series	unn. C [wt.-%]	norm. C [wt.-%]	Atom. C [at.-%]	Error [%]
O	8	K-series	55.78	51.50	65.79	9.6
Si	14	K-series	23.45	21.65	15.76	2.9
Al	13	K-series	19.24	17.76	13.46	2.3
K	19	K-series	6.04	5.58	2.92	1.1
Fe	26	K-series	1.90	1.75	0.64	1.2
Mg	12	K-series	1.78	1.65	1.39	0.5
Ca	20	K-series	0.11	0.10	0.05	0.3
C	6	K-series	0.00	0.00	0.00	0.0

Total: 108.3 %

Fig. 12.6 Energy-dispersive X-ray spectroscopy analyses made at the proboscis tip of *Mureropodia* (area of Fig. 12.5e with muscle fibers). Note the presence of carbon in a fiber (spectrum 1), which is absent in the surrounding matrix (2)

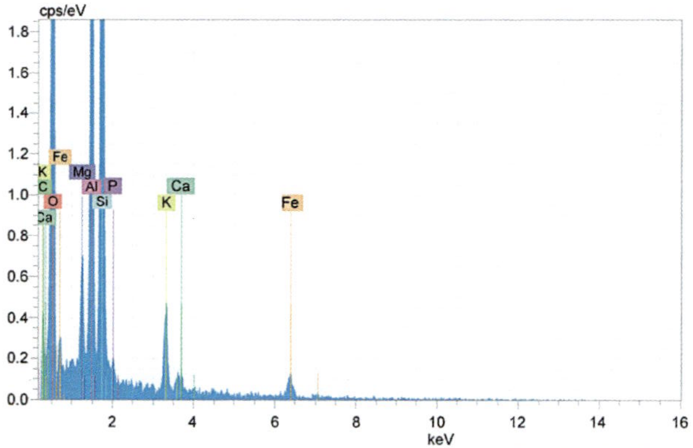

Spectrum: 1

HV: 15.0 kV

El	AN	Series	unn. C [wt.-%]	norm. C [wt.-%]	Atom. C [at.-%]	Error [%]
O	8	K-series	59.94	53.14	66.97	10.3
Si	14	K-series	24.36	21.60	15.50	3.2
Al	13	K-series	15.32	13.59	10.15	2.0
Fe	26	K-series	4.32	3.83	1.38	2.4
K	19	K-series	3.34	2.96	1.53	0.8
Mg	12	K-series	3.31	2.94	2.44	0.7
C	6	K-series	0.91	0.81	1.36	0.4
P	15	K-series	0.70	0.62	0.40	0.4
Ca	20	K-series	0.60	0.53	0.27	0.4
		Total:	112.8 %			

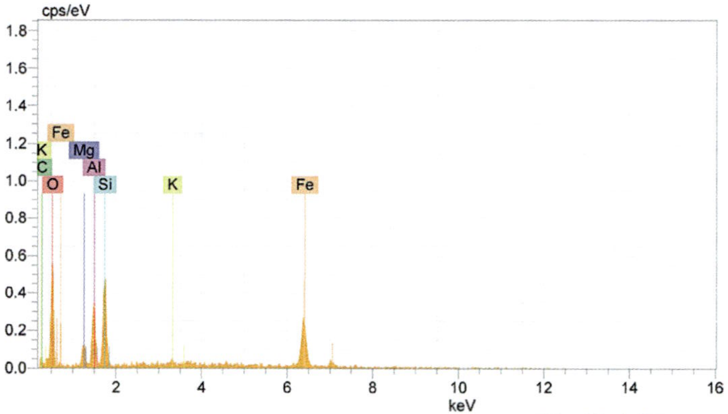

Spectrum: 2

HV: 15.0 kV

El	AN	Series	unn. C [wt.-%]	norm. C [wt.-%]	Atom. C [at.-%]	Error [%]
O	8	K-series	20.85	31.54	49.13	4.8
Fe	26	K-series	16.50	24.97	11.14	4.5
Si	14	K-series	14.53	21.98	19.50	3.6
Al	13	K-series	10.04	15.19	14.03	2.6
Mg	12	K-series	3.68	5.57	5.71	1.7
K	19	K-series	0.49	0.74	0.47	0.8
C	6	K-series	0.00	0.00	0.01	0.0
		Total:	66.1 %			

Fig. 12.7 Energy-dispersive X-ray spectroscopy analyses of the proboscis tip of *Mureropodia* (area of Fig. 12.5d with polygonal cuticular areoles). Note the presence of carbon and phosphorus in the microplate (spectrum 1), which are absent from the bottom layer of the microplate and the surrounding matrix (2)

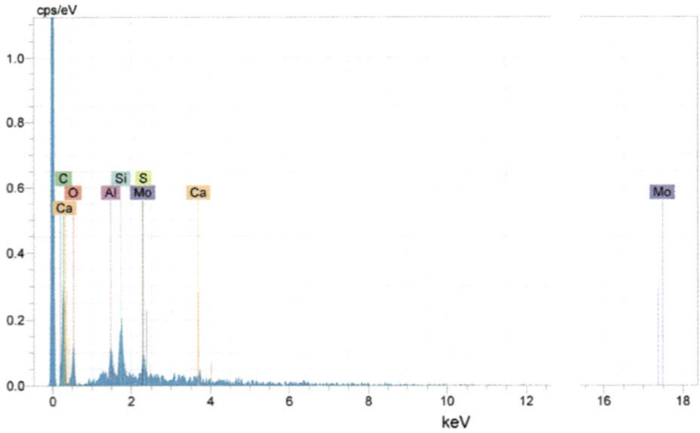

Spectrum: 1

HV: 15.0 kV

El	AN	Series	unn. C [wt.-%]	norm. C [wt.-%]	Atom. C [at.-%]	Error [%]
O	8	K-series	45.48	49.57	43.97	17.0
C	6	K-series	41.65	45.39	53.63	15.8
Si	14	K-series	2.16	2.35	1.19	0.8
S	16	K-series	1.06	1.15	0.51	0.8
Al	13	K-series	1.02	1.12	0.59	0.6
Mo	42	L-series	0.20	0.22	0.03	0.5
Ca	20	K-series	0.19	0.21	0.08	0.5
		Total:	91.8 %			

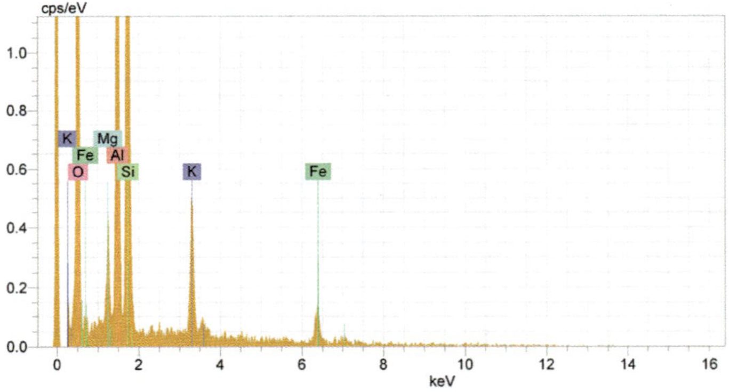

Spectrum: 2

HV: 15.0 kV Pulsd.: 1.23 kcps

El	AN	Series	unn. C [wt.-%]	norm. C [wt.-%]	Atom. C [at.-%]	Error [%]
O	8	K-series	47.92	47.14	62.32	8.8
Si	14	K-series	23.05	22.67	17.07	3.3
Al	13	K-series	17.23	16.95	13.28	2.5
K	19	K-series	5.29	5.20	2.81	1.2
Fe	26	K-series	5.15	5.07	1.92	2.3
Mg	12	K-series	3.03	2.98	2.59	0.8
		Total:	101.7 %			

Fig. 12.8 Energy-dispersive X-ray spectroscopy analyses made on a carbonaceous pectinate sclerite (that of Fig. 12.5g, j) and the matrix at the proboscis tip of *Mureropodia*. Note the very high contents of carbon and the presence of sulfur and molybdenum in the sclerite (spectrum 1), all of which are absent in the surrounding matrix (2)

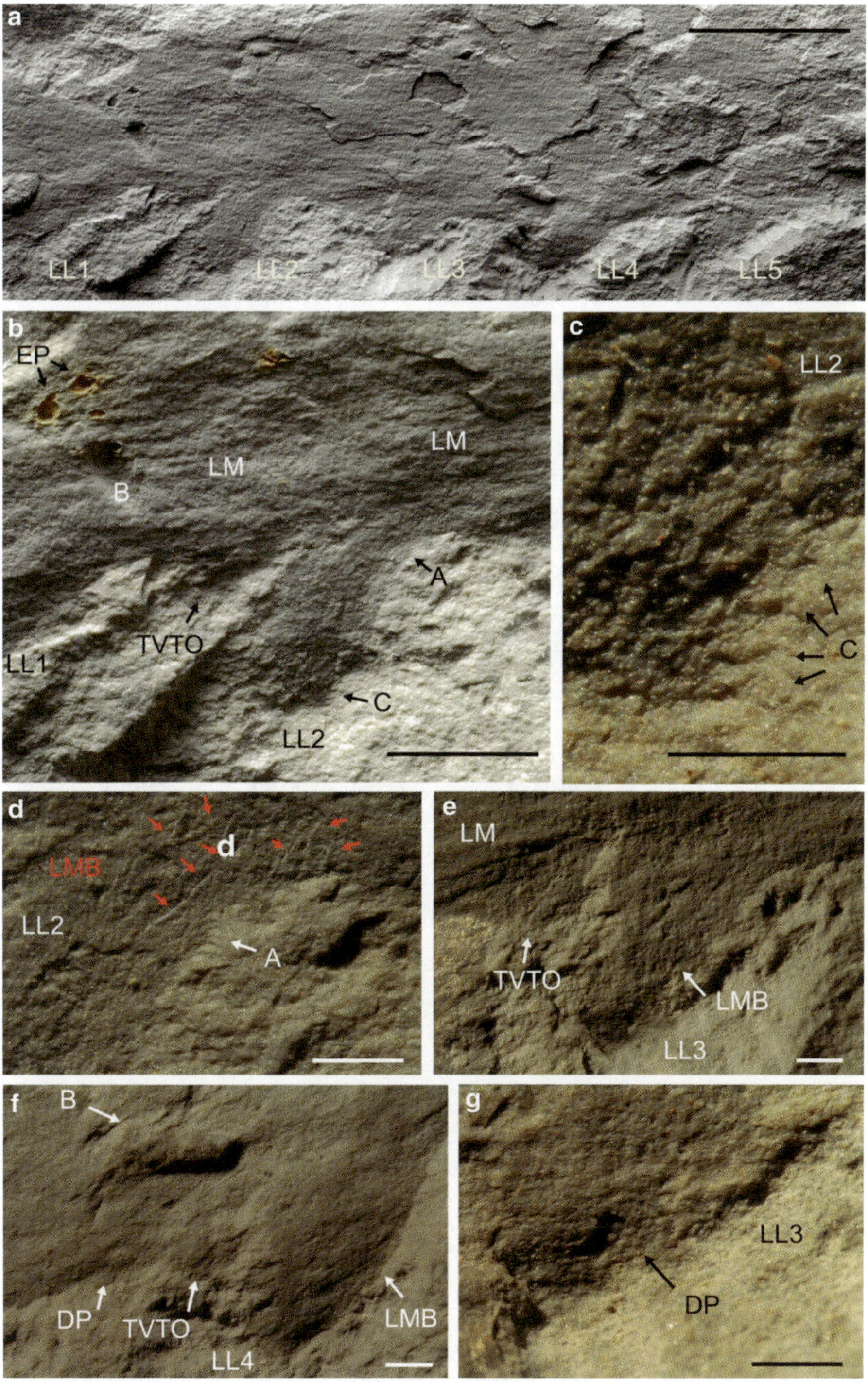

Fig. 12.9 *Mureropodia apae* gen. and sp. nov., MPZ 2009/1241. (**a**) trunk and the five left lobopods (*LL1–LL5*). Ammonium chloride coating (very low-angle lighting from N). (**b**) Anterior

similar outgrowths (but bigger) are at the tip and ventral surface of the proboscis (Fig. 12.3a, b).

Triangular ventral trunk outgrowths develop just ahead of each limb from LL2 to LL5 (plus the non-preserved LL6). They show thicker longitudinal ridges and give the ventral side of the trunk the appearance to be tilted rearward on those places (Figs. 12.4f and 12.9a, b, e, f). Some dorsal outgrowths of the trunk are observed, imparting the fossil a rugged appearance (Figs. 12.3e and 12.4a–c).

Dermal papillae are preserved on some areas of the proboscis (Fig. 12.3d), lobopods (Fig. 12.9f, g), and the trunk (mainly ventral area) (Fig. 12.5f). They range from 0.100 to 0.250 mm across, and are rounded to elliptical in outlines. Smaller papillae (0.040–0.070 mm across) are present at the proboscis tip.

Remarks. The flattened compression shows proboscis and, thus, is interpreted as the anterior part of a xenusian body including the head and anterior trunk area. The anterior–posterior orientation of the body is defined in accordance with *Xenusion*, *Aysheaia pedunculata*, and *Paucipodia* models where the position of terminal mouth on the tip of anterior proboscis is better supported (Whittington 1978; Dzik and Krumbiegel 1989; Ramsköld 1992; Hou et al. 2004).

The proboscis however is not a simple tubular organ alike those of *Xenusion* and *Paucipodia*. A long slender, slightly curling, outgrowth although resembling a muscle bundle is thicker than any muscle bundles of this fossil and runs off the limits of the body. Besides, it bears minute papillae, which are absent from the muscle bundles. By both its position and shape, the outgrowth resembles antenniform frontal appendages of some other xenusians, namely, *Hadranax* (Budd and Peel 1998, Pl. 1) and *Luolishania* (Ma et al. 2009, Fig. 3K).

It seems that the proboscis is sharply truncated anteriorly being cut by a burrow. However, such an interpretation is doubtful because the "burrow" delimits a lower triangular area of the proboscis only. This area bears facet-like polygonal microplates, which resemble as cuticular microplates of onychophorans as well as areoles of nematomorphs (Robson 1964, Fig. 2; De Villalobos and Zanca 2001, Fig. 2D; Poinar et al. 2004, Figs. 3 and 4). The pore canals penetrating the frontal part of the proboscis can be interpreted as sensory structures alike those of tardigrades and various extinct and extant vermiform ecdysozoans (Crowe et al. 1970; Malakhov and Adrianov 1995). Pectinate structures observing on the tip of the proboscis can be interpreted as pharyngeal teeth alike those of cephalorhynch worms (Malakhov and Adrianov 1995) and some xenusians (Liu et al. 2006b).

Fig. 12.9 (continued) part of the trunk with longitudinal muscle bundles and lobopods LL1 and LL2 (with telescopic structure, claws, and appendicules preserved). (**c**) Detail of "b"; tip of LL2 with claws and interwoven muscle fibers. (**d**) Detail of "d" with lobopod muscle system and appendicules (note the long fiber running from the body wall down into the limb). (**e**) Thick longitudinal muscle fibers in the ventral trunk outgrowth ahead of LL3. (**f**) Oblique view of left lobopod 4. (**g**) Detail of "e" showing dermal papillae. Very low-angle lighting from N (**a–c**) and NW (**d–g**). Scale bar = 1 cm (**a**), 5 mm (**b**), 1 mm (**c–g**). A appendicules, B microburrow (void), *C* claws, *DP* dermal papillae, *EP* edrioasteroid plates, *LL1–LL5* left lobopods, *LM* longitudinal muscle bundles, *LMB* lobopod muscle bundles, *TVTO* triangular ventral trunk outgrowths

Another interesting feature of *Mureropodia* is fine regular reticulated pattern of its surface including the trunk (Fig. 12.4c–e), the proboscis (Fig. 12.5e), and lobopods (Figs. 12.4f and 12.9e, f). Some cephalorhynch worms, namely, priapulids and nematomorphs, possess a regular reticulate system of collagen fibers underlying their cuticles, which provides some flexibility of the integument (Malakhov and Adrianov 1995; Poinar 1999). However, this system has a clear cross-hatching orientation rather than transverse and parallel to the body surface one. On the contrary, distinct transverse (circular) and longitudinal fiber orientation is observed in a dermatomuscular sac of cephalorhynch worms as well as some Cambrian xenusians. Fine transverse striation covering the thinnest muscle fibers of *Mureropodia* is typical of myofibrils in *Pambdelurion* (Budd 1998). *Pambdelurion* (Budd 1998) and *Kerygmachela* (Budd 1999) from Sirius Passet, *Paucipodia* from Chengjiang (Hou et al. 2004), and Sinsk xenusian from the Siberian Platform (Zhuravlev 2005) show well-expressed peripheral circular musculature consisting of fiber bundles of 0.008–0.010 mm (in *Pambdelurion*) to 0.065–0.070 mm (in *Kerygmachela*) across. Although the thickness of transverse and longitudinal ridges in *Mureropodia* (0.075–0.150 mm) is over this size range, their very pattern fits closely to a dermatomuscular sac structure. The size difference can be resulted from a different taphonomic pathway: if muscle fibers in the Sirius Passet Lagerstätte were replaced by silica and those in the Sinsk Lagerstätte by a phosphate mineral, soft tissues of Murero fossils were replicated with clay minerals, the crystals of which were enlarged later on due to incipient anchimetamorphism. Even thicker longitudinal ridges of *Mureropodia* (0.25–0.50 mm), which are seen on its anterior part, can represent muscle bundles (Fig. 12.3a, c). In modern priapulids and fossil palaeoscolecidans, a retractor–protractor muscle system of reversible proboscis is present (Malakhov and Adrianov 1995; Zhuravlev et al. 2011 in press). These muscles form thick bundles covered with collagenous envelope. On the contrary, muscles which are preserved at the very tip of the proboscis are much thinner than any other muscles of *Mureropodia* (Fig. 12.5e). Possibly, these were muscles regulating movement of antenniform appendages.

Peripheral muscular sacs are, probably, observed in lobopods of *Mureropodia*, allowing them a telescopic movement. Besides, lobopods reveal a limb musculature consisting of thick bundles of muscle both in the triangular ventral outgrowths and also running down into the limb, possibly, from the body wall (Fig. 12.9d). The triangular ventral outgrowths themselves can represent second, smaller, branches of appendages alike those of the Ordovician lobopodian from the Soom Shale (Whittle et al. 2009, text-fig. 2). These branches are better expressed in front of L2–L5 (Figs. 12.4b and 12.9a, b, e, f).

Noteworthy, that claws of *Mureropodia* match closely to those of modern onychophorans (e.g., Morera-Brenes and Monge-Nájera 2010, Fig. 8). At the same time, they resemble a widespread lowermost Cambrian phosphatic microfossil *Mongolodus* (e.g., Esakova and Zhegallo 1996, Pl. 4, Figs. 4–6; Steiner et al. 2007, Fig. 4G), which is commonly assigned to protoconodonts (Bengtson et al. 1990; Esakova and Zhegallo 1996). If our suggestion will be confirmed by histological comparative studies, the fossil record of xenusians would be continued to the basal Cambrian, thus, adding further 20 million years to existing body fossil record of this group (Fig. 12.1).

Noteworthy, that a structure similar in size and morphology to claws are observed around the midpoint of the proboscis (Fig. 12.5i, l). This claw-like structure is also comparable with sensory–locomotory organs of cephalorhynchs, which are scalids (Malakhov and Adrianov 1995). Such scalids would be useful if the proboscis is retractile. This suggestion is supported by the observation on the presence of retractor–protractor muscle system (see above).

Some features of the fossil are difficult for an interpretation. Thus, outgrowths, which are observed at the tip and ventral surface of the proboscis (Fig. 12.3a, b), can be either fragments of loose muscle bundles pressed out the proboscis due to taphonomic processes or genuine appendicules alike those of *Onychodictyon*. A rugged relief of the dorsal surface may appear either by preservation of the tips of the right limbs or, simply, by features of postmortem body distortion. A position of the mouth opening can be suggested only because there are two probable openings on the tip of the proboscis. The first of them is located terminally on the tip, the second one occurs antero-ventrally (Fig. 12.3a, c).

The fossil reveals several full relief knobs and voids on its surface. Some stellate voids certainly represent edrioasteroid plates, which although being dissolved still keep imprints of the stereomic structure (Figs. 12.3 and 12.9b). Other voids and knobs are elongate and show faint transversal ridges; they resemble microburrows (0.50–1.45 mm across, up to 2.80 mm in length) and their restriction to the fossil area can be explained by scavenging activity (Figs. 12.3 and 12.4a, b, e, f, 12.9a, b).

12.3 General Morphology of the Xenusia

Each new Cambrian xenusian possesses a number of features that differs it easily from other lobopodians. This is why, current systematics of the Xenusia counts for three orders and eight families. Probably, the systematics does not reflect principal features of xenusians well enough. For instance, an establishment of the order Scleronychophora Hou and Bergström (1995) to bring together all xenusian taxa bearing sclerites is hardly plausible because roughly the same morphological area of sclerites is typical of Cambrian cephalorhynchs of the class Palaeoscolecida (Zhuravlev et al. 2011 in press).

More fruitful would be subdivision of xenusian taxa into groups in accordance with an expression of the head tagmosis following Chen (2009). Three groups are distinguished by this feature. The first group includes proboscis-bearing xenusians lacking any appendage differentiation: *Xenusion*, *Microdictyon*, *Paucipodia*, *Diania*, and *Mureropodid* (Chen et al. 1989, 1995; Dzik and Krumbiegel 1989; Hou et al. 2004; Liu et al. 2011). The second group comprises xenusians having an elongated head (shorten proboscis) and anterior appendages resembling slightly degenerated trunk limbs: *Luolishania* [+*Miraluolishania*], *Cardiodictyon*, *Onychodictyon*, *Hallucigenia*, *Aysheaia*, and *Facivermis* (Whittington 1978; Hou and Chen 1989; Hou et al. 1991; Ramsköld 1992; Ramsköld and Chen 1998; Bergström and Hou 2001; Liu et al. 2006a, 2008a; Ma et al. 2009). The third group consists of xenusians revealing advanced tagmosis – head appendages are modified and branched, eyes may be present, the proboscis is reduced to a mouth cone located ventro–anteriorly: *Hadranax*, *Megadictyon*, *Jianshanopodia*, *Pambdelurion*, and *Kerygmachela* (Budd 1997, 1998, 1999; Budd and Peel 1998; Liu et al. 2006b, 2007; Schoenemann et al. 2009).

All other features, such as trunk annulation, presence of sclerites and trunk outgrowths, number of limbs, their length and stiffness, and number and arrangement of terminal claws, are highly variable and do not form any set of characters matching exactly the head tagmosis expression. For instance, *Diania*, which possesses the most elaborated sclerotized legs, has a simple head bearing proboscis (Liu et al. 2011). *Facivermis*, *Pambdelurion*, and *Kerygmachela* are listed among xenusians because each of them possesses a number of characters typical of the Xenusia including lobopods, terminal mouth, and absence of even incipient articulation (arthrodization).

In general, the xenusian ground plan includes a vermiform body, segmented with exception for *Mureropodia*; a proboscis, which, if retractile, bears possible pharyngeal teeth and scalids, reducing to a mouth cone in advanced forms; paired lobopod appendages equipped with terminal claws; a straight digestive tract with both a terminal mouth and an anus. Other features can be extrapolated from the anatomy of better preserved specimens. Xenusians certainly belong to ecdysozoans because possible exuvia are known from *Aysheaia*? *prolata*, *Xenusion*, and *Hadranax* (Robison 1985; Dzik and Krumbiegel 1989; Budd and Peel 1998). Superimposed sclerite duplication in *Microdictyon* and *Quadratopora*, when the sclerites display identical morphology but different size, is also indicative for molting process (Chen et al. 1995; Zhang and Aldridge 2007). These exuviated integuments consist of relatively rigid, annulated, multilayered cuticle commonly incorporating large net-like plates and spines (*Microdictyon*, *Cardiodictyon*, *Onychodictyon*, *Hallucigenia*, *Luolishania*) and polygonal cuticular patterns or areoles (*Orstenotubulus* and *Mureropodia*) (Bengtson et al. 1986; Maas et al. 2007; Zhang and Aldridge 2007; Fig. 12.5d herein). The polygonal patterns resemble, in both size and appearance, cuticular areoles of some cephalorhynchs (nematomorphs), while large plates are comparable with sclerites of the Palaeoscolecida. The only three-dimensionally preserved phosphatized microspecies *Orstenotubulus evamuellerae* also demonstrates spines, sensory papillae, and a single ventrally preterminal gonopore (Maas et al. 2007) and *Mureropodia* preserves sensitive pore canals, which are features widely distributed among cephalorhynchs, tardigrades, and onychophorans. A nervous system probably consisted of a circumpharyngeal brain ring and a ventral cord with paired ganglia. The latter, possibly, has been documented in *Paucipodia* (Hou et al. 2004), while circumpharyngeal brain ring is implied by analogy with tardigrades and cephalorhynchs having terminal mouth. Probable pharingeal teeth are detected in *Paucipodia* and *Jianshanopodia* (Hou et al. 2004; Liu et al. 2006b), while jaws are definitely absent from any xenusians (Ramsköld and Chen 1998; Liu et al. 2008a). Axial paired reniform midgut diverticulae with an internal structure of submillimetric lamellae are distinguished in *Pambdelurion*, *Kerygmachela*, *Megadictyion*, and *Jianshanopodia* (Budd 1998, 1999; Liu et al. 2006b, 2007). Similar structures were retained in anomalocaridids, stem-group euarthropods, and stem-group cephalorhynchs (Butterfield 2002; Zhuravlev et al. 2011 in press). Peripheral circular and longitudinal cross-striated musculature is distinguishable in addition to which diagonally oriented fibers might be developed as evident in *Pambdelurion*, *Kerygmachela* (Budd 1998, 1999), and *Paucipodia* (Hou et al. 2004). Peripheral circular muscular sac is also well expressed in phosphatised xenusians from the lower Cambrian Sinsk, and lower Silurian Waukesha Lagerstätten (Wilson et al. 2004; Zhuravlev 2005), while in *Mureropodia* it is preserved being replaced by clay minerals (Figs. 12.4 and 12.6–12.8). Such a dermomuscular sac surrounded a spacious body cavity of a probably coelomic type because coelom is preserved in onychophorans and euarthropods (during embryogenesis and as sacculi in adults), and in priapulid *Meiopriapulis* (around the foregut) (Störch et al. 1989; Eriksson et al. 2003; Schmidt-Rhaesa 2006), and, thus, coelom can be a plesiomorphic state of ecdysozoans. An approximately centrally positioned intestine running along the whole length of the body is visible in *Cardiodictyon* (Hou et al. 1991), *Paucipodia* (Hou et al. 2004), *Megadictyon* (Liu et al. 2007), *Onychodictyon ferox* and *O. gracilis* (Liu et al. 2008a), and the Soom Shale lobopodian (Whittle et al. 2009). Such a position of intestine implies a development of supportive dorsoventral mesenteries, thus fitting the presence of coelom.

What *Mureropodia* adds to our knowledge about xenusians is unusual ratio of the limb length to the body width (0.26–0.92). Among all other xenusians, this coefficient is well above 1.0, for instance, 1.4–2.9 in *Cardiodictyon catenulum*, 2.0–2.3 in *Microdictyon sinicum*, 2.0–2.9 in *Luolishania longicruris*, 4.2 in *Hallucigenia fortis* (based on measurements of trunk width and limb length in Monge-Nájera and Hou 2000), 1.3–2.5 in *Paucipodia inermis* (based on Hou et al. 2004), and 1.5 in *Hadranax augustus* (based on Budd and Peel 1998). *Mureropodia* could use the limbs for anchoring the body by terminal claws either to the substrate or even within the sediment during wormlike crawling rather than for walking on the bottom surface. The absence of any rigid integument structures (sclerites, annulations), and the presence of a vermiform dermomuscular sac and even, probably, a retractile proboscis fit the suggestion that *Mureropodia* moved by peristaltic locomotion rather than by walking legs. Such living conditions, probably, were not artificial for xenusians as some of them possessed appendages pointed laterally (*Orstenotubulus, Luolishania*), some others had appendages extremely long as shown above for a normal walking locomotion and *Facivermis* born appendages on the anterior body area only. Budd (2001b) suggested that in long-legged xenusians large sclerites served for the attachment of muscle bundles and this enforced muscle system allowed them to walk. However, *Paucipodia* displays long limbs but lacks sclerites (Hou et al. 2004), and many Cambrian legless vermiform ecdysozoans possess large sclerites (Han et al. 2007). It seems that vermiform peristaltic locomotion fits better to the morphology of long-legged xenusians. The limbs themselves could be alternatively used to push the body forward while limb appendicules and claws helped to anchor it to the substrate

(Maas et al. 2007; Ma et al. 2009). This does not mean that all xenusians were crawling animals. Some of them possessing an advanced muscle system of both peripheral and skeletal muscle as was shown for *Pambdelurion* (Budd 1998), and those having robust sclerotized appendages as was revealed in *Diania* (Liu et al. 2011) were walking epifaunal species.

12.4 Conclusions

In summary, the early Cambrian – middle Silurian Xenusia were a highly morphologically heterogeneous group, which includes vermiform lobopod-bearing animals with diverse cuticular elements, different muscular body systems, and variously differentiated head region. The xenusians crawling with their limbs along the sediment surface, may gave rise to four morphofunctional lineages, namely, to stem-group cephalorhynchs or cycloneuralians (*via Mureropodid* and *Facivermis*-type forms) by adaptation for burrowing with proboscis; to tardigrades (*via Hadranax*-like forms) by adaptation for interstitial habitat; to onychophorans (*via Jianshanopodia*-type forms) by adaptation to walking lifestyle with muscular appendages; and to anomalocaridids (*via Kerygmachela*-type forms) by adaptation to swimming with lateral flaps in the pelagic realm (Fig. 12.10). Stem-group euarthropods could have further originated from either onychophorans or anomalocaridids, or even, if polyphyletic, from both these groups. The direct descendents of xenusians still retain their features during embryonic development at least. Thus, studies of embryology and brain anatomy of present-day onychophorans revealed that their mouth has been originally terminal and the "labrum" is developed as a muscular outgrowth from the pharynx and, thus, is not homologous with arthropod labrum (Eriksson et al. 2003); antennae, jaws, and slime papillae were originated from legs (Mayer and Koch 2005). Actually, this scenario of evolutionary events was predicted by Hutchinson (1969). Similar studies of tardigrades have shown a circumpharyngeal brain morphology closely resembling the brain of vermiform ecdysozoans (Zantke et al. 2008). Besides, both onychophorans and tardigrades lack a ventral nervous ladder (Mayer and Harzsch 2007; Zantke et al. 2008). Noteworthy, that first pentastomids were lobopod-bearing animals (Waloszek et al. 2006), and early pycnogonids had long tubular proboscises (Charbonnier et al. 2007). Although according to some authors (Williamson 2009a, b), certain insect larvae are produced by hybridization with onychophorans, they merely recapitulate some features of their far ancestors. Molecular data grouping tardigrades and nematodes within the Ecdysozoa (Dunn et al. 2008) merely confirm that these clades originated from the same ancestral stock, which is xenusians.

Thus, the Xenusia represented the common ancestral group from which all other ecdysozoan clades originated. What is even more important is that slightly paraphrasing Darwin's words, "geological research has revealed the former

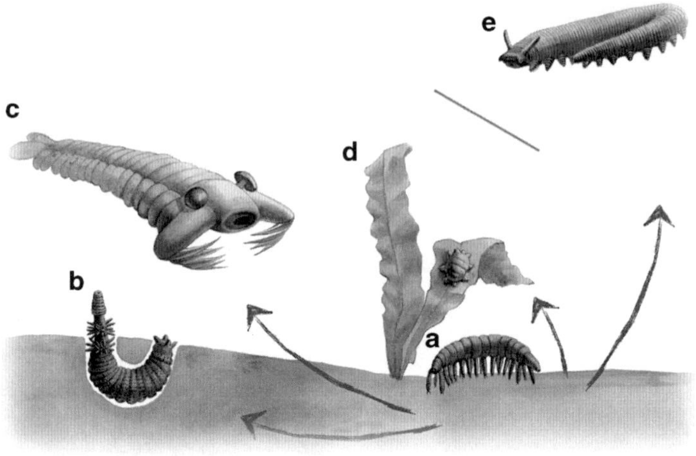

Fig. 12.10 Generalized reconstructions of early Cambrian xenusian (**a**), cephalorhynch worm (**b**), freely swimming anomalocaridid (**c**), dwarfed tardigrade (**d**), and present-day terrestrial onychophoran (**e**) (Drawing by Anastasia Besedina)

existence of infinitely numerous gradations, as fine as existing varieties," but still are silent about saltational and sporadic processes like hybridogenesis of remote forms.

Acknowledgments We thank sincerely the Association pour l'Étude de l'Évolution Biologique, and Prof. Pierre Pontarotti (Université de Provence, Marseilles), organizers of the 14th Evolutionary Biology Meeting at Marseilles (France), for the invitation to participate in the meeting and to publish this paper. This is a contribution to the projects: Consolíder CGL2006–12975/BTE ("MURERO"; Ministerio de Educación y Ciencia-FEDER–EU, Spain), Grupo Consolidado E–17 ("Patrimonio y Museo Paleontológico"; Gobierno de Aragón), ACI2009–1307 (Ministerio de ciercia e Innovación, Spain), and IGCP 587 ("Of Identity, Facies and Time, the Ediacaran (Vendian) Puzzle"). JAGV received financial support from the Ministerio de Ciencia e Innovación of Spain ("Juan de la Cierva" contract, ref. JCI-2009-05319). AZ benefited from the grants MI042/2006, Departamento de Ciencia, Tecnología y Universidad (Gobierno de Aragón) and CB 3/08 Programa Europa XXI de Estancias de Investigación (CAI–CONAI + D) 2008. Ms I. Pérez Urresti (MEC-European Social Fund-Universidad de Zaragoza) and Mrs Anastasia Besedina assisted with some of the drafting, and we are grateful for their contribution. Mr Ignacio Tacchini Ciudad (Instituto de Carboquímica, CSIC, Zaragoza) and Ms Ana Cristina Gallego Benedicto (Servicios de Apoyo a la Investigación, Universidad de Zaragoza) took the SEM micrographs.

References

Adoutte A, Balavoine G, Lartillot N, Lespinet O, Prud'homme B, de Rosa R (2000) The new animal phylogeny: reliability and implications. Proc Natl Acad Sci USA 97:4453–4456

Bauluz B, Fernández-Nieto C, González López JM (1998) Diagenesis – very low-grade metamorphism of clastic Cambrian and Ordovician sedimentary rocks in the Iberian Range (Spain). Clay Miner 33:373–394

Bauluz B, Peacor DR, González López JM (2000) Transmission electron microscopy study of illitization in pelites from the Iberian Range, Spain: layer-by-layer replacement? Clays Clay Miner 48:374–384

Bengtson S, Matthews SC, Missarzhevsky VV (1986) The Cambrian netlike fossil *Microdictyon*. In: Hoffman A, Nitecki MH (eds) Problematic fossil taxa. Oxford University Press, New York; Clarendon Press, Oxford, pp 97–115

Bengtson S, Conway Morris S, Cooper BJ, Jell PA, Runnegar B (1990) Early Cambrian fossils from South Australia. Mem Assoc Australas Palaeontols 9:1–364

Bergström J, Hou X (2001) Cambrian Onychophora or xenusians. Zool Anz 240:237–245

Budd GE (1997) Stem group arthropods from the Lower Cambrian Sirius Passet fauna of North Greenland. In: Fortey RA, Thomas RH (eds) Arthropod relationships. Chapman and Hall, London, pp 125–138

Budd GE (1998) Arthropod body-plan evolution in the Cambrian with an example from anomalocaridid muscle. Lethaia 31:197–210

Budd GE (1999) The morphology and phylogenetic significance of *Kerygmachela kierkegaardi* Budd (Buen Formation, Lower Cambrian, N Greenland). Trans R Soc Edinb Earth Sci 89:249–290

Budd GE (2001a) Tardigrades as 'stem-group arthropods': the evidence from the Cambrian fauna. Zool Anz 240:265–279

Budd GE (2001b) Why are arthropods segmented? Evol Dev 3:332–342

Budd GE, Peel JS (1998) A new xenusiid lobopod from the Early Cambrian Sirius Passet fauna of North Greenland. Palaeontology 41:1201–1213

Budd GE, Telford MJ (2009) The origin and evolution of arthropods. Nature 457:812–817

Butterfield NJ (2002) *Leanchoilia* guts and the interpretation of three-dimensional structures in Burgess Shale-type fossils. Paleobiology 28:155–171

Charbonnier S, Vannier J, Riou B (2007) New sea spiders from the Jurassic La Voulte-sur-Rhône Lagerstätte. Proc R Soc Lond B 274:2555–2561

Chen J (2009) The sudden appearance of diverse animal body plans during the Cambrian explosion. Int J Dev Biol 53:733–751

Chen J, Zhou G (1997) Biology of the Chengjiang biota. In: Chen J, Cheng Y, Iten HV (eds) The Cambrian explosion and the fossil record. National Museum of Natural Sciences, Taichung, pp 11–105

Chen J, Hou X, Lu H (1989) Early Cambrian nettled scale-bearing worm-like sea animal. Acta Palaeontol Sin 28:1–16 (in Chinese)

Chen J, Zhou G, Ramsköld L (1995) The Cambrian lobopodian, *Microdictyon sinicum*. Bull Natl Mus Nat Sci (Taichung, Taiwan) 5:1–93

Conway Morris S (1977) A new metazoan from the Cambrian Burgess Shale of British Columbia. Palaeontology 20:623–640

Conway Morris S, Robison RA (1986) Middle Cambrian priapulids and other soft-bodied fossils from Utah and Spain. Univ Kansas Paleontol Contrib 117:1–22

Crowe JH, Newell IM, Thomson WW (1970) *Echiniscus viridus* (Tardigrada): fine structure of the cuticle. Trans Am Microsc Soc 89:316–325

Daley AC, Budd GE, Caron J-B, Edgecombe GD, Collins D (2009) The Burgess Shale anomalocaridid *Hurdia* and its significance for early euarthropod evolution. Science 323: 1597–1600

Darwin C (1859) The origin of species by means of natural selection or the preservation of favoured races in the struggle for life, 6th edn. 1872. A Mentor Book. New York, Scarborough, Ontario

De Villalobos C, Zanca F (2001) Scanning electron microscopy and intraspecific variation of *Chordodes festae* Camerano, 1897 and *C. peraccae* (Camerano, 1894) (Nematomorpha: Gordioidea). Syst Parasitol 50:117–125

Delle Cave L, Simonetta AM (1975) Notes on the morphology and taxonomic position of *Aysheaia* (Onychophora?) and of *Skania* (undetermined phylum). Monitore Zool Ital (Nov Ser) 9:67–81

Dies Álvarez ME (2004) Bioestratigrafía y Paleoecología de la Formación Valdemiedes (límite Cámbrico Inferior-Medio) en las Cadenas Ibéricas. Tesis Doctoral, Universidad de Zaragoza, Zaragoza, p 147

Dunn CW, Hejnol A, Matus DQ, Pang K, Browne WE, Smith SA, Seaver E, Rouse GW, Obst M, Edgecombe GD, Sørensen MV, Haddock SHD, Schmidt-Rhaesa A, Okusu A, Kristensen RM, Wheeler WC, Martindale MQ, Giribet G (2008) Broad phylogenomic sampling improves resolution of the animal tree of life. Nature 452:745–749

Dzik J, Krumbiegel G (1989) The oldest 'onychophoran' *Xenusion*: a link connecting phyla? Lethaia 22:169–181

Eriksson BJ, Tait NN, Budd GE (2003) Head development in the onychophoran *Euperipatoides kanangrensis* with particular reference to the central nervous system. J Morphol 255:1–23

Esakova NV, Zhegallo EA (1996) Biostratigraphy and fauna of the Lower Cambrian of Mongolia. Nauka, Moscow, in Russian

Gámez Vintaned JA (1995) Nuevo hallazgo de un anélido (?) paleoscolécido en el Cámbrico Medio de Murero (Cadena Ibérica Occidental, NE de España). In: Gámez Vintaned JA, Liñán E (eds) Memorias de las IV Jornadas Aragonesas de Paleontología: "La expansión de la vida en el Cámbrico". Libro homenaje al Prof. Klaus Sdzuy. Institución "Fernando el Católico", Zaragoza, pp 205–218

Gámez Vintaned JA, Mayoral E (1995) Paleoicnología del Grupo Mesones (Cámbrico Inferior-Medio) en Murero (Cadena Ibérica Occidental, NE de España). In: Gámez Vintaned JA, Liñán E (eds) Memorias de las IV Jornadas Aragonesas de Paleontología: "La expansión de la vida en el Cámbrico". Libro homenaje al Prof. Klaus Sdzuy. Institución "Fernando el Católico", Zaragoza, pp 219–252

Gámez Vintaned JA, Liñán E, Zhuravlev AYu, Bauluz B, Gozalo R, Zamora S, Esteve J (2009a) The preservation of the Cambrian Murero biota in the Mesones Group, Cadenas Ibéricas, Spain. In: Smith MR, O'Brien LJ, Caron J-B (eds) International Conference on the Cambrian Explosion. Walcott 2009. Abstract Volume. The Burgess Shale Consortium, Toronto, pp 32–33

Gámez Vintaned JA, Schmitz U, Liñán E (2009b) Upper Vendian–lowest Ordovician sequences of the western Gondwana margin, NE Spain. In: Craig J, Thurow J, Thusu B, Whitham A, Abutarruma Y (eds) Global Neoproterozoic petroleum systems: the emerging potential in North Africa, vol 326. Geological Society of London Special Publications, Bath, pp 231–244

García-Bellido DC, Gozalo R, Chirivella Martorell JB, Liñán E (2007) The demosponge genus *Leptomitus* and a new species from the Middle Cambrian of Spain. Palaeontology 50:467–478

Gozalo R (1995) El Cámbrico de las Cadenas Ibéricas. In: Gámez Vintaned JA, Liñán E (eds) Memorias de las IV Jornadas Aragonesas de Paleontología: "La expansión de la vida en el Cámbrico". Libro homenaje al Prof. Klaus Sdzuy. Institución "Fernando el Católico", Zaragoza, pp 137–167

Gozalo R, Liñán E (1988) Los materiales hercínicos de la Cordillera Ibérica en el contexto del Macizo Ibérico. Estud Geol 44(5–6):399–404

Gozalo R, Liñán E, Dies ME (2003) Intraspecific dimorphism in an evolutionary series of paradoxidids from the Middle Cambrian of Murero, Spain. In: Lane PD, Siveter DJ, Fortey RA (eds) Trilobites and their relatives (proceedings of Oxford conference 2001). Spec Pap Palaeontol 70:141–156

Gozalo R, Chirivella Martorell JB, Esteve J, Liñán E (2011, in press) Proposal of correlation between the base of Drumian Stage and the base of middle Caesaraugustan Stage in the Iberian Chains (NE Spain). Bull Geosci 86(4)

Han J, Liu J, Zhang Z, Zhang X, Shu D (2007) Trunk ornament on the palaeoscolecid worms *Cricocosmia* and *Tabelliscolex* from the Early Cambrian Chengjiang deposits of China. Acta Palaeontol Pol 52:423–431

Harvey THP, Dong X, Donoghue PCJ (2010) Are palaeoscolecids ancestral ecdysozoans? Evol Dev 12(2):177–200

Hou X, Bergström J (1995) Cambrian lobopodians – ancestors of extant onychophorans? Zool J Linn Soc 114:3–19

Hou X, Chen J (1989) Early Cambrian arthropod-annelid intermediate animal, *Luolishania* gen. nov. from Chengjiang, Yunnan. Acta Palaeontol Sin 28:205–213 (in Chinese)
Hou X, Ramsköld L, Bergström J (1991) Composition and preservation of the Chengjiang fauna – a Lower Cambrian soft-bodied biota. Zool Scr 20:395–411
Hou X, Ma X, Zhao J, Bergström J (2004) The lobopodian *Paucipodia inermis* from the Lower Cambrian Chengjiang fauna, Yunnan, China. Lethaia 37:235–244
Hutchinson GE (1969) *Aysheaia* and general morphology of the Onychophora. Am J Sci 267:1062–1066
Liñán E (2003) The Lower and Middle Cambrian biota of Murero. In: Alcalá L (ed) European Palaeontological Association-Workshop 2003, Exceptional Preservation, Teruel, Spain. Fundación Conjunto Paleontológico de Teruel, Teruel - Museo Nacional de Ciencias Naturales (CSIC), Madrid, pp 19–23
Liñán E, Gozalo R (1986) Trilobites del Cámbrico Inferior y Medio de Murero (Cordillera Ibérica). Mem Mus Paleontol Univ Zaragoza 2:1–104
Liñán E, Fernández-Nieto C, Gámez JA, Gozalo R, Mayoral E, Moreno-Eiris E, Palacios T, Perejón A (1993) Problemática del límite Cámbrico Inferior-Medio en Murero (Cadenas Ibéricas, España). Rev Españ Paleontol Nº Extr:26–39
Liñán E, Gámez Vintaned JA, Gozalo R (1996) Hallazgo de una biota tipo Burgess Shale en el Cámbrico Inferior tardío de Murero (Zaragoza). In: Palacios T, Gozalo R (eds) Comunicaciones de las XII Jornadas de Paleontología: Badajoz, 30 de octubre-2 de noviembre de 1996. Universidad de Extremadura, Servicio de Publicaciones, Cáceres, 1996, pp 77–78
Liñán E, Gozalo R, Dies Álvarez ME, Gámez Vintaned JA, Mayoral E, Chirivella Martorell JB, Esteve J, Zamora S, Zhuravlev AYu, Andrés, JA (2008) Fourth International Trilobite Conference Trilo 08 Toledo, Spain, 2008. Post-Conference Field Trip. Lower and Middle Cambrian trilobites of selected localities in Cadenas Ibéricas (NE Spain), Universidad de Zaragoza, Zaragoza, p. 52
Liu J, Han J, Simonetta AM, Hu S, Zhang Z, Yao Y, Shu D (2006a) New observations of the lobopod-like worm *Facivermis* from the early Cambrian Chengjiang Lagerstätte. Chin Sci Bull 51:358–363
Liu J, Shu D, Han J, Zhang Z, Zhang X (2006b) A large xenusiid lobopod with complex appendages from the Lower Cambrian Chengjiang Lagerstätte. Acta Palaeontol Pol 51:215–222
Liu J, Shu D, Han J, Zhang Z, Zhang X (2007) Morpho-anatomy of the lobopod *Magadictyon* cf. *haikouensis* from the early Cambrian Chengjiang Lagerstätte, South China. Acta Zool (Stockholm) 88:279–288
Liu J, Shu D, Han J, Zhang Z, Zhang X (2008a) The lobopod *Onychodictyon* from the Lower Cambrian Chengjiang Lagerstätte revisited. Acta Palaeontol Pol 53:285–292
Liu J, Shu D, Han J, Zhang Z, Zhang X (2008b) Origin, diversification, and relationships of Cambrian lobopods. Gondwana Res 14:277–283
Liu J, Steiner M, Dunlop JA, Keupp H, Shu D, Ou Q, Han J, Zhang Z, Zhang X (2011) An armoured Cambrian lobopodian from China with arthropod-like appendages. Nature 470:526–530
Ma X, Hou X, Bergström J (2009) Morphology of *Luolishania longicruris* (Lower Cambrian, Chengjiang Lagerstätte, SW China) and the phylogenetic relationships within lobopodians. Arthropod Struct Dev 38:271–291
Maas A, Mayer G, Kristensen RM, Waloszek D (2007) A Cambrian micro-lobopodian and the evolution of arthropod locomotion and reproduction. Chin Sci Bull 52:3385–3392
Malakhov VV, Adrianov AV (1995) Cephalorhyncha – a new phylum of the animalia kingdom. KMK Scientific Press, Moscow, in Russian; English summary
Mayer G, Harzsch S (2007) Immunolocalization of serotonin in Onychophora argues against segmented ganglia being an ancestral feature of arthropods. BMC Evol Biol 7:118–125
Mayer G, Koch M (2005) Ultrastructure and fate of nephridial anlagen in the antennal segment of *Epiperipatus biolleyi* (Onychophora, Peripatidae) – evidence for the onychophoran antennae being modified legs. Arthropod Struct Dev 34:471–480

Monge-Nájera J, Hou X (2000) Disparity, decimation and the Cambrian "explosion": comparison of early Cambrian and Present faunal communities with emphasis on velvet worms (Onychophora). Rev Biol Trop 48:333–351
Morera-Brenes B, Monge-Nájera J (2010) A new giant species of placented worm and the mechanism by which onychophorans weave their nets (Onychophora: Peripatidae). Rev Biol Trop 58:1127–1142
Poinar G Jr (1999) *Palaeochordes protus* n.g., n.sp. (Nematomorpha, Chordodidae), parasites of a fossil cockroach, with a critical fossil hairworms and helminths examination of other of extant cockroaches (Insecta: Blattaria). Invertebr Biol 118:109–115
Poinar G Jr, Rykken J, LaBonte J (2004) *Parachordodes tegonotus* n.sp. (Gordioidea: Nematomorpha), a hairworm parasite of ground beetles (Carabidae: Coleoptera), with a summary of gordiid parasites of carabids. Syst Parasitol 58:139–148
Pompeckj JF (1927) Ein neues Zeugnis uralten Lebens. Paläontol Z 9:287–313
Ramsköld L (1992) Homologies in Cambrian Onychophora. Lethaia 25:443–460
Ramsköld L, Chen J (1998) Cambrian lobopodians: morphology and phylogeny. In: Edgecombe GD (ed) Arthropods fossils and phylogeny. Columbia University Press, New York, pp 107–150
Robison RA (1985) Affinities of *Aysheaia* (Onychophora), with description of new Cambrian species. J Paleontol 59:226–235
Robison RA (1991) Middle Cambrian biotic diversity: examples from four Utah Lagerstätten. In: Simonetta AM, Conway Morris S (eds) The early evolution of Metazoa and the significance of problematic taxa. Cambridge University Press, Cambridge, pp 77–98
Robson EA (1964) The cuticle of *Peripatopsis moseleyi*. Q J Microsc Sci 105:281–299
Schmidt-Rhaesa A (2006) Perplexites concerning the Ecdysozoa: a reply to Pilato et al. Zool Anz 244:205–208
Schoenemann B, Liu JN, Shu DG, Han J, Zhang ZF (2009) A miniscule optimized visual system in the Lower Cambrian. Lethaia 42:265–273
Steiner M, Li G, Qian Y, Zhu M, Erdtmann B-D (2007) Neoproterozoic to early Cambrian small shelly fossil assemblages and a revised biostratigraphic correlation of the Yangtze Platform (China). Palaeogeogr Palaeoclimatol Palaeoecol 254:67–99
Storch V, Higgins RP, Morse P (1989) Ultrastructure of the body wall of *Meiopriapulus fijiensis* (Priapulida). Trans Am Microsc Soc 108(4):319–331
Tarlo LBH (1967) *Xenusion* – onychophoran or coelenterate? Mercian Geol 2:97–99
Van Roy P, Orr PJ, Botting JP, Muir LA, Vinther J, Lefebvre B, el Hariri K, Briggs DEG (2010) Ordovician faunas of Burgess Shale type. Nature 465:215–218
von Bitter PH, Purnell MA, Tetreault DK, Stott CA (2007) Eramosa Lagerstätte – exceptionally preserved soft-bodied biotas with shallow-marine shelly and bioturbating organisms (Silurian, Ontario, Canada). Geology 35:879–882
Walcott CD (1911) Cambrian geology and paleontology. II. – Middle Cambrian annelids. Smithson Misc Collects 57(5):109–144
Waloszek D, Repetski JE, Maas A (2006) A new Late Cambrian pentastomid and a review of relationships of this parasitic group. Trans R Soc Edinb Earth Sci 96:163–176
Whittington HB (1978) The lobopod animal *Aysheaia pedunculata* Walcott, Middle Cambrian, Burgess Shale, British Columbia. Philos Trans R Soc Lond B 284:165–197
Whittle RJ, Gabbott SE, Aldridge RJ, Theron J (2009) An Ordovician lobopodian from the Soom Shale Lagerstätte, South Africa. Palaeontology 52:561–567
Williamson DI (2009a) Caterpillars evolved from onychophorans by hybridogenesis. Proc Natl Acad Sci USA 106:19901–19905
Williamson DI (2009b) Reply to Giribet: Caterpillars evolved from onychophorans by hybridogenesis. Proc Natl Acad Sci USA 106:E132
Wilson HM, Briggs DEG, Mikulic DG, Kluessendorf J (2004) Affinities of the Lower Silurian Waukesha 'myriapod'. Geol Soc Am Abstr Programs 36:525

Zamora S, Gozalo R, Liñán E (2009) Middle Cambrian gogiid echinoderms from Northeast Spain: taxonomy, palaeoecology, and palaeogeographic implications. Acta Palaeontol Pol 54:253–265

Zantke J, Wolff C, Scholtz G (2008) Three-dimensional reconstruction of the central nervous system of *Macrobiotus hufelandi* (Eutardigrada, Parachela): implications for the phylogenetic position of Tardigrada. Zoomorphology 127:21–36

Zhang X, Aldridge RJ (2007) Development and diversification of trunk plates of the Lower Cambrian lobopodians. Palaeontology 50:401–415

Zhao Y, Zhu M, Babcock LE, Yuan J, Parsley RL, Peng J, Yang X, Wang Y (2005) Kaili Biota: a taphonomic window on diversification of metazoans from the basal Middle Cambrian: Guizhou, China. Acta Geol Sin 79:751–765

Zhuravlev AYu (2005) Tardipolypodians. In: Ponomarenko AG (ed) Unique Sinsk localities of early Cambrian organisms (Siberian platform). Nauka, Moscow, pp 56–61 (in Russian)

Zhuravlev AYu, Gámez Vintaned JA, Liñán E (2011, in press) The Palaeoscolecida and the evolution of the Ecdysozoa. Palaeontogr Can 31

Part IV
Genome Evolution

Chapter 13
Genomic Perspectives on the Long-Term Absence of Sexual Reproduction in Animals

Etienne G.J. Danchin, Jean-François Flot, Laetitia Perfus-Barbeoch, and Karine Van Doninck

Abstract Sexual reproduction, the exchange and recombination of genetic material between different individuals, is commonly viewed as one of the most important sources of genomic diversity in animals. This genomic diversity is subject to natural selection and, consequently, the fittest genomes relative to the environment survive and persist. According to this vision, the absence of sexual reproduction in animals is believed to inexorably lead to an evolutionary dead end as asexual animals become unable to adapt to changing environmental conditions. Yet, several animal lineages suspected to have been reproducing exclusively asexually for millions of years actually survived environmental changes and are not necessarily restricted to specialized ecological niches. The sources of genomic variations that have contributed to the evolutionary success and persistence of these lineages is currently unknown. Here we will review and discuss these known cases of long-term survival of asexually reproducing animal lineages with a focus on recent genomic findings.

13.1 Introduction

The most common reproductive mode throughout the animal tree of life is sexual reproduction, in which meiosis and fertilization occur sequentially. It is also the most widespread both in terms of species and phyla. One reason commonly cited for its evolutionary success is the series of advantages associated with sexual

E.G.J. Danchin • L. Perfus-Barbeoch
INRA, CNRS, Université de Nice-Sophia Antipolis, UMR 1301, 400 route des Chappes, B.P. 167, F-06903 Sophia-Antipolis Cedex, France
e-mail: etienne.danchin@sophia.inra.fr

J.-F. Flot • K. Van Doninck
University of Namur (FUNDP), Unit of Research in Organism Biology (URBO), 61 rue de Bruxelles, B-5000 Namur, Belgium

P. Pontarotti (ed.), *Evolutionary Biology – Concepts, Biodiversity, Macroevolution and Genome Evolution*, DOI 10.1007/978-3-642-20763-1_13,

reproduction. These advantages comprise higher genome plasticity through mixing and recombination of different haplotypes expected to provide better adaptability (Weismann 1886; Burt 2000), as well as the possibility to overcome invasion of deleterious mutations within genomes (Muller 1932; Muller 1964). Therefore, it is believed that sexually reproducing species are more likely to survive environmental changes and persist over long evolutionary time periods, whereas species reproducing exclusively asexually cannot persist in the long term and represent evolutionary dead ends. Indeed, the absence of recombination and mixing reduces possibilities for an overall genetic variability and there is an ongoing accumulation of deleterious mutations. Most examples of strictly asexually reproducing animals belong to short emerging branches dispersed among clades of sexually reproducing taxa in the animal tree of life (Butlin 2002). Additionally, these recently emerging asexual lineages generally possess only part of the genotypic diversity and occupy a more restricted ecological niche than their sexual ancestors (Doncaster et al. 2000; Janko et al. 2008). This dominance of sexual animals supports the view that sexual reproduction confers advantages over asexuality. In apparent contradiction with this view, a few animal lineages are suspected to have survived in the absence of sexual reproduction for several millions of years. How they survived environmental changes and persisted in the long term in the absence of known mechanisms to generate genetic diversity remains unexplained. In this chapter, we define as ancient asexuals, animal lineages that have been surviving in the absence of sexual reproduction for at least 10 million years. To our knowledge, four such animal lineages have been reported so far and are considered as putative ancient asexuals (Fig. 13.1, Table 13.1): two groups of arthropods (darwinulid ostracods and oribatid mites), one nematode lineage (the tropical root-knot nematodes), and the notorious bdelloid rotifers considered by Maynard Smith as being "something of an evolutionary scandal" (Maynard Smith 1986). The whole genome of the root-knot nematode *Meloidogyne incognita* has recently been sequenced and annotated (Abad et al. 2008). This nematode reproduces without meiosis and without sex and the availability of the genome sequence of its sexual sister species *Meloidogyne hapla* (Opperman et al. 2008) will allow a detailed examination of the genomic consequences of the loss of sex. In parallel, the genome sequence of the bdelloid rotifer *Adineta vaga*, an ancient asexual, is currently underway, and it will be interesting to compare it with the genome of a monogonont rotifer (a closely related group that reproduces both sexually and asexually) in order to identify the genomic signatures of a complete loss of sexuality. Comparison of these two genomic models (nematodes and rotifers) that have survived for a long time in the absence of sexual reproduction will allow determining whether some genomic singularities are shared between different lineages of asexually reproducing animals. To our knowledge, no genome sequence data is currently available for the two arthropod examples (oribatid mites and darwinulid ostracods), but their genomes will be worth exploring in the future to fill the current lack of model and large-scale data for animal species considered as ancient asexuals.

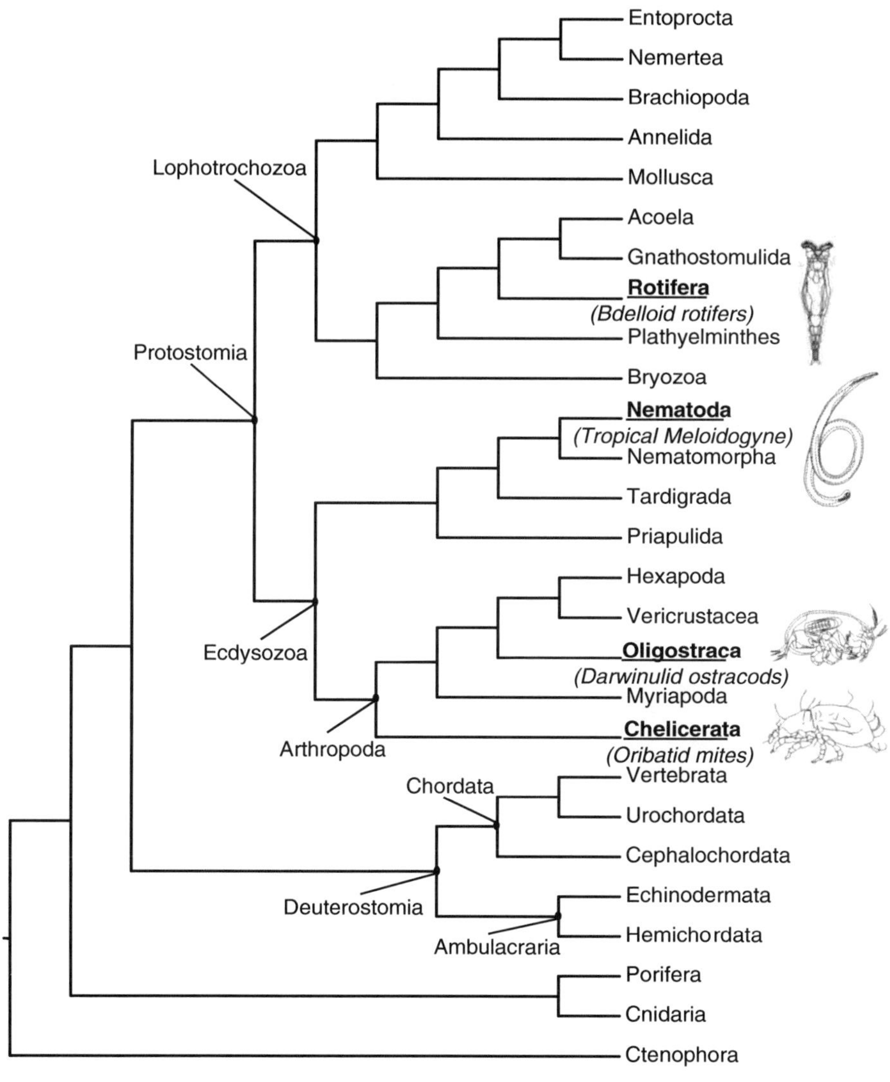

Fig. 13.1 Phylogenetic position of ancient asexuals in the animal tree of life. This schematic representation of the phylogeny of animal species is based on a phylogenomic analysis performed on 71 animal taxa based on 150 different genes (Dunn et al. 2008). The phylogenetic position of the different groups within the Arthropoda lineage is based on a focused phylogeny performed on 75 arthropod species and covering 62 genes (Regier et al. 2010). The main animal divisions are indicated at corresponding nodes. Clades that contain the four animal lineages considered as ancient asexuals in the present chapter are *underlined* and in *bold*

Table 13.1 Characteristics of four known lineages estimated to have survived in absence of sexual reproduction for more than 10 million years

Characteristics	**Darwinulid ostracods**	**Oribatid mites**	**Root-knot nematodes**	**Bdelloid rotifers**
Number of living species	35	10 000	80	460
Habitat	Aquatic	Soil, trees, aquatic	Soil, parasite of plant roots	Freshwater , semi-terrestrial (mosses, lichens)
Phylogenetic group	Arthropoda	Arthropoda	Nematoda	Rotifera
Asexual species				
Model species	*D. stevensoni* *V. cornelia*	*P. peltifer* *M. nasalis*	*M. incognita* *M. arenaria* *M. javanica*	*A. vaga* *P. roseolata*
Parthenognesis	Mitotic or meiotic?	Meiotic with inverted meiosis sequence	Mitotic => female clones	Mitotic => female clones
Males	Virtually absent, only 3 *V. cornelia* live males (functional ?)	Very rare and not functional.	Rare, not functional.	No
Genomic singularities associated with ancient asexual status				
Presence of highly-diverged gene copies	No	No	Yes	Yes
Polyploidy/Hybridization	Not described	Not described	Yes Hybridization or two ancient parental haplotypes	Yes WGD or Hybridization => degenerate tetraploid
Homogenization mechanism, low divergence level	Yes, efficient DNA repair?	Yes, efficient DNA repair?	Not described	Yes, gene conversion?
Ancientness of asexuality: Datation source				
Age of asexuality	~25/200 million years	~100/200 million years	~17/43/80 million years	~35/40 million years
Fossil	Yes	No	No	Yes
Emergence of obligate asexual reproduction	Single?	Multiple	Multiple	Single?
Nucleotide divergence	Low at the nuclear level.	High divergence observed in the COI mitochondrial gene	Low at the mitochondrial level, high at the nuclear level	High divergence observed in the HSP82 region
Genomic discrepancies with prediction for ancient asexual status				
Presence of retrotransposons	Yes	?	Yes	Yes
Sexual/Amphimictic relative species				
Facultative / cyclical parthenogenesis	None	Yes and cases of reversion towards sexuality	Clade II Meloidogyne	monogonont rotifers
Obligate sexuals	None	-	*M. microtyla* *M. carolinensis* *M. macrotyla*	-

13.2 An Overview of Putative Ancient Asexual Animals

As mentioned in the introduction, a number of animal lineages appear to contradict the common view that species unable to reproduce sexually represent evolutionary dead ends. Each of these lineages is considered to have survived in the absence of sexual reproduction for more than 10 million years (Neiman et al. 2009). Here, we present the four known such lineages and the associated evidence supporting their long-term abandon of sexual reproduction (Fig. 13.1, Table 13.1).

Frequently, obligate asexuality in animals is inferred from the absence of males, their extreme rarity, or their sterility in extant populations, whereas the ancientness of asexuality is usually assessed by examining the fossil record, by comparing the divergence level between nuclear or mitochondrial genes in asexual and sexual lineages or by estimating the genetic divergence within individuals (the so-called Meselson effect). In this last case, high allelic sequence divergence observed within asexual individuals (Birky 1996) is interpreted as long-term evolution in the absence of sexual recombination because of the independent accumulation of allelic mutations. However, it has been shown that ameiotic recombination such as gene conversion occurs in asexual species and may reduce heterozygosity (Omilian et al. 2006): consequently, the absence of Meselson effect cannot be considered as a proof of sexual reproduction.

Evolutionary theory also predicts that ancient asexuals should contain few or no functional retrotransposons because sex, despite facilitating the spread of these elements within populations, also limits their intragenomic proliferation. In asexuals, the uncontrolled multiplication of deleterious retrotransposons could predictably lead to their extinction (Arkhipova and Meselson 2005b). Another evolutionary expectation in species that have abandoned meiosis and fertilization a long time ago is that genes involved specifically in those processes should have accumulated a high number of deleterious mutations and become nonfunctional. Although methods such as "meiosis detection toolkit" have been proposed to assess ancientness of asexuality (Schurko and Logsdon 2008), it can still be argued that these genes might have been co-opted for other processes unrelated to meiosis and sexual reproduction.

13.2.1 Darwinulid Ostracods

Darwinulid ostracods are small, bivalve crustaceans for which a rich fossil record is available. The darwinulids are exclusively non-marine brooders and only 35 living species have been described (Martens et al. 1998). Researchers generally agree that no traces of male darwinulid ostracods have been found in the fossil record since 65–100 million years ago; and although putative males have been reported for the species *Darwinula stevensoni* for periods comprised between 200 and 100 million years, they actually turned out to be females according to a more detailed analysis (Martens et al. 2003). Based on this long-term absence of observed males, darwinulid ostracods appear to have been surviving without sexual reproduction for at least 200 million years. However, three males of the darwinulid species *Vestalenula cornelia* have recently been found and described and it is not clear whether they represent nonfunctional male relicts or whether they actually participate in rare sexual reproduction (Smith et al. 2006). To date, no copulation in *V. cornelia* has been observed and sperm could not be found in the three males nor in sympatric females (Schön et al. 2009). At the molecular level, analysis of three nuclear regions of a darwinulid species revealed no Meselson effect. Indeed, the level of nucleotide divergence within and between individuals was low compared to a related fully sexual lineage (Schön and Martens 2003). Whether this is due to rare cases of sexual reproduction, to gene conversion, or to a particularly efficient DNA repair mechanism remains to be elucidated. The genome of a darwinulid species was screened for the presence of RT-encoding non-LTR retrotransposons (LINEs)and two novel families were characterized, one being apparently functional, but their role and position have not yet been determined (Schön and Arkhipova 2006).

Until now, the only evidence for the ancient asexual status of darwinulid ostracods is the absence or extreme rarity of males. Schön et al. (2009) suggested that the "model-species" *D. stevensoni* is the most likely candidate to be a true ancient asexual. Indeed, no recent or fossil male has been found since at least 25 million years and this species appears to feature genetic mechanisms that homogenize their genome and maintain their general-purpose genotype (Schön et al. 2009). Such a generalized genotype appears to allow survival in a wide range of ecological conditions (Van Doninck et al. 2002; Van Doninck et al. 2003), as further detailed in Sect. 13.3.1.

13.2.2 Oribatid Mites

Oribatid mites are a species-rich group with around 10,000 currently described species inhabiting soils, trees, and aquatic habitats. They are small arthropods belonging to the Acari, and parthenogenesis, a mode of asexual reproduction, is supposed to have emerged multiple times independently in this clade. Although asexuality is the most frequent reproductive mode in the oribatid phylum, reversion toward sexuality has been reported to occur in some species of this group (Domes et al. 2007). The exact mode of asexual reproduction is not clear yet but a recent review suggests terminal fusion automixis with holokinetic chromosomes and inverted meiosis sequence (Heethoff et al. 2009). In oribatid mites, the hypothesis of an obligate asexual reproduction is not supported by the absence of males but rather by their rarity and sterility. Also supporting this idea is the apparent absence of cyclical parthenogenesis in these species (Palmer and Norton 1991), and the balanced sex ratio observed in sexual lineages (Heethoff et al. 2007). Concerning the ancientness of asexuality, divergence levels in the mitochondrial COI gene between and within clades of *Platynothrus peltifer* suggest that asexual reproduction is at least 100 million years old for this species (Heethoff et al. 2007). Another fully asexual species in this group, *Mucronothrus nasalis*, is thought to be 200 million years old (Hammer and Wallwork 1979). Similarly to darwinulid ostracods, no Meselson effect has been identified in oribatid mites and analyses of nuclear regions also suggest homogenization mechanisms in the absence of sexual recombination (Schaefer et al. 2006). This low divergence at the nuclear level contrasts with the high divergence level observed in the mitochondrial gene COI, used as an indication of ancientness. The presence of active retrotransposons within their genome has not yet been screened.

13.2.3 Root-Knot Nematodes

Root-knot nematodes (*Meloidogyne* genus) comprise ca. 80 described species, dwell in soil, and parasitize plant roots. Phylogenetic analyses have shown that in this lineage, asexual reproduction through obligate mitotic parthenogenesis has emerged at least two times independently (Holterman et al. 2009). These lineages have thus not only abandoned sexual reproduction but also meiotic division. The *Meloidogyne* clade I contains tropical root-knot nematodes (e.g., *M. incognita*, *M. arenaria*, *M. javanica*) that are considered as ancient mitotic parthenogenetic species (De Ley et al. 2002; Castagnone-Sereno 2006; Holterman et al. 2009). Clade II, its most closely related clade (Holterman et al. 2009) is essentially composed of facultative meiotic parthenogenetic species like *M. hapla* but also comprise two species described as obligate sexuals (*M. microtyla* and *M. spartinae*). Clade III, which holds an outgroup position relative to clades I–II, contains species that have all been described as facultative meiotic parthenogens (e.g., *M. chitwoodii*, *M. fallax*), except *M. oryzae* that is considered as a mitotic parthenogenetic species (Holterman et al. 2009). The common assumption that tropical *Meloidogyne* species (*M. incognita*, *M. arenaria*, *M. javanica*) are obligate parthenogens is not based on the absence of males. Indeed, males are observed in these tropical root-knot nematodes but they are rare and are assumed not to contribute genetically to the offspring (Castagnone-Sereno 2006). Furthermore, meiosis has never been observed in these nematodes and offspring results from mitotic division from the female germline, thus giving rise to clones (Van der Beek et al. 1998). High frequency of polyploidy, aneuploidy, and variable chromosome number within one species have all been reported in strict parthenogenetic *Meloidogyne* (Sasser and Carter 1985; Castagnone-Sereno 2006). Such observations are commonly viewed as indicative of frequent asexual reproduction although not necessarily of obligate asexuality.

On the basis of a comparative analysis of enzymatic profiles in the genus *Meloidogyne*, Esbenshade and Triantaphyllou (1987) estimated the last common ancestor of tropical mitotic parthenogenetic (strictly asexual) and of facultative meiotic parthenogenetic (able to reproduce sexually) nematodes to be ca. 43 million years old. Based on the phylogenetic tree presented by these authors, the age of the last common ancestor of mitotic species is estimated to be ca. 17 million years old. Another analysis, based on the comparison of the level of divergence of the mtDNA between mitotic and meiotic RKN, suggests that the last common ancestor of these two lineages may be as old as 80 million years (Hugall et al. 1997). In both analyses, the authors acknowledge that these divergence times could be overestimated due to an AT-rich composition or other biases. Anyhow, taking these potential biases into account, the last common ancestor of the obligate asexual root-knot nematodes is unlikely to be more recent than several millions of years.

A recent analysis offers a different interpretation of the evidences for long-term asexual reproduction in *Meloidogyne*. In a phylogenetic analysis of various nuclear genes, Lunt (2008) confirms the occurrence of large divergence in allelic sequences in tropical *Meloidogyne*, as expected in an asexually reproducing species due to

Meselson effect, and shows that the alleles do not cluster according to recognized morphological species in phylogenies. However, due to high similarity at the mtDNA level between different tropical apomictic (strict parthenogen) species, the author interprets the observed high allelic sequence divergence as the result of past interspecific hybridizations rather than as Meselson effect.

13.2.4 Bdelloid Rotifers

Bdelloid rotifers are common microinvertebrates inhabiting freshwater environments and semiterrestrial habitats such as mosses, lichens, and temporary pools. Bdelloidea, in which more than 460 morphospecies have been described, is the only class of the phylum Rotifera composed entirely of obligate parthenogenetic species (Segers 2007; Segers 2008). Despite much observation since the eighteenth century, neither males nor vestigial male structures have ever been observed in bdelloid rotifers. However, males have been clearly identified in monogonont rotifers, the sister class reproducing by cyclical parthenogenesis (Velázquez-Rojas et al. 2002; Leasi et al. 2010). The absence of males and the fact that single females can be reared in laboratories to produce "female" clones have led to the hypothesis that bdelloid rotifers are indeed asexual. Hsu (1956a, b) studied bdelloid oogenesis and demonstrated that oocytes are produced without any chromosome pairing or reduction in chromosome number and that after two mitotic divisions, one egg and two polar bodies are produced. These cytological results indicate the absence of meiosis and hence a reproduction by obligate mitotic parthenogenesis in bdelloid rotifers.

The presence of bdelloid fossils in old amber dated 35–40 million years (Waggoner and Poinar 1993) indicates that bdelloid rotifers originated more than 40 million years ago. Another signature of the ancient asexual status of bdelloid rotifers is that, unlike monogonont rotifers and other tested eukaryotic animals, bdelloids seem to lack high-copy number retrotransposons within their genome (Arkhipova and Meselson 2000). Commonly, those elements will propagate within the genome and if specific meiotic mechanisms are absent to control their proliferation; their unchecked invasion will lead to the extinction of the lineage (Arkhipova and Meselson 2000). Therefore, asexuals can only persist if vertically transmitted deleterious elements are maintained at a low level or are absent within their genome, a situation observed in bdelloids.

High levels of allelic divergence in the *hsp82* region were first reported by Mark Welch and Meselson (2000), suggesting that Meselson effect, an accumulation of mutations between former alleles that may lead to functional divergence, did occur in bdelloid rotifers. However, more recent studies of both the *hsp82* and histone regions of bdelloid genomes have demonstrated that they are in fact degenerate tetraploids, resulting either from an ancient whole genome duplication (autotetraploidization) or an interspecies hybridization (allotetraploidization) (Mark Welch et al. 2008; Hur et al. 2009; Van Doninck et al. 2009). Consequently, their genome is structured as two colinear pairs of genomic regions corresponding to two ancient lineages (A and B). The two lineages A and B have only few genes in common and these genes present a high level of nucleotide divergence. Within each lineage, the divergence between copies is low although a few gene copies diverge by as much as 20% (Ks value), indicating that over time, in the absence of recombination or homogenization mechanisms, synonymous divergence accumulates (Mark Welch et al. 2009).

13.3 The Challenges of Long-Term Asexuality

13.3.1 Adaptability Without Sex

One argument to explain why sexual reproduction is the most widely represented reproductive mode in animals is that it allows better adaptation (Weismann 1886; Burt 2000). By allowing mixis between arrangements of alleles (haplotypes), sexual reproduction produces at each generation new combination of alleles that can provide a selective advantage. An allele may turn out to be advantageous only when expressed together with other alleles in a combined effect. In asexual species, such a combined advantageous effect is substantially less likely to occur as emergence of a new mutation is restricted to one individual and its offspring and has no chance to mix with mutations that occurred independently in other individuals. Similarly, in diploid species, if a mutation provides an advantage only when it is present in the homozygote state, this mutation has to occur twice and independently in the two former alleles of an asexual lineage (Kirkpatrick and Jenkins 1989), unless gene conversion using the "advantageous" gene as template occurs (Mandegar and Otto 2007). Another aspect is that an advantageous mutation cannot spread easily in populations of asexual species. Indeed, assuming that such a mutation has occurred in one individual, it can only be transferred to its own offspring, but to be spread in the population, it has to be by competitive replacement of the offspring of other individuals (that may bear other mutations that would have been beneficial in different conditions). Again fixation of an advantageous mutation is supposed to be much longer and difficult in asexual populations than in sexual ones, according to the Fisher-Muller accelerated evolution theory (Fisher 1930). Intuitively, we may postulate that sexual lineages possess a better adaptation potential to environmental or ecological changes than asexuals. Furthermore, if the asexual lineage emerged from an individual genotype of the source sexual lineage, it probably possesses only a reduced frozen subset of the whole pool of genetic diversity present in the source population(Vrijenhoek and Parker 2009). In such a case, asexuals must occupy more restricted ecological niches than their sexual relatives. However, if the asexual lineage possesses a more versatile genotype than the source sexual lineage(s), as a result of hybridization for example, the asexuals may present a broader niche than their sexual relatives. In all cases, in the absence of sexual recombination, parthenogenetic species lack an important mechanism of genotypic plasticity. Thus, while they may have an adaptive advantage in a relatively stable environment due to a frozen efficient genotype and a reproductive efficiency since males are not produced, they appear clearly disadvantaged in cases of multiple environmental changes as it is expected for lineages that have been surviving for long evolutionary periods. As counterintuitive as it may appear, the four asexual animal lineages discussed here exhibit a wide geographical distribution.

In darwinulid ostracods (all asexuals), species that are ubiquitous and cosmopolitan, like *Darwinula stevensoni*, seem to contain a general-purpose genotype (GPG), i.e., a genotype providing a broad environmental tolerance (Van Doninck et al. 2002, 2003). Unlike sexuals, for which selection acts over individual genes, in asexuals such as dawinulids the unit of selection appears to be the complete genome. As a consequence, natural selection over time can favor clones with a wide tolerance (see Vrijenhoek and Parker 2009) and once such a GPG clone evolved, it can be maintained because the absence of recombination will avoid breaking up those well-adapted genotypes. How such a GPG evolved in the darwinulid ostracods is not known but the cosmopolitan species clearly have a wide tolerance of variations in abiotic factors whereas endemic darwinulids exhibit a narrow tolerance (Van Doninck et al. 2003).

In oribatid mites, an analysis of the ecological distribution according to the reproductive mode has been performed to test several predicates of differences between sexual and asexual lineages (Cianciolo and Norton 2006). No evidence for difference in ecological niche breadth between sexual and asexual lineages could be found. Another postulate commonly held is that the frequency of asexual lineages should be negatively correlated to the biological diversity in an ecological niche. The same analysis showed no negative correlation between these two features. Thus, no significant difference could be found in ecological pattern or niche breadth between sexual and asexual lineages of oribatid mites.

In root-knot nematodes, apomictic (asexual, clade I) species have a broader host spectrum as well as a wider geographical and ecological distribution than their amphimictic (sexual, clade II) relatives (Triantaphyllou 1985; Castagnone-Sereno 2006). This observation is in total contradiction with the postulated better adaptability of sexual species. In the particular case of plant parasites, this apparent evolutionary success of asexual lineages may be related to their competitive advantage due to the "twofold cost of sex" in a relatively stable and uniform environment as recently proposed for agricultural pests (Hoffmann et al. 2008). However, this argument only holds partially in root-knot nematodes as amphimictic competitors of apomictic species are able to perform facultative meiotic parthenogenesis, and are thus not completely subject to the twofold cost of sex. Furthermore, if these tropical root-knot nematodes have actually been surviving for millions of years without sex, they predated the development of agriculture and must have survived in competition with sexual relatives in unstable environments. Hence, other mechanisms of currently unknown nature may provide these obligate parthenogenetic nematodes with a competitive advantage.

Bdelloid rotifers are common micro-invertebrates inhabiting freshwater environments but also temporary habitats that dry out frequently. They are able to colonize such environments because they are both asexual and desiccation resistant. Asexuality allows individual bdelloids to colonize empty patches or to reestablish a population after experiencing severe bottlenecks. Desiccation resistance enables bdelloids to inhabit environments that are prone to desiccation and to easily disperse as dried propagules (Ricci 1998). Indeed, many bdelloid species are cosmopolitan, exhibiting a worldwide distribution (Fontaneto et al. 2008) and reaching a surprisingly high level of diversity at local scale (Fontaneto et al. 2006). Moreover, a recent study by Wilson and Sherman (2010) showed that bdelloids can eliminate a lethal fungal parasite by drying out completely for a prolonged period and escape by wind dispersal. Therefore, the combination of asexuality and desiccation seems to allow bdelloids to thrive in unstable environments and to compete with biotic factors. Finally, research by Fontaneto et al. (2007) demonstrated that bdelloids have been able to diversify into distinct evolutionary entities, successfully adapted to specific niches, in the absence of sexual reproduction. Thus, evolution and speciation appear to have proceeded unimpeded by this group's lack of sexuality.

From the four examples discussed in this section, there is presently no evidence that asexually reproducing animal species present a narrower geographical distribution or a more restricted (specialized) ecological niche than their sexual relatives (if any). In contrast, some asexuals even present a wider distribution than their sexual relatives. Specific mechanisms of yet unknown nature or peculiar genomic structures observed in these organisms may be related to their adaptability despite the lack of sexual recombination.

13.3.2 Rates of Evolution and Clonal Decay

According to Muller's ratchet theory (Muller 1964), strictly asexual lineages should undergo "clonal decay" and disappear within a few thousand years. Therefore, the persistence of some lineages for much longer periods of time without sex contradicts this model and we expect those lineages to have a low rate of mutation accumulation (maybe due to a particularly efficient DNA repair system). Supporting this hypothesis, darwinulid ostracods (Schön et al. 1998) and oribatid mites (Schaefer et al. 2006) have been shown to undergo slower rates of molecular evolution than their sexual relatives. Thus, at least two out of the four examples of ancient asexuals presented here exhibit relatively slower rates of evolution, but, is that true for the other examples?

In bdelloid rotifers, the rate of accumulation of potentially deleterious mutations appears to be higher than in their closest sexual relatives the monogonont rotifers (Barraclough et al. 2007). Other works comparing bdelloids and monogononts showed slightly higher rates of non-synonymous substitutions and slightly lower rates of synonymous substitutions in bdelloids as compared with monogononts

(Mark Welch and Meselson 2001). However, it should be noted that in the case of bdelloid rotifers, this apparently high rate of mutation accumulation is probably related to the tetraploid genome structure (made of two colinear pairs of chromosomes with a high level of divergence between pairs). Within one colinear pair, lower levels of divergence have been shown, including tracts of near identity that may result from homogenization events such as gene conversion (Mark Welch et al. 2009). These homogenizing events are suspected to occur during rounds of desiccation and recovery that involve DNA double stand break repair.

In the root-knot nematode *M. incognita*, no precise evaluation of the rate of mutation accumulation has been conducted so far to our knowledge. However, an analysis of the internal transcribed spacers (ITS) of nuclear ribosomal genes showed an extremely high heterogeneity of sequences within apomictic (obligate asexuals) *Meloidogyne* species whereas the heterogeneity was virtually absent in sexually reproducing *Meloidogyne* (Hugall et al. 1999). Nevertheless, as for bdelloid rotifers, this apparent high divergence at the genetic level has to be put in parallel with a peculiar genomic structure. Indeed, in *M. incognita*, most of the genome is present as two highly diverged copies (~8% divergence at the nucleotide level) that may represent former allelic regions or the result of an interspecies hybridization. This feature, not observed in the close relative facultative sexual species *Meloidogyne hapla*, is discussed further in Sect. 13.4.1. Concerning possible mechanisms of homogenization such as gene conversion, none have been revealed so far in apomictic *Meloidogyne*.

Overall, no clear tendency regarding the rate of mutation accumulation appears to emerge in the considered ancient asexual animal lineages. Half of the examples show higher rates of mutation accumulation than their sexual relatives, though it is certainly related to peculiar genomic structures, while the other half exhibit lower rates. There is no clear evidence for a positive or negative correlation between asexual reproduction and the rate of accumulation of potentially deleterious mutation when considering these examples. Assuming that all examples truly represent ancient strict asexuals, we cannot argue in that case that sexual reproduction provides an evolutionary advantage in terms of resistance to the accumulation of potentially deleterious mutations.

13.4 Genomic Consequences of Long-Term Asexuality

No whole-genome sequence for an obligate asexually reproducing animal was available until recently, with the publication in 2008 of the genome of the tropical root-knot nematode *Meloidogyne incognita* (Abad et al. 2008). Interestingly, the genome of a facultative sexually reproducing relative was published a few months later the same year (Opperman et al. 2008) and comparison of these two genomes will allow identifying genomic marks of long-term asexual reproduction. Genome sequence data is also emerging in bdelloid rotifers as the genome of *Adineta vaga*, a long-term obligate asexual, is currently being sequenced and assembled. We will thus focus here on the genomic singularities that emerged from the analysis of the genome sequence of *M. incognita* and from the preliminary assembly of the genome of *A. vaga*.

13.4.1 Insights from the Genome of M. incognita

The *M. incognita* genome, sequenced using a whole-genome shotgun strategy and assembled with Arachne (Jaffe et al. 2003) yielded 2,817 supercontigs. The size of the assembly, totaling 86 Mb, is almost twice the size (between 47 and 51 Mb) that had been estimated experimentally using flow cytometry approach (Leroy et al. 2003). Interestingly, an all-against-all comparison of supercontigs revealed that the genome of *M. incognita* is mainly composed of pairs of homologous yet divergent copies. The average divergence level at the nucleotide level observed between two homologous pairs is ca. 8% (Abad et al. 2008). For comparison, the average level of nucleotide divergence between individuals within an animal species is usually below 2% and higher levels of dissimilarity is considered as an indication of speciation (Birky et al. 2005). Highly divergent pairs in the genome of *M. incognita* cannot be interpreted as a mixture of individuals from different lineages or populations since the sequenced material results from repeated infections from the clonal offspring of a single female. Thus, this high divergence level can be considered to occur within one individual. The highly divergent pairs can represent former allelic regions or they can be the result of hybridization between two sexual progenitors from distinct but closely related species (Triantaphyllou 1985; Castagnone-Sereno 2006; Lunt 2008). In both cases, long-term absence of sexual recombination may have allowed mutations to accumulate independently and persist to reach the currently observed high divergence level as proposed under the "Meselson effect" model (Mark Welch et al. 2004). These features, associated to the relatively high frequency of observed polysomy and aneuploidy, are compatible with a strictly mitotic parthenogenetic reproductive mode. The peculiar genomic structure observed in *M. incognita*, composed of pairs of highly diverged regions, is not found in its facultative sexual relative *M. hapla*. Indeed, this species, able to do meiosis, harbors a small genome (54 Mb, the smallest so far for an animal) totally conform to the predicted size based on flow cytometry experiments and no trace of pairs of diverged regions has been found (Opperman et al. 2008; Bird et al. 2009). To evaluate whether the peculiar genome structure observed in *M. incognita* had consequences at the protein level, predicted proteins were grouped in cluster of at least 95% identical sequences, using the program CD-HIT (Li and Godzik 2006). The results of this clustering showed that more than 69% of protein sequences were more than 5% divergent to any other. This indicates that the observed 8% average divergence at the nucleotide level between pairs of similar genomic regions include non-synonymous substitutions that may be associated to functional divergence between gene copies.

In the case of root-knot nematodes, it appears that one remarkable consequence of the long-term absence of sexual reproduction and meiotic recombination is a genome constituted of a juxtaposition of pairs of homologous but divergent copies that might represent former paternal and maternal haplotypes. These divergent copies include genes that encode proteins divergent enough to potentially support subfunctionalization or neofunctionalization events. It will be necessary to check whether similar peculiar genomic structures are observed in other obligate asexual root-knot nematodes as well as in other species that have abandoned sexual reproduction a long time ago in order to find out whether this might represent a general genomic signature of long-term asexual reproduction.

The only other feature that emerged as an idiosyncrasy in the genome of *M. incognita* as compared to those of *M. hapla* and other nematodes was the proportion of the genome covered by repetitive elements, including transposable elements. More than 36% of the *M. incognita* genome is covered by such elements (Abad et al. 2008). This is substantially higher than for other nematodes, as only 17% were reported in *M. hapla* (Opperman et al. 2008); 16.5% and 22%, respectively, reported for *C. elegans* and *C. briggsae* (Stein et al. 2003), 17% in *P. pacificus* (Dieterich et al. 2008), and between 12% and 15% in *B. malayi* (Ghedin et al. 2007). Whether some of these transposable elements are active and potentially play a role in the plasticity of the genome of *M. incognita*, as previously suggested (Castagnone-Sereno 2006), remains to be determined.

13.4.2 Emerging Results from the Adineta vaga *Genome Project*

Initial investigations of parts of the genome of bdelloid rotifers appeared to match what was expected for ancient obligate asexuals: high intraindividual divergence (ca. 15% at nucleotide level) between what was believed to be ancient allelic sequences (Mark Welch and Meselson 2000), and an apparent lack of retrotransposons (Arkhipova and Meselson 2000). Later on, however, the picture started to change completely: the highly divergent gene copies co-occurring in *Philodina roseola* were found to be actually ohnologs (Mark Welch et al. 2008), i.e., paralogs resulting from complete genome duplication (Wolfe 2000), whereas the level of divergence between ancient alleles was markedly lower (ca. 3%) and not very different from the range observed in sexually reproducing species such as *Ciona savingnyi* (Small et al. 2007). Subsequent observations in *Adineta vaga* (another bdelloid species) confirmed this result (Hur et al. 2009). Recently, a wide diversity of transposable elements was found in *Adineta vaga* near chromosome ends (Arkhipova and Meselson 2005a), most of them inactivated or decaying but some still apparently active. Another unexpected finding at the telomeric regions of *Adineta vaga* was the discovery of abundant horizontally transferred genes (Gladyshev et al. 2008). Similarly, a high number of genes acquired via horizontal transfer was found in the genome of the root-knot nematode *M. incognita* but also in its facultative sexual relative *Meloidogyne hapla* as well as in many other plant-parasitic nematodes, including obligate sexuals (Danchin et al. 2010).

Therefore, this feature may not be indicative of the absence of sexual reproduction and it would be interesting to check whether similar abundance of genes acquired by lateral transfer is found in sexual relatives of bdelloid rotifers such as monogononts.

Since all these results were obtained from the analysis of a few selected genomic fragments (fosmids), an international consortium decided to sequence the complete genome of *Adineta vaga* in order to check the generality of these observations and conduct more in-depth analyses. Sequencing was performed using mostly paired-end pyrosequencing (Margulies et al. 2005), but the assembly proved challenging due to the medium-range heterozygosity of this genome, a problem also encountered in other whole-genome shotgun sequencing projects such as the ascidians *Ciona savignyi* and *Ciona intestinalis* (Vinson et al. 2005; Kim et al. 2007; Small et al. 2007). While very divergent sequences assemble separately and very similar ones are fused during assembly process, intermediate-level heterozygosity result in incomplete fusion that makes a reference sequence particularly difficult to produce.

Although the assembly and annotation of the complete genome sequence of *Adineta vaga* is still in progress, the first preliminary results from this project seem to confirm the absence of the Meselson effect in this species. Indeed, the average divergence level between former allelic regions within a colinear pair is around 3% over the whole genome of *Adineta vaga*. This figure appears surprisingly low for an organism whose last genomic homogenization through meiosis is supposed to have occurred several millions years ago. In comparison, the average divergence level between homologous regions that might represent former alleles in *Meloidogyne incognita* reaches 8%. Either *Adineta* does actually perform meiosis, albeit rarely (and a search for meiosis-related genes in the complete genome sequence will be required to bring a definitive answer to this question), or there must be some other mechanism acting to homogenize ancient alleles and prevent their divergence. The alternation of desiccation and rehydration phases in the life cycle of bdelloid rotifers (Gilbert 1974) may provide such a mechanism. As shown by experiments on *Deinococcus radiodurans* (Mattimore and Battista 1996), desiccation usually results in DNA double-strand breaks. In eukaryotes, double-strand breaks are repaired through heteroduplex formation (Resnick 1976), which, in turn, often leads to gene conversion (Bishop et al. 1987), i.e., the copying of one region of a chromosome over the homologous region of another chromosome, thus resulting in sequence homogenization. Moreover, such repair mechanism only works if two homologous regions are not too divergent. Hence, bdelloid rotifer that would have accumulated a large amount of divergence between homologs would probably not survive desiccation, as their damaged DNA could not be repaired. At the present time, however, the only experimental evidence for the occurrence of gene conversions in bdelloid rotifer comes from the isolation and sequencing of *hsp70*- and histone-containing fosmids in *A. vaga* and *P. roseola* that revealed several tracks of sequence identity or near-identity between ancient alleles (Hur et al. 2009). Therefore, this result will have to be confirmed and quantified at the genome scale to find out whether gene conversion really plays a role in limiting the divergence between homologs in bdelloid rotifers.

13.5 Concluding Remarks

Strict asexuality and its ancientness in animal species both remain difficult to establish. The four animal lineages we have described in the present chapter represent, to our knowledge, the most plausible ancient asexual candidates. However, for none of these lineages, ancient asexuality can be stated in an absolutely incontestable manner. The evidences used to indicate asexuality rely on the absence of current observation of sexual-specific features such as males, meiosis or fertilization, while support from the fossil record or divergence level between individuals are used to state age of asexuality. The extensive study of the genomes of presumed long-term asexuals, showing, e.g., that key genes involved in sexual reproduction or meiosis are absent, may represent the most solid evidence in the near future. Considering that the lineages presented here are most probably ancient asexuals, several peculiarities and features can be sorted out. Recent genomic data for bdelloid rotifers and root-knot nematodes suggest that a peculiar genomic structure in which at least part of the genes are present in divergent copies may support functional divergence and provide a genetic pool for adaptation. These singular genomic structures may represent partial alternatives to sexual reproduction as a source of genomic plasticity necessary for adaptation in changing environment. On the other hand, in darwinulid ostracods, it has been proposed that a more or less fixed general-purpose genotype has allowed these species to survive in a variety of environments.

Another constraint linked to the absence of sexual recombination is that deleterious mutations are not eliminated as rapidly and tend to accumulate if no alternative elimination mechanism exists. In oribatid mites and in darwinulid ostracods, it has been observed that asexual lineages present lower rates of accumulation of mutations than their sexual relatives, possibly due to particularly efficient DNA repair mechanisms. In bdelloid rotifers, a homogenization mechanism, possibly via gene conversion during DNA repair after desiccation, has been proposed to maintain a low level of divergence between gene copies within a colinear pair while allowing high divergence between copies in different pairs. Overall, it appears that ancient asexuals may have evolved substitutes to sexual reproduction that would allow their genomes to adapt to environmental changes while maintaining a low level of potentially deleterious mutations. With the first genome for an animal reproducing strictly without sex recently available and the forthcoming release of another such genome, we are currently at the dawn of the genomic era for asexual animals. Moreover, comparative analysis with genomes of close sexual relatives will probably shed light on new features in the genomes of asexual species that might represent signatures of the long-term absence of sexual reproduction.

References

Abad P, Gouzy J et al (2008) Genome sequence of the metazoan plant-parasitic nematode *Meloidogyne incognita*. Nat Biotechnol 8:909–915

Arkhipova I, Meselson M (2000) Transposable elements in sexual and ancient asexual taxa. Proc Natl Acad Sci USA 26:14473–14477

Arkhipova I, Meselson M (2005a) Deleterious transposable elements and the extinction of asexuals. Bioessays 1:76–85

Arkhipova IR, Meselson M (2005b) Diverse DNA transposons in rotifers of the class Bdelloidea. Proc Natl Acad Sci USA 33:11781–11786

Barraclough TG, Fontaneto D et al (2007) Evidence for inefficient selection against deleterious mutations in cytochrome oxidase I of asexual bdelloid rotifers. Mol Biol Evol 9:1952–1962

Bird DM, Williamson VM et al (2009) The genomes of root-knot nematodes. Annu Rev Phytopathol 47:333–351

Birky CW Jr (1996) Heterozygosity, heteromorphy, and phylogenetic trees in asexual eukaryotes. Genetics 1:427–437

Birky CW, Wolf C et al (2005) Speciation and selection without sex. Hydrobiologia 1:29–45

Bishop DK, Williamson MS et al (1987) The role of heteroduplex correction in gene conversion in *Saccharomyces cerevisiae*. Nature 6128:362–364

Burt A (2000) Sex, recombination, and the efficacy of selection – was Weismann right? Evolution 2:337–351

Butlin R (2002) The costs and benefits of sex: new insights from old asexual lineages. Nat Rev Genet 4:311–317

Castagnone-Sereno P (2006) Genetic variability and adaptive evolution in parthenogenetic root-knot nematodes. Heredity 4:282–289

Cianciolo JM, Norton RA (2006) The ecological distribution of reproductive mode in oribatid mites, as related to biological complexity. Exp Appl Acarol 1:1–25

Danchin EG, Rosso MN et al (2010) Multiple lateral gene transfers and duplications have promoted plant parasitism ability in nematodes. Proc Natl Acad Sci USA 41:17651–17656

De Ley IT, De Ley P et al (2002) Phylogenetic analyses of *Meloidogyne* small subunit rDNA. J Nematol 4:319–327

Dieterich C, Clifton SW et al (2008) The *Pristionchus pacificus* genome provides a unique perspective on nematode lifestyle and parasitism. Nat Genet 10:1193–1198

Domes K, Norton RA et al (2007) Reevolution of sexuality breaks Dollo's law. Proc Natl Acad Sci USA 17:7139–7144

Doncaster CP, Pound GE et al (2000) The ecological cost of sex. Nature 6775:281–285

Dunn CW, Hejnol A et al (2008) Broad phylogenomic sampling improves resolution of the animal tree of life. Nature 7188:745–749

Esbenshade PR, Triantaphyllou AC (1987) Enzymatic relationships and evolution in the genus *Meloidogyne* (Nematoda: Tylenchida). J Nematol 1:8–18

Fisher RA (1930) The genetical theory of natural selection. Clarendon, Oxford

Fontaneto D, Ficetola GF et al (2006) Patterns of diversity in microscopic animals: are they comparable to those in protists or in larger animals? Glob Ecol Biogeogr 2:153–162

Fontaneto D, Herniou EA et al (2007) Independently evolving species in asexual bdelloid rotifers. PLoS Biol 4:e87

Fontaneto D, Barraclough TG et al (2008) Molecular evidence for broad-scale distributions in bdelloid rotifers: everything is not everywhere but most things are very widespread. Mol Ecol 13:3136–3146

Ghedin E, Wang S et al (2007) Draft genome of the filarial nematode parasite *Brugia malayi*. Science 5845:1756–1760

Gilbert JJ (1974) Dormancy in rotifers. Trans Am Microsc Soc 4:490–513

Gladyshev EA, Meselson M et al (2008) Massive horizontal gene transfer in bdelloid rotifers. Science 5880:1210–1213

Hammer M, Wallwork JA (1979) A review of the world distribution of oribatid mites (Acari: Cryptostigmata) in relation to continental drift. Biol Skr Dan Vid Selsk 22:1–31

Heethoff M, Domes K et al (2007) High genetic divergences indicate ancient separation of parthenogenetic lineages of the oribatid mite *Platynothrus peltifer* (Acari, Oribatida). J Evol Biol 1:392–402

Heethoff M, Norton RA et al (2009) Parthenogenesis in oribatid mites (Acari, Oribatida): evolution without sex. In: Schön I, Martens K, Dijk P (eds) Lost sex. Springer, Dordrecht, pp 241–257

Hoffmann AA, Reynolds KT et al (2008) A high incidence of parthenogenesis in agricultural pests. Proc Biol Sci 1650:2473–2481

Holterman M, Karssen G et al (2009) Small subunit rDNA-based phylogeny of the Tylenchida sheds light on relationships among some high-impact plant-parasitic nematodes and the evolution of plant feeding. Phytopathology 3:227–235

Hsu WS (1956a) Oogenesis in the Bdelloidea rotifer, *Philodina roseola*. Cellule 57:283–296

Hsu WS (1956b) Oogenesis in *Habrotrocha tridens* (Milne). Biol Bull 3:364–374

Hugall A, Stanton J et al (1997) Evolution of the AT-rich mitochondrial DNA of the root knot nematode, *Meloidogyne hapla*. Mol Biol Evol 1:40–48

Hugall A, Stanton J et al (1999) Reticulate evolution and the origins of ribosomal internal transcribed spacer diversity in apomictic *Meloidogyne*. Mol Biol Evol 2:157–164

Hur JH, Van Doninck K et al (2009) Degenerate tetraploidy was established before bdelloid rotifer families diverged. Mol Biol Evol 2:375–383

Jaffe DB, Butler J et al (2003) Whole-genome sequence assembly for mammalian genomes: Arachne 2. Genome Res 1:91–96

Janko K, Drozd P et al (2008) Clonal turnover versus clonal decay: a null model for observed patterns of asexual longevity, diversity and distribution. Evolution 5:1264–1270

Kim JH, Waterman MS et al (2007) Diploid genome reconstruction of *Ciona intestinalis* and comparative analysis with *Ciona savignyi*. Genome Res 7:1101–1110

Kirkpatrick M, Jenkins CD (1989) Genetic segregation and the maintenance of sexual reproduction. Nature 6222:300–301

Leasi F, Fontaneto D et al (2010) Phylogenetic constraints in the muscular system of rotifer males: investigation on the musculature of males versus females of *Brachionus manjavacas* and *Epiphanes senta* (Rotifera, Monogononta). J Zool 2:109–119

Leroy S, Duperray C et al (2003) Flow cytometry for parasite nematode genome size measurement. Mol Biochem Parasitol 1:91–93

Li W, Godzik A (2006) Cd-hit: a fast program for clustering and comparing large sets of protein or nucleotide sequences. Bioinformatics 13:1658–1659

Lunt DH (2008) Genetic tests of ancient asexuality in root knot nematodes reveal recent hybrid origins. BMC Evol Biol 8:194

Mandegar MA, Otto SP (2007) Mitotic recombination counteracts the benefits of genetic segregation. Proc Biol Sci 1615:1301–1307

Margulies M, Egholm M et al (2005) Genome sequencing in microfabricated high-density picolitre reactors. Nature 7057:376–380

Mark Welch DB, Meselson M (2000) Evidence for the evolution of bdelloid rotifers without sexual reproduction or genetic exchange. Science 5469:1211–1215

Mark Welch DB, Meselson MS (2001) Rates of nucleotide substitution in sexual and anciently asexual rotifers. Proc Natl Acad Sci USA 12:6720–6724

Mark Welch DB, Cummings MP et al (2004) Divergent gene copies in the asexual class Bdelloidea (Rotifera) separated before the bdelloid radiation or within bdelloid families. Proc Natl Acad Sci USA 6:1622–1625

Mark Welch DB, Mark Welch JL et al (2008) Evidence for degenerate tetraploidy in bdelloid rotifers. Proc Natl Acad Sci USA 13:5145–5149

Mark Welch DB, Ricci C et al (2009) Bdelloid rotifers: progress in understanding the success of an evolutionary scandal. In: Schön I, Martens K, Dijk P (eds) Lost sex. Springer, Dordrecht, pp 259–279

Martens K, Horne DJ et al (1998) Age and diversity of non-marine ostracods. In: Martens K (ed) Sex and parthenogenesis: evolutionary ecology of reproductive modes in non-marine ostracods. Backhuys Publishers, Leiden, pp 37–55
Martens K, Rossetti G et al (2003) How ancient are ancient asexuals? Proc Biol Sci 1516:723–729
Mattimore V, Battista JR (1996) Radioresistance of *Deinococcus radiodurans*: functions necessary to survive ionizing radiation are also necessary to survive prolonged desiccation. J Bacteriol 3:633–637
Maynard Smith J (1986) Contemplating life without sex. Nature 6095:300–301
Muller HJ (1932) Some genetic aspects of sex. Am Nat 703:118–138
Muller HJ (1964) The relation of recombination to mutational advance. Mutat Res 106:2–9
Neiman M, Meirmans S et al (2009) What can asexual lineage age tell us about the maintenance of sex? Ann NY Acad Sci 1168:185–200
Omilian AR, Cristescu ME et al (2006) Ameiotic recombination in asexual lineages of Daphnia. Proc Natl Acad Sci USA 49:18638–18643
Opperman CH, Bird DM et al (2008) Sequence and genetic map of *Meloidogyne hapla*: A compact nematode genome for plant parasitism. Proc Natl Acad Sci USA 39:14802–14807
Palmer SC, Norton RA (1991) Taxonomic, geographic and seasonal distribution of thelytokous parthenogenesis in the Desmonomata (Acari: Oribatida). Exp Appl Acarol 1:67–81
Regier JC, Shultz JW et al (2010) Arthropod relationships revealed by phylogenomic analysis of nuclear protein-coding sequences. Nature 7284:1079–1083
Resnick MA (1976) The repair of double-strand breaks in DNA: a model involving recombination. J Theor Biol 1:97–106
Ricci C (1998) Anhydrobiotic capabilities of bdelloid rotifers. Hydrobiologia 387/388:321–326
Sasser JN, Carter CC (1985) Overview of the international *Meloidogyne* project 1975–1984. In: Sasser JN, Carter CC (eds) An advance treatise on *Meloidogyne*, I: biology and control. North Carolina State University Graphics, Raleigh, pp 19–24
Schaefer I, Domes K et al (2006) No evidence for the 'Meselson effect' in parthenogenetic oribatid mites (Oribatida, Acari). J Evol Biol 1:184–193
Schön I, Arkhipova IR (2006) Two families of non-LTR retrotransposons, Syrinx and Daphne, from the Darwinulid ostracod, Darwinula stevensoni. Gene 2:296–307
Schön I, Martens K (2003) No slave to sex. Proc R Soc Lond B Biol Sci 1517:827–833
Schön I, Butlin RK et al (1998) Slow molecular evolution in an ancient asexual ostracod. Proc R Soc Lond B Biol Sci 1392:235–242
Schön I, Rossetti G et al (2009) Darwinulid ostracods: ancient asexual scandals or scandalous gossip? In: Schön I, Martens K, van Dijk P (eds) Lost sex. The Evolutionary Biology of Parthenogenesis Springer, Dordrecht, Heidelberg, London, New York, pp 217–240
Schurko AM, Logsdon JM Jr (2008) Using a meiosis detection toolkit to investigate ancient asexual "scandals" and the evolution of sex. Bioessays 6:579–589
Segers H (2007) Annotated checklist of the rotifers (Phylum Rotifera), with notes on nomenclature, taxonomy and distribution. Zootaxa 1564:1–104
Segers H (2008) Global diversity of rotifers (Rotifera) in freshwater. Hydrobiologia 1:49–59
Small K, Brudno M et al (2007) A haplome alignment and reference sequence of the highly polymorphic *Ciona savignyi* genome. Genome Biol 3:R41
Smith RJ, Kamiya T et al (2006) Living males of the 'ancient asexual' Darwinulidae (Ostracoda: Crustacea). Proc Biol Sci 1593:1569–1578
Stein LD, Bao Z et al (2003) The genome sequence of *Caenorhabditis briggsae*: a platform for comparative genomics. PLoS Biol 2:E45
Triantaphyllou AC (1985) Cytogenetics, cytotaxonomy and phylogeny of root-knot nematodes. In: Sasser JN, Carter CC (eds) An advance treatise on *Meloidogyne*, 1. North Carolina State University Graphics, Raleigh, pp 113–126
Van der Beek JG, Los JA et al (1998) Cytology of parthenogenesis of five *Meloidogyne* species. Fundam Appl Nematol 4:393–399

Van Doninck K, Schön I et al (2002) A general purpose genotype in an ancient asexual. Oecologia 132(2):205–212

Van Doninck K, Schön I et al (2003) Ecological strategies in the ancient asexual animal group Darwinulidae (Crustacea, Ostracoda). Freshw Biol 8:1285–1294

Van Doninck K, Mandigo ML et al (2009) Phylogenomics of unusual histone H2A variants in bdelloid rotifers. PLoS Genet 3:e1000401

Velázquez-Rojas CA, Santos-Medrano GE et al (2002) Sexual reproductive biology of *Platyias quadricornis* (Rotifera: Monogononta). Int Rev Hydrobiol 1:97–105

Vinson JP, Jaffe DB et al (2005) Assembly of polymorphic genomes: algorithms and application to *Ciona savignyi*. Genome Res 8:1127–1135

Vrijenhoek RC, Parker ED Jr (2009) Geographical parthenogenesis: general purpose genotypes and frozen niche variation. In: Schön I, Martens K, van Dijk P (eds) Lost sex. The Evolutionary Biology of Parthenogenesis Springer, Dordrecht, Heidelberg, London, New York, pp 99–131

Waggoner BM, Poinar GO (1993) Fossil habrotrochid rotifers in Dominican amber. Experientia 4:354–357

Weismann A (1886) Die Bedeutung der sexuellen Fortpflanzung für die Selektions-Theorie. Verlag von Gustav Fischer, Jena

Wilson CG, Sherman PW (2010) Anciently asexual bdelloid rotifers escape lethal fungal parasites by drying up and blowing away. Science 5965:574–576

Wolfe K (2000) Robustness – it's not where you think it is. Nat Genet 1:3–4

Chapter 14
Evolutionary Constraint on DNA Shape in the Human Genome

Thomas D. Tullius, Stephen C.J. Parker, and Elliott H. Margulies

Abstract In the age of genomics, DNA is depicted as a string of letters. While this is a useful device for representing the information in a genome, the molecular nature of DNA is obscured. Proteins cannot actually "read" DNA letters – they discriminate between DNA binding sites via molecular recognition, which is sensitive to DNA structure. Since shape is essential to DNA's biological function, we hypothesized that natural selection can act to preserve DNA shape without maintaining the exact sequence of nucleotides. To test this hypothesis, we developed a DNA structure database, ORChID, and used it to map structural variation throughout the human genome. We then devised a computational algorithm, Chai, to detect evolutionary constraint on DNA shape. We found that Chai regions correlate better with experimental functional elements than do genomic regions that are sequence-constrained. Our results support the hypothesis that DNA shape can be a substrate for natural selection.

T.D. Tullius
Department of Chemistry, Boston University, Boston, MA 02215, USA

Program in Bioinformatics, Boston University, Boston, MA 02215, USA
e-mail: tullius@bu.edu

S.C.J. Parker
Program in Bioinformatics, Boston University, Boston, MA 02215, USA

Genome Informatics Section, Genome Technology Branch, National Human Genome Research Institute, National Institutes of Health, Bethesda, MD 20892, USA
e-mail: stephen.parker@nih.gov

E.H. Margulies
Genome Informatics Section, Genome Technology Branch, National Human Genome Research Institute, National Institutes of Health, Bethesda, MD 20892, USA
e-mail: elliott@nhgri.nih.gov

P. Pontarotti (ed.), *Evolutionary Biology – Concepts, Biodiversity, Macroevolution and Genome Evolution*, DOI 10.1007/978-3-642-20763-1_14,

14.1 Introduction

Since the discovery of the genetic code, in which triplets of nucleotides represent the amino acids that make up proteins, the medium through which evolutionary selection operates has been taken to be the sequence of nucleotides in a genome. It seems clear how a random mutation in a gene might affect fitness: a mutation could be neutral, deleterious, or beneficial to the fitness of the organism. Deleterious or beneficial mutants would then be subject to either purifying or positive selection, respectively.

However, if we look a bit closer, we see that such a mechanism for evolutionary selection presumes that the only important attribute of a genome is the sequence of nucleotides (A, C, G, and T), which eventually is expressed as the sequence of a protein. The simplicity and universal nature of the three-letter genetic code has led to the abstraction of representing a genome by a string of letters. There is no doubt that the simplification of a very long biopolymer into a string has made computational analysis of genomes feasible.

But when the human genome was sequenced, it was found that less than 2% of the genome codes for proteins. The vast majority of the genome, therefore, is not made up of genetic-code triplets. Within the 98% of non-coding sequence is housed the elements that control how the genome functions. How does evolutionary selection operate on the non-coding parts of the genome? In this chapter, we present a new framework for understanding evolutionary constraint in the non-coding functional regions of the human genome. We have found that the shape and structure of DNA, and not just the sequence of nucleotide letters, can be under evolutionary constraint.

14.2 The ENCODE Project

Our work began as part of the Encyclopedia of DNA Elements (ENCODE) Project (ENCODE Consortium 2004). This large international collaboration, organized and supported by the National Human Genome Research Institute of the National Institutes of Health, aims to uncover all of the functional elements in the human genome. These functional elements include transcription factor binding sites, deoxyribonuclease I (DNase) hypersensitive sites that are associated with active chromatin, promoters, enhancers, histone epigenetic modifications, origins of replication, and so on. As part of ENCODE, a computational pipeline was developed to integrate multiple datasets to determine "signatures" of various functional elements in the genome.

14.2.1 Comparative Genomics and ENCODE

The ENCODE Project also makes use of comparative genomics to pinpoint non-coding functional elements in the human genome (Margulies et al. 2007). This approach is based on the plausible expectation that genomic regions that are conserved among distantly related species are highly likely to be functional. For the Pilot phase of the ENCODE Project, which focused on 1% of the human genome (30 Mb), the ENCODE consortium constructed multispecies alignments of the orthologous genome sequences of 28 vertebrates. Four different computational methods were then used to detect sequences in this 1% of the human genome that are under evolutionary constraint. These methods produced a set of constrained sequences that represented approximately 5% of the ENCODE Pilot Project regions (ENCODE Consortium 2007). Since most protein-coding sequences are identified as constrained by these methods, this analysis suggests that around 3% of non-coding sequences are constrained and therefore likely to be functional.

The ENCODE Pilot Project's next step was to compare experimentally annotated functional regions with regions identified as constrained by evolution. In essence, two questions were asked: (1) Are constrained sequences functional? (2) Are functional sequences constrained? Of course, most expected that these two questions would give the same answer: (almost) all constrained sequences are functional, and (almost) all functional sequences are constrained.

In Fig. 14.1, we depict four possible scenarios that might result from comparing experimental functional annotations with constrained regions, as was done in the ENCODE Pilot Project: (a) there is a substantial overlap between constrained and functional sequences; (b) a region is identified as constrained, but no function is detected there by experiment; (c) a small constrained region is found within a larger functional region; (d) an experiment identifies a function in this region, but no constraint is detected.

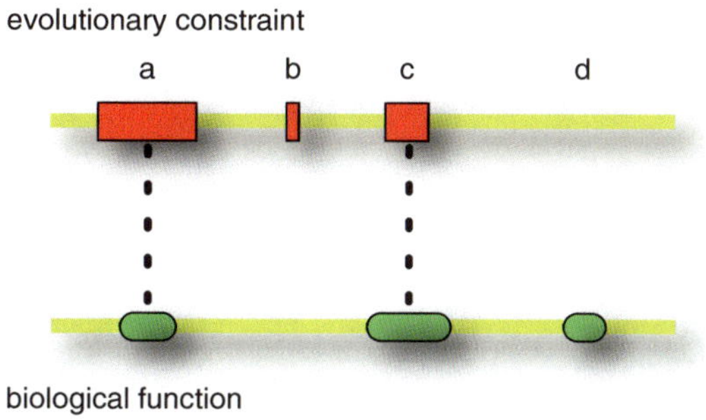

Fig. 14.1 Possible outcomes in comparing evolutionary constraint and biological function in the genome. *Horizontal lines* represent a particular segment of the human genome. On this segment are mapped regions identified as evolutionarily constrained in sequence (*boxes*, *top*), and regions found by experiment to be biologically functional (*ovals*, *bottom*). When these two maps are compared, four of the possible outcomes are depicted: (**a**), evolutionary constraint and biological function overlap; (**b**) a region identified as constrained is not found by experiment to be functional; (**c**) a constrained region overlaps only part of a larger region that is demonstrated to be functional; and (**d**) biological function is found in a region of the genome that is not identified as being evolutionarily constrained

The first three scenarios can readily be understood: (a) is the most likely expectation, (b) would result from the limited number of experiments (transcription factors or cell types studied, for example) that are included in the ENCODE Pilot Project datasets, and (c) is the consequence of the low resolution of some experimental annotations. Scenario (d), however, is not what would be expected, and if found to be widespread would question the assumption that evolutionary constraint is always associated with function.

The results of the comparative genomics analysis were perhaps the most surprising of the ENCODE Pilot Project (ENCODE Consortium 2007). These are the answers that were found for the two questions posed above: (1) 40% of constrained sequences have no identified function in the ENCODE Pilot Project experimental datasets; (2) only 50% of most experimentally defined non-coding functional elements are identified as constrained across all mammals. These functional elements include 5′ and 3′ untranslated regions (UTRs), DNase hypersensitive sites, FAIRE sites (nucleosome-free regions), and regulatory factor binding regions (RFBRs).

As pointed out above, the lack of function exhibited by a large fraction of constrained sequences (question (1)) is most likely the result of the limited set of experiments included in the ENCODE Pilot Project. One might confidently expect this fraction to increase as new assays, cell types, experimental conditions, and so on are integrated into the analysis. And in fact, one result of the Pilot Project was the demonstration that extension of ENCODE to the whole human genome is feasible, and so we are now in the midst of the full ENCODE Project. Parallel projects focused on model organisms (*Drosophila*, *C. elegans*, and mouse) also have been initiated in recent years (the modENCODE Projects) (Celniker et al. 2009). So a cornucopia of whole-genome functional data is imminent, on a much wider variety of cell types and organisms (modENCODE Consortium 2010; Gerstein et al. 2010). These data will be freely available to all interested scientists.

The real surprise of the ENCODE Pilot Project was the remarkably low fraction of functional sequences that were identified as being constrained (question (2)). In the paper reporting on the Pilot Project, one of the possibilities mentioned by the ENCODE consortium to explain this result was that there exists a neutral pool of biologically functional elements that do not confer a selective advantage to the organism, and that this neutral pool may be larger than previously supposed (ENCODE Consortium 2007).

14.3 Is There Evolutionary Constraint on DNA Shape?

When we thought about these results, we wondered whether there might be another reason for the lack of constraint exhibited by many functional elements. Our idea was that *nucleotide-sequence-based constraint* might not be the only way that evolutionary selection could work on non-coding functional sequences in a genome. If another property of DNA is essential for some biological function, but this property is not strictly tied to nucleotide sequence identity, methods that assess constraint only on the basis of nucleotide sequence may not assign these elements as being under evolutionary selection.

More specifically, our hypothesis was that standard nucleotide sequence-based analyses of the genome might miss signals that are encoded by DNA shape. After all, a protein that controls genome function cannot "read" DNA letters to locate its specific binding site. Proteins interact with the genome by exploiting intermolecular forces, which ultimately are a function of molecular shape. And molecular shape, while sequence-dependent, is not sequence specific. That is, different DNA sequences can in principle give rise to similar DNA shapes.

So, if DNA shape is the feature that imparts function to a region in the genome, then it is possible that the nucleotide sequence of the region could vary as mutations accumulate, but function could be maintained if these sequence changes did not affect shape. A method that relies on sequence identity to score evolutionarily conserved genomic regions would not assign such a region as being constrained, even though the functionally relevant feature (shape) actually is maintained (conserved) through a set of species.

14.3.1 Mapping DNA Shape at Single-Nucleotide Resolution

We were well equipped to investigate this idea, because we had already produced a genomic dataset in which an important property of DNA did not map perfectly to nucleotide sequence. As our contribution to the ENCODE Pilot Project, we had generated a high-resolution map of DNA structural variation throughout the human genome (Greenbaum et al. 2007).

14.3.1.1 Hydroxyl Radical as a Chemical Probe of DNA Structure

Our work took advantage of the chemistry of the hydroxyl radical (•OH) (Tullius 1987). This reactive free radical makes a single-nucleoside gap in the DNA backbone by abstracting a hydrogen atom from a deoxyribose residue (Pogozelski and Tullius 1998). Because of its high reactivity, the hydroxyl radical is very nonselective in its reactions with the DNA backbone, and so every nucleotide is susceptible to attack.

We measure the extent of cleavage at each nucleotide in a DNA molecule by electrophoretic separation of the reaction products. This method is capable of resolving DNA strands that differ in length by only one nucleotide, so we are able to obtain structural data on DNA at single-nucleotide resolution. While the extent of cleavage is nearly equal at each nucleotide in a duplex DNA molecule, we do observe relatively small but highly reproducible differences in cleavage that reflect the sequence-dependent structural variation of DNA (Tullius and Dombroski 1985; Price and Tullius 1992).

We showed directly that hydroxyl radical cleavage is related to DNA shape through deuterium kinetic isotope effect experiments (Balasubramanian et al. 1998), in which we chemically substituted deuterium for hydrogen in the deoxyribose residues of DNA. We found that the solvent-accessible surface area of a given hydrogen atom governs the extent of its reaction with the hydroxyl radical. This is what would be expected for a highly reactive free radical for which the rate of its chemical reaction with another species is controlled by diffusion (i.e., nearly every collision of the hydroxyl radical with another molecule results in a reaction). If the surface area presented by a particular deoxyribose is low, because of local shape variation of the DNA molecule (e.g., a narrow minor groove), the extent of reaction with the hydroxyl radical will be low, and we will observe less cleavage.

14.3.2 DNA Shape and Biological Function

The idea that proteins recognize DNA based on shape and not nucleotide sequence has gained substantial momentum in recent years (Tullius 2009; Rohs et al. 2010). A pioneering study showed that the heterodimer of the *Drosophila* Hox protein Sex combs reduced (Scr) and its cofactor Extradenticle (Exd) distinguishes between two closely related DNA binding sites by differences in the shape of the DNA minor groove (Joshi et al. 2007). A wide range of experiments, including x-ray crystallography, Monte Carlo simulation, electrostatics calculations, and in vitro and in vivo molecular biological assays, was used to show that Scr inserts an arginine residue into an especially narrow region of the DNA minor groove in the paralog-specific binding site. A physical basis for the specificity of this interaction came from Poisson–Boltzmann calculations of the electrostatic potential in the minor groove of the DNA molecule. These calculations demonstrated that a narrow minor groove has a more negative electrostatic potential, and so would interact more favorably with a positively charged arginine residue from the Scr protein.

In a recent study, these same authors showed that shape-specific recognition of DNA by protein is a widely occurring phenomenon (Rohs et al. 2009). They surveyed the high-resolution, three-dimensional structures of a large number of DNA–protein complexes obtained from the Protein Data Bank, and found that there were many previously unrecognized examples of DNA-binding proteins that insert arginine residues into narrow regions of the DNA minor grove. They also showed that sequences with a narrow minor groove most often consist of short runs of adenine nucleotides. This form of recognition was even seen in structures of the nucleosome core particle. Other investigators showed that the *E. coli* Fis protein specifically binds to narrow minor groove regions in DNA (Stella et al. 2010), providing further evidence that DNA shape recognition is a general feature of protein–DNA complexes.

14.4 ORChID: A Database of DNA Structure

Our previous work showed that experimentally determining the hydroxyl radical cleavage pattern is an effective way to map structural variation for DNA molecules in solution (Price and Tullius 1992). DNA molecules up to several hundred base pairs in length may be studied conveniently. But in our work for the ENCODE Project, in which we planned to map DNA structural variation throughout the human genome, experimentally measuring hydroxyl radical cleavage patterns for 3 billion base pairs would be a daunting task. Instead, we took another approach, summarized in Fig. 14.2 (Greenbaum et al. 2007).

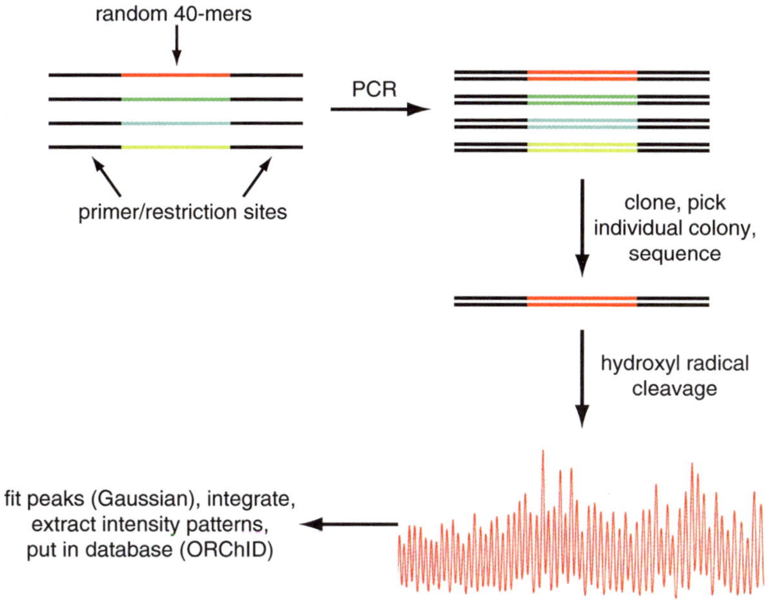

Fig. 14.2 Construction of ORChID, a database of DNA structural patterns. A library of DNA molecules, each consisting of a central 40-nucleotide random sequence flanked by common sequences, was synthesized. PCR was used to generate the complementary strand of each library member. The library of DNA duplexes was cloned in a plasmid and used to transform *E. coli*. Individual bacterial colonies, each of which harbored a different member of the library, were picked. A library member was sequenced, and the DNA molecule was subjected to hydroxyl radical cleavage. The cleavage pattern consists of a series of peaks on an electrophoretogram. The peaks making up the cleavage pattern were integrated to determine the extent of hydroxyl radical-induced cleavage at each nucleotide, a measure of the local variation in shape of the DNA molecule. Finally, the cleavage patterns were housed in a DNA structural database, ORChID

14.4.1 A Library of Experimental Hydroxyl Radical Cleavage Patterns

We began by synthesizing a library of DNA molecules 158 nucleotides long, each one containing a 40-nucleotide random segment in the center. The common flanking nucleotides in each member of the library were used to prime the polymerase chain reaction (PCR) and to serve as internal standards for data normalization. We next used PCR to synthesize the complementary strands of the library members, cloned the double-stranded DNA molecules into *E. coli*, picked individual colonies each harboring one library member, and sequenced the inserts. We then used PCR, with one primer labeled at its 5′ terminus with a fluorescent tag, to produce a singly labeled duplex DNA molecule with one library sequence at the center. We performed the hydroxyl radical cleavage reaction on individual library members, and separated the cleavage products on an automated slab gel electrophoresis device designed for DNA sequencing. As DNA fragments emerged from the bottom of the electrophoresis device, the fluorescence was monitored. The fluorescence intensity of each emerging peak provided a quantitative measure of the extent of cleavage that occurred at each nucleotide in the DNA molecule. Finally, the peaks in the fluorescence data trace were integrated and the intensity patterns were deposited in a custom-constructed relational database, **O**H Radical **C**leavage **I**ntensity **D**atabase (ORChID). This database is accessible to interested scientists at http://dna.bu.edu/orchid/.

One of our first analyses of the experimental cleavage patterns in ORChID focused on comparing the relative sizes of the “spaces” of DNA nucleotide sequence and structure. In other words, how unique is a particular DNA backbone shape? Can different DNA sequences adopt similar structures? We found that this is indeed the case.

When we clustered experimental cleavage patterns of trinucleotides, we found that sets of different trinucleotide sequences can yield very similar cleavage patterns. We extended this analysis by searching ORChID for longer DNA sequences that are similar in cleavage pattern but divergent in nucleotide sequence (Greenbaum et al. 2007). For example, we found a pair of 10-mers with no sequence identity (0 base pairs in common) but with highly similar cleavage patterns (Pearson’s $R = 0.94$). Similarly, two 20-mers with only 10% sequence identity had cleavage patterns that exhibited a correlation coefficient of 0.81. These observations supported our idea that “structure space” is smaller than “sequence space,” and that structure could be maintained without strict conservation of sequence.

14.4.2 Computational Prediction of Hydroxyl Radical Cleavage Patterns

Once we had accumulated a sufficient number of experimental cleavage patterns in ORChID, we developed a computational algorithm based on a sliding tetranucleotide window that was capable of predicting the hydroxyl radical cleavage pattern for any input DNA sequence of any length with high accuracy (Greenbaum et al. 2007). Leave-one-out cross-validation studies showed that the mean correlation coefficient (*R* value) between the predicted cleavage pattern of a 40-nucleotide DNA sequence compared to the experimental cleavage pattern was 0.88, demonstrating the excellent correspondence between experimental and predicted cleavage patterns.

We used the ORChID server to predict the hydroxyl radical cleavage pattern, first for the 30 Mb of the ENCODE Pilot Project regions of the human genome, and more recently for the entire genome. We have deposited the ORChID patterns in the UC Santa Cruz genome browser (http://genome.ucsc.edu/), where they are available for anyone to use. As an example, the ORChID pattern for a specific segment of the human genome is shown in Fig. 14.3, at three genomic scales.

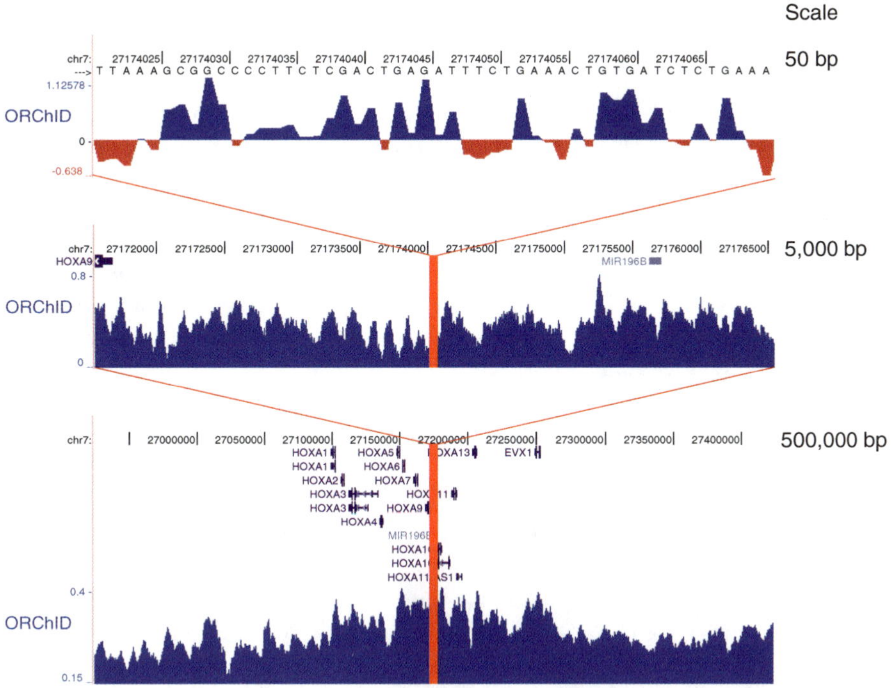

Fig. 14.3 DNA structural profile of the human genome. A 500-Kb region (*bottom*) of the human genome is shown as depicted in the UCSC genome browser. The region highlighted is the ENCODE target region ENm010 (the *HoxA* cluster). The same data are shown at two additional resolutions: 5 Kb (*middle*) and 50 bp (*top*). The "ORChID" track represents the predicted structural profile, based on the hydroxyl radical cleavage pattern, for the given region. Note the variability in the structural profile that is apparent at each scale

14.5 Chai: A Computational Method to Detect Evolutionary Constraint on DNA Shape

Having established that ORChID can be used to map DNA shape variation throughout the human genome at high resolution, we next embarked on testing our idea that DNA shape is under evolutionary selection (Parker et al. 2009). To do this we adapted methods that had been used by the ENCODE Project to assess evolutionary constraint on nucleotide sequence in the human genome (Margulies et al. 2007). In brief, these methods require multispecies alignments, and a way to take into account phylogenetic distance to properly weight observed sequence similarities among species. The basic idea is that nucleotide sequence identity among species that are distant from each other in the phylogenetic tree is more significant in indicating evolutionary constraint. For example, because human and chimpanzee have nearly identical genome sequences, it is not possible to use sequence identity to infer evolutionary constraint in a particular genomic region. But if a region is identical in sequence (or nearly so) between human and chicken, then it is much more likely that this region is under evolutionary constraint.

For our analysis we adapted the binCons algorithm (Margulies et al. 2003), one of the four methods that had been used in the ENCODE Pilot Project to map evolutionary constraint in the human genome (Margulies et al. 2007). As input we used the 36-vertebrtate genome alignment that had been constructed for the ENCODE Pilot Project. The first step in our method was to use ORChID to compute hydroxyl radical cleavage patterns for all of the aligned genomic sequences. We then calculated the Euclidean distance between cleavage patterns, as a quantitative measure of the similarity of DNA shape between aligned regions of the genomes that we studied. Finally, we developed an algorithm, based on binCons, that weighted similarity in cleavage pattern by phylogenetic distance to detect regions in the human genome that exhibit signs of evolutionary selection for structure. Our computational approach is named Chai.

14.5.1 Sequence Constraint Versus Structure Constraint in the Human Genome

For comparison, we also applied the binCons algorithm to the same set of alignments we used for our Chai analysis. In Fig. 14.4, we show a genome browser shot that depicts these alignments for a segment of human chromosome 7. At the top of the figure we plot the Chai score that is calculated for each nucleotide, with higher scores indicating a greater likelihood of evolutionary constraint on structure. Below the raw Chai scores, and just above the alignments, we compare the genomic regions that are marked by binCons and Chai as likely to be under evolutionary constraint. While some regions are found to be both sequence- and structure-constrained, there are numerous additional regions that are detected only by Chai.

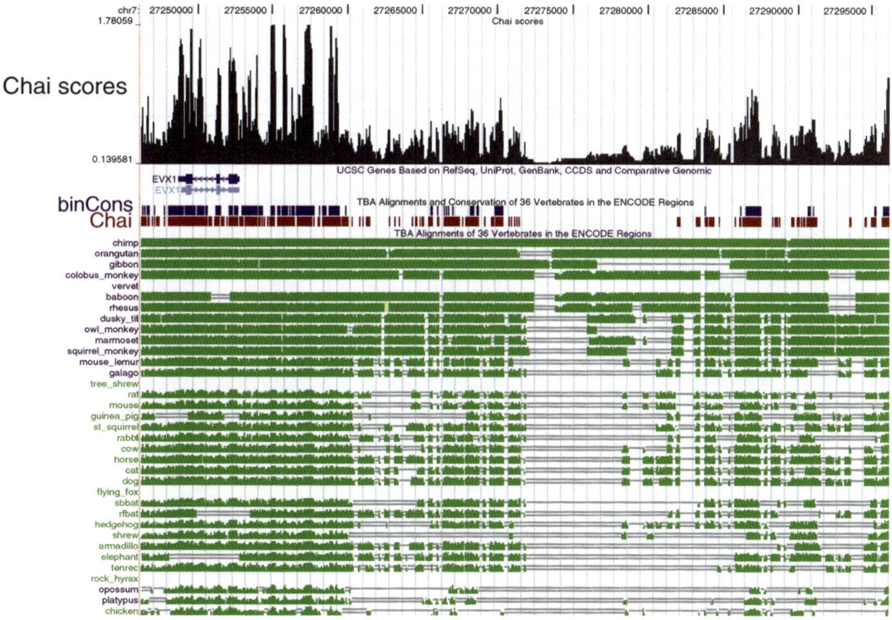

Fig. 14.4 Chai scores and regions in the human genome. This figure shows a UCSC genome browser screenshot zoomed in on the 3′ end of the human *Hox A* cluster (for which ORChID patterns are depicted in Fig. 14.3). The "Chai scores" track shows the score produced by the Chai algorithm, where higher scores represent regions with higher levels of constraint. The "binCons" and "Chai" tracks show distinct regions that are identified as evolutionarily constrained by the binCons and Chai algorithms, respectively. Note that there are several Chai regions with no overlapping binCons region. Below the binCons and Chai tracks, the similarity of orthologous sequence from different species relative to the human reference is shown. The height of the bar graph for each species is proportional to the similarity to the human sequence at that position

Analysis of Chai and binCons regions throughout the ENCODE Pilot Project regions of the human genome showed that 6.7% of bases are sequence-constrained, and 12% are structure-constrained (at a 5% false discovery rate).

Direct comparison of Chai and binCons regions in the 30 Mb of the human genome that were studied in the ENCODE Pilot Project demonstrated that nearly all regions that are identified as sequence-constrained by binCons also are scored as structure-constrained by Chai (1.8 Mb). This makes sense, since identical sequences will yield identical structures. We also find that an additional 1.8 Mb are structure-constrained but not sequence-constrained.

Returning to our initial hypothesis, we tested whether genomic regions that are under evolutionary constraint for structure are associated with experimentally annotated functional elements. Recall that a big surprise of the ENCODE Pilot Project was that only around half of experimentally determined functional elements were found to overlap with sequence-constrained regions.

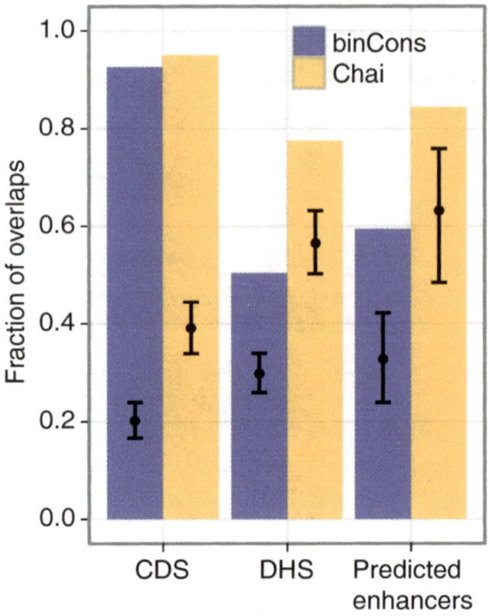

Fig. 14.5 Chai regions are enriched for non-coding functional regions. Plotted is the fraction of different types of genomic regions that overlap binCons- and Chai-detected regions. *Black points* are the mean of a null distribution constructed using the GSC method (ENCODE Consortium 2007). Error bars represent 95% confidence intervals. *CDS* coding sequences, *DHS* DNase hypersensitive sites. (This plot was adapted from Fig. 2c in Parker et al. (2009))

In Fig. 14.5, we show the overlap of three types of functional elements (coding sequences, DNase hypersensitive sites, and predicted enhancers) with binCons (sequence-constrained) regions and Chai (structure-constrained) regions. As expected, nearly all coding sequences overlap both types of constrained regions, since protein-coding genes are highly conserved in sequence throughout vertebrates, which also implies that they are conserved in structure.

Of much greater interest is our observation that functional elements (like the enhancers and DNase hypersensitive sites that are shown in Fig. 14.5) have a much greater overlap with structure-constrained regions than they do with sequence-constrained regions. More than 80% of these two functional elements occur in structure-constrained regions of the genome, compared to less than 60% that are found in sequence-constrained regions (Fig. 14.5). This result demonstrates that structure-constrained regions are not just randomly distributed around the human genome. Instead, the structure-constrained territory of the genome preferentially harbors genomic sequences that are involved in genome function.

14.6 Concluding Remarks

Recent studies clearly show that the shape of DNA is important for conferring function to non-coding regions of the genome. Combining the ORChID DNA structure prediction method with the Chai evolutionary conservation method allows for the detection of functional non-coding regions that were previously undetectable by conventional sequence comparisons. These results are important because they suggest a new way to interpret how information is encoded in genomic sequences.

We note that it is likely that not all of the structure-informed constrained regions in the human genome have been discovered. The Chai algorithm relies on existing multispecies alignments that were created based on primary DNA sequence. Because the optimal DNA structural alignment may not be the same as the optimal sequence alignment in all cases, there is an inherent bias in using existing sequence alignments. The development of new computational tools to align large regions of multiple genomes using DNA structural information will be an important future step.

Incorporating local DNA structural information into genomic analyses is a promising new direction for genomics. The largest contribution will most likely be in understanding how functional non-coding information is encoded within a genome. Perhaps new technologies that enable high-resolution profiling of protein–DNA interactions in vivo will aid in deciphering the base preferences and binding affinities for the many different regulatory factors that interact with DNA. Integration of these types of data with DNA sequence and structural profiles within and between species may ultimately lead to a set of rules that govern protein–DNA interactions. Even though such an ambitious goal may not be realized soon, it is clear that incorporating DNA structural profiles – instead of relying solely on four "letters" to represent structural variations within the double helix – will help propel the field forward.

Acknowledgments This work was funded by a grant to T.D.T. from the National Human Genome Research Institute (NHGRI) of the NIH (R01 HG003541). E.H.M. was supported by the Intramural Research Program of the NHGRI, NIH. S.C.J.P. was the recipient of a National Academies Ford Foundation Dissertation Fellowship.

References

Balasubramanian B, Pogozelski W, Tullius T (1998) DNA strand breaking by the hydroxyl radical is governed by the accessible surface areas of the hydrogen atoms of the DNA backbone. Proc Natl Acad Sci USA 95:9738–9743

Celniker SE, Dillon LAL, Gerstein MB, Gunsalus KC, Henikoff S, Karpen GH, Kellis M, Lai EC, Lieb JD, Macalpine DM, Micklem G, Piano F, Snyder M, Stein L, White KP, Waterston RH, modENCODE Consortium (2009) Unlocking the secrets of the genome. Nature 459:927–930

ENCODE Consortium (2004) The ENCODE (ENCyclopedia of DNA Elements) project. Science 306:636–640

ENCODE Consortium (2007) Identification and analysis of functional elements in 1% of the human genome by the ENCODE pilot project. Nature 447:799–816

Gerstein MB, Lu ZJ, Van Nostrand EL, Cheng C, Arshinoff BI, Liu T, Yip KY, Robilotto R, Rechtsteiner A, Ikegami K, Alves P, Chateigner A, Perry M, Morris M, Auerbach RK, Feng X, Leng J, Vielle A, Niu W, Rhrissorrakrai K, Agarwal A, Alexander RP, Barber G, Brdlik CM, Brennan J, Brouillet JJ, Carr A, Cheung M-S, Clawson H, Contrino S, Dannenberg LO, Dernburg AF, Desai A, Dick L, Dose AC, Du J, Egelhofer T, Ercan S, Euskirchen G, Ewing B, Feingold EA, Gassmann R, Good PJ, Green P, Gullier F, Gutwein M, Guyer MS, Habegger L, Han T, Henikoff JG, Henz SR, Hinrichs A, Holster H, Hyman T, Iniguez AL, Janette J, Jensen M, Kato M, Kent WJ, Kephart E, Khivansara V, Khurana E, Kim JK, Kolasinska-

Zwierz P, Lai EC, Latorre I, Leahey A, Lewis S, Lloyd P, Lochovsky L, Lowdon RF, Lubling Y, Lyne R, Maccoss M, Mackowiak SD, Mangone M, Mckay S, Mecenas D, Merrihew G, Miller DM, Muroyama A, Murray JI, Ooi S-L, Pham H, Phippen T, Preston EA, Rajewsky N, Ratsch G, Rosenbaum H, Rozowsky J, Rutherford K, Ruzanov P, Sarov M, Sasidharan R, Sboner A, Scheid P, Segal E, Shin H, Shou C, Slack FJ, Slightam C, Smith R, Spencer WC, Stinson EO, Taing S, Takasaki T, Vafeados D, Voronina K, Wang G, Washington NL, Whittle CM, Wu B, Yan K-K, Zeller G, Zha Z, Zhong M, Zhou X, Ahringer J, Strome S, Gunsalus KC, Micklem G, Liu XS, Reinke V, Kim SK, Hillier LW, Henikoff S, Piano F, Snyder M, Stein L, Lieb JD, Waterston RH (2010) Integrative analysis of the *Caenorhabditis elegans* genome by the modENCODE Project. Science 330:1775–1787

Greenbaum JA, Pang B, Tullius T (2007) Construction of a genome-scale structural map at single-nucleotide resolution. Genome Res 17:947–953

Joshi R, Passner JM, Rohs R, Jain R, Sosinsky A, Crickmore MA, Jacob V, Aggarwal AK, Honig B, Mann RS (2007) Functional specificity of a Hox protein mediated by the recognition of minor groove structure. Cell 131:530–543

Margulies E, Blanchette M, Haussler D, Green ED (2003) Identification and characterization of multi-species conserved sequences. Genome Res 13:2507–2518

Margulies EH, Cooper G, Asimenos G, Thomas DJ, Dewey CN, Siepel A, Birney E, Keefe D, Schwartz AS, Hou M, Taylor J, Nikolaev S, Montoya-Burgos JI, Löytynoja A, Whelan S, Pardi F, Massingham T, Brown JB, Bickel P, Holmes I, Mullikin JC, Ureta-Vidal A, Paten B, Stone EA, Rosenbloom KR, Kent WJ, Bouffard GG, Guan X, Hansen NF, Idol JR, Maduro VVB, Maskeri B, McDowell JC, Park M, Thomas PJ, Young AC, Blakesley RW, Muzny DM, Sodergren E, Wheeler DA, Worley KC, Jiang H, Weinstock GM, Gibbs RA, Graves T, Fulton R, Mardis ER, Wilson RK, Clamp M, Cuff J, Gnerre S, Jaffe DB, Chang J, Lindblad-Toh K, Lander ES, Hinrichs A, Trumbower H, Clawson H, Zweig A, Kuhn RM, Barber GP, Harte R, Karolchik D, Field MA, Moore RA, Matthewson CA, Schein JE, Marra MA, Antonarakis SE, Batzoglou S, Goldman N, Hardison R, Haussler D, Miller W, Pachter L, Green ED, Sidow A (2007) Analyses of deep mammalian sequence alignments and constraint predictions for 1% of the human genome. Genome Res 17:760–774

modENCODE Consortium (2010) Identification of functional elements and regulatory circuits by *Drosophila* modENCODE. Science 330:1787–1797

Parker SCJ, Hansen L, Abaan HO, Tullius TD, Margulies EH (2009) Local DNA topography correlates with functional noncoding regions of the human genome. Science 324:389–392

Pogozelski W, Tullius T (1998) Oxidative strand scission of nucleic acids: routes initiated by hydrogen abstraction from the sugar moiety. Chem Rev 98:1089–1108

Price M, Tullius T (1992) Using hydroxyl radical to probe DNA structure. Methods Enzymol 212:194–219

Rohs R, West SM, Sosinsky A, Liu P, Mann RS, Honig B (2009) The role of DNA shape in protein-DNA recognition. Nature 461:1248–1253

Rohs R, Jin X, West SM, Joshi R, Honig B, Mann RS (2010) Origins of specificity in protein-DNA recognition. Annu Rev Biochem 79:233–269

Stella S, Cascio D, Johnson RC (2010) The shape of the DNA minor groove directs binding by the DNA-bending protein Fis. Genes Dev 24:814–826

Tullius T (1987) Chemical "snapshots" of DNA: using the hydroxyl radical to study the structure of DNA and DNA-protein complexes. Trends Biochem Sci 12:297–300

Tullius T (2009) DNA binding shapes up. Nature 461:1225–1226

Tullius T, Dombroski B (1985) Iron(II) EDTA used to measure the helical twist along any DNA molecule. Science 230:679–681

Chapter 15
Evolution of Fungi and Their Respiratory Metabolism

Marina Marcet-Houben and Toni Gabaldón

Abstract The oxidative phosphorylation (OXPHOS) pathway plays a central role in the energetic metabolism of aerobic organisms. Despite such centrality, this pathway has not remained unaltered through evolution, and variations of it, including its complete loss, can be found in organisms adapted to different ecological niches. Fungi, a eukaryotic group of species with a high metabolic diversity, represent an ideal phylum in which to study the evolutionary plasticity of the OXPHOS pathway from a phylogenomics perspective. With more than 100 completely sequenced genomes, and thanks to recent progress in elucidating their evolutionary relationships, fungal species have served to reveal the evolutionary mechanisms that underlie the evolution of the core respiratory pathways. In this chapter, we review recent progress toward the characterization of OXPHOS components in fungi and in understanding their evolution. A special focus is devoted to the history of duplications that the multi-protein complexes in OXPHOS have experienced.

15.1 Introduction

The fungal kingdom, one of the eukaryotic groups with the highest number of fully sequenced genomes (http://www.genomesonline.org), comprises a large diversity of species, including mushrooms, yeasts, and molds. The exact number of species is unknown but estimates set this value around 1.5 million (Hawksworth 1991), being 700,000 a conservative, lower estimate (Schmit and Mueller 2007). Therefore, only between 5% and 10% of the total diversity of fungi has been characterized so far (Mueller and Schmit 2007). Fungi are ubiquitous and are able to colonize a broad

M. Marcet-Houben • T. Gabaldón
Comparative Genomics Group, Bioinformatics and Genomics Programme, Centre for Genomic Regulation (CRG), UPF, Doctor aiguader, 88. 08003, Barcelona, Spain
e-mail: mmarcet@crg.es

P. Pontarotti (ed.), *Evolutionary Biology – Concepts, Biodiversity, Macroevolution and Genome Evolution*, DOI 10.1007/978-3-642-20763-1_15,

diversity of habitats. This high ecological plasticity has been driven by parallel changes in their basic metabolism, resulting in a broad diversity of metabolic capacities found across fungal lineages.

15.1.1 Adaptation of the Respiratory Pathway in Fungal Species

While most fungal species are aerobic, some lineages have adopted alternative modes of respiration. A clear example is, for instance, the adaptation of some yeast species to fermentation and to anaerobic lifestyles. Adaptation to anaerobic environments can sometimes be very high. An early study performed by Visser et al. (1990) found that some fermentative strains were able to grow even under strict anaerobic conditions (i.e., *Candida tropicalis* or *Saccharomyces cerevisiae*). Although in most cases growth rates were impaired, species closely related to *S. cerevisiae,* were able to maintain near optimal growth rates. While all species that have the ability to grow under anaerobic conditions are fermentative, the opposite is not true. Species such as *Debaryomyces hansenii* or *Pachysolen tannophilus* are fermentative and yet unable to grow under strict anaerobic conditions. Filamentous fungi also have mechanisms to obtain energy when oxygen is limited. Two such mechanisms are nitrate respiration (also known as denitrification) and ammonia fermentation. These mechanisms have been thoroughly studied, for instance, in the filamentous fungi *Fusarium oxysporum* (Takaya 2009). Under limited oxygen conditions, *F. oxysporum* obtains energy from the denitrification pathway by reducing nitrate to nitrous oxide. When the amount of oxygen decreases further, *F. oxysporum* switches to ammonia fermentation, reducing nitrate to ammonia.

Two extreme adaptations to anaerobic conditions are illustrated by two groups of basal fungi: Chitrids and Microsporidians. Chitrids are anaerobic fungal species that contain hydrogenosomes instead of regular mitochondria (Voncken et al. 2002). Although initially considered to be different organelles, it is now clearly established that hydrogenosomes evolved from mitochondria (Hackstein et al. 2006). Hydrogenosomes are able to use protons as electron acceptors and generate hydrogen, acetate, and carbon dioxide. On the other hand, Microsporidia contain yet another form of highly derived mitochondria, the so-called mitosomes. Mitosomes are organelles devoid of DNA and which completely lost all components of the OXPHOS pathway (Burri et al. 2006). It is likely that these species only have the ability to produce energy via glycolysis and that they prefer to import ATP from the host (Williams et al. 2010). Remarkably, genome analysis of the microsporidian *Enterocytozoon bieneusi* shows a complete lack of the glycolitic pathway in this intracellular parasite, illustrating extreme dependence on energy provided by the host (Keeling et al. 2010).

There is plenty of evidence that all fungi have a mitochondria or related organelle and even if the ability for aerobic respiration has been lost in some of the early diverging fungal groups, it is safe to assume that the last common ancestor of fungi was an aerobic organism with a fully functional OXPHOS pathway that has since undergone changes in a lineage specific manner, thereby enabling the adaptation of fungi to a diverse variety of ecological niches.

15.1.2 Clinical Relevance of Fungal Respiratory Metabolism

Numerous fungal species are pathogenic. They can be commensals, becoming pathogenic when the host is immunocompromised (e.g., *Candida glabrata* or *Candida albicans*), or they can reside in the environment and only become pathogenic when they enter in contact with their host (as it is the case with *Histoplasma capsulatum* or *Cryptococcus neoformans*). The inner tissues of the hosts are usually oxygen-poor, therefore adaptation to anaerobic lifestyles as discussed above can be particularly helpful for survival of the fungal pathogen during infection. This adaptation is extreme when the mode of parasitism is intracellular. Microsporidia are obligate intracellular parasites that are unable to survive outside their host unless it is in the form of spores. Such strict host-dependent lifestyle does not require aerobic respiration, and as a consequence these species have lost the entire OXPHOS pathway.

Adaptation to poor oxygen levels in plant and animal pathogens should be parallel to adaptation to high oxidative stress. Although this may seem contradictory, it is explained by the fact that oxidative burst, the rapid increase of reactive oxygen compounds, is a common defense mechanism against infection. The objective of this defense mechanism is to degrade pathogenic organisms that enter certain plant tissues or that have been internalized by phagocytes in animals. As a defense, many fungi have an alternative respiratory mechanism. The alternative oxidase bypasses complexes III and IV and greatly reduces the reactive oxygen compounds generated by the fungal cell during respiration. This, coupled with enzymes that are able to reduce the reactive oxygen species generated by the host, increases the chance of survival of fungi during infection. For instance, *Aspergillus fumigatus* (Magnani et al. 2008) is able to survive the host's defense mechanisms and continue with the infection once it has escaped the macrophages.

15.1.3 Industrial Relevance of Fungal Respiratory Metabolism

Fungi have been traditionally used in the elaboration of food and beverages such as bread, cheese, wine, sake, or beer. Similarly, fungi were the source for the first antibiotics (penicillin). Nowadays they are still used in industries related to these and other products such as enzymes and acids, and it is clear that the interest in the biotechnological potential of fungi continues to grow. Some of the species that are more widely used in industry are the filamentous fungi *Aspergillus* or *Penicillium* and the yeasts *S. cerevisiae* and *Pichia pastoris* (Li et al. 2010). For instance, citric acid is used widely in the food and pharmaceutical industries. Before 1923, this compound was produced from lemons. Nowadays, this practice has been replaced by the use of *Aspergillus niger* to synthesize citric acid. Fungi are also often used for the synthesis of antibiotics. One example is cyclosporin, which can be isolated from *Tolypocladium inflatum* (Suvase et al. 2010). This compound was first used as an antifungal, but later, it was shown to possess immunosuppressive activity. Cyclosporin A is currently the most widely used drug for preventing rejection of human organ transplants.

In a biotechnological setting, fungal species are often placed in bioreactors and subjected to over-oxygenation. This can lead to oxidative stress, defined as the overproduction of reactive oxygen species (ROS), which cells cannot defend against with their antioxidant defenses. ROS is generated in mitochondria during the process of aerobic respiration. Fungal species under oxidative stress conditions may suffer morphological changes, a slow growth rate, a low substrate consumption rate, low protein, and ATP content. This last consequence can be attributed to the switch from normal respiration to alternative respirations. In such situations, alternative pathways can substitute parts of the electron transport chain, lowering the production of ROS compounds. However, these alternative routes are unable to translocate protons, therefore the amount of ATP derived from the OXPHOS pathway is severely reduced.

15.1.4 The Oxidative Phosphorylation Pathway

The OXPHOS pathway is used by aerobic organisms to obtain energy. It can be found in all the domains of life with very few exceptions. In most eukaryotic organisms, the pathway is formed by five multi-subunit complexes that work coordinately to produce energy in the form of ATP (Joseph-Horne et al. 2001). The four first complexes (NADH: ubiquinone oxidoreductase, succinate dehydrogenase, ubiquinol cytochrome c reductase, and cytochrome c oxidase) in the pathway form the electron transport chain. This chain starts with the transference of electrons from NADH or $FADH_2$ to ubiquinone, a task carried out by complexes I and II. The next step, performed by complex III, transfers the electrons from ubiquinol to cytochrome c. Finally, complex IV transfers the electrons to molecular oxygen. Electron transfer in complexes I, III and IV, is coupled with the translocation of protons across the inner mitochondrial membrane, creating a proton gradient. The energy produced during the release of the proton gradient is used by complex V in order to produce ATP. The pathway is located in the mitochondrial inner membrane, and, at least under certain circumstances, the complexes are associated into supramolecular structures called respirasomes (Wittig et al. 2006).

Some of the subunits of the OXPHOS pathway are encoded in the mitochondrial genome. Phylogenetic studies have tried to establish the origin of the different components of the pathway and while some nuclearly encoded subunits have a clear mitochondrial origin, the phylogenies of other genes are much more complex (Gabaldon and Huynen 2003, 2004; Gabaldon et al. 2005).

15.1.5 Alterations to the Main Pathway in Fungi

Beyond the common electron transport chain, in plants, fungi and several protists, there are alternative pathways that can bypass some steps. These alternative pathways have conferred a better adaptation for some fungal species and produced fail-safe mechanisms that act when the organism is under stress or toxic conditions.

15.1.5.1 Alternative NADH Dehydrogenase

In contrast to mammals, plants, fungi, and bacteria can bypass complex I with alternative types of NADH dehydrogenases (Kerscher 2000). The main functional difference between alternative NADH dehydrogenases and complex I is that the electron transport to ubiquinone by alternative NADH dehydrogenases is not coupled to proton translocation. In addition, they use FAD or FMN as prosthetic groups and work either as monomers or homodimers.

Alternative NADH dehydrogenases can act either alone, substituting complex I, or in tandem with the regular complex. In *S. cerevisiae*, for instance, complex I has been lost and in its place three different alternative NADH dehydrogenases can be found. One of them faces the matrix while the others face the cytoplasm and compensate for the absence of the malate/aspartate shuttle in this microorganism. In baker's yeast, the alternative NADH dehydrogenases are single polypeptides of 53 or 58 kDa (external and internal, respectively) and are devoid of iron–sulfur centers (Helmerhorst et al. 2002).

These three alternative NADH dehydrogenases can also be found in *Neurospora crassa*, which, unlike baker's yeast, does contain complex I. It has been suggested that in this case, the alternative dehydrogenases are used to prevent the over-reduction of electron-transport carriers and the production of reactive oxygen species. Numerous studies have been performed in order to elucidate the function of these enzymes in *N. crassa*. For instance, it was seen that the internal NADH dehydrogenase, while not essential for growth or sexual development, affected the germination of ascospores and conidia (Duarte et al. 2003).

Numerous human mitochondrial diseases are associated with defects in the functionality of complex I such as Leigh syndrome or Parkinson's disease (Yagi et al. 2006). In recent years, the inner alternative NADH of *S. cerevisiae* (NDI1) has been considered as a potential treatment for those diseases. It has been shown in mouse models that once NDI1 is imported into the cells it can be expressed and is functionally active, and that it remains active for at least several years (Marella et al. 2009). Thanks to the great structural differences between NDI1 and complex I, the former is only one peptide while complex I is formed by around 40 different subunits, this enzyme is totally insensitive to the compounds that can damage complex I.

15.1.5.2 Alternative Oxidase

Alternative oxidases can be found in numerous fungal species, higher plants, algae, and some protozoa, though its distribution within these groups is patchy. Their main function is to serve as an alternative pathway for the completion of the electron transport chain. They bypass complexes III and IV by transferring electrons from ubiquinol to the final acceptor, O_2. They are located in the inner membrane of mitochondria and confer resistance against toxic compounds such as cyanide, for which complex IV is sensitive.

As with alternative NADH dehydrogenases, AOX are unable to translocate protons when electrons are transferred, therefore no energy is produced in this step. It would be extremely unlikely to find alternative NADH dehydrogenases and AOX working together as that would imply a futile cycle. For this reason, it is not surprising to observe that fungal species that lack complex I have also lost the alternative oxidase.

Surprisingly, functional copies of AOX have been found in some anaerobic Microsporidians (Williams et al. 2010). These enzymes have a high quinol oxidase activity and it is thought that they are used in order to lower the reducing potential created through continuous use of glycolysis. Even so, the distribution of AOX in Microsporidians remains patchy and is missing in completely sequenced genomes such as *Encephalitozoon cuniculi*, *E. bieneusi*, or *Nosema ceranae*.

15.2 Establishing the Evolutionary Framework: The Fungal Species Tree

In order to trace the evolution of OXPHOS components across sequenced fungal species, we first need to establish a reliable scenario depicting their evolutionary relationships. In other words, we need to reconstruct a fungal species tree as well as to evaluate the level of confidence of the inferred lineages. Efforts to reconstruct fully resolved phylogenies have been much influenced by the availability of completely sequenced genomes. In particular, methods based on tree concatenation and supertree strategies (Delsuc et al. 2005) have been extensively used at different taxonomic levels (Snel et al. 2005; Wolf et al. 2001). The large amount of available sequence data for fungal species has opened the doors for the use of phylogenomics to address the reconstruction of the still elusive fungal species tree.

Several species trees have been reconstructed over the last few years based on completely sequenced genomes (Marcet-Houben and Gabaldon 2009; Fitzpatrick et al. 2006; Wang et al. 2009). These trees, while slightly different in the distribution of some species, mostly support a similar topology (Marcet-Houben and Gabaldon 2009).

In Fig. 15.1, we present the largest tree that has been reconstructed to date based on phylogenomics data. The tree represents the evolution of 102 different fungal

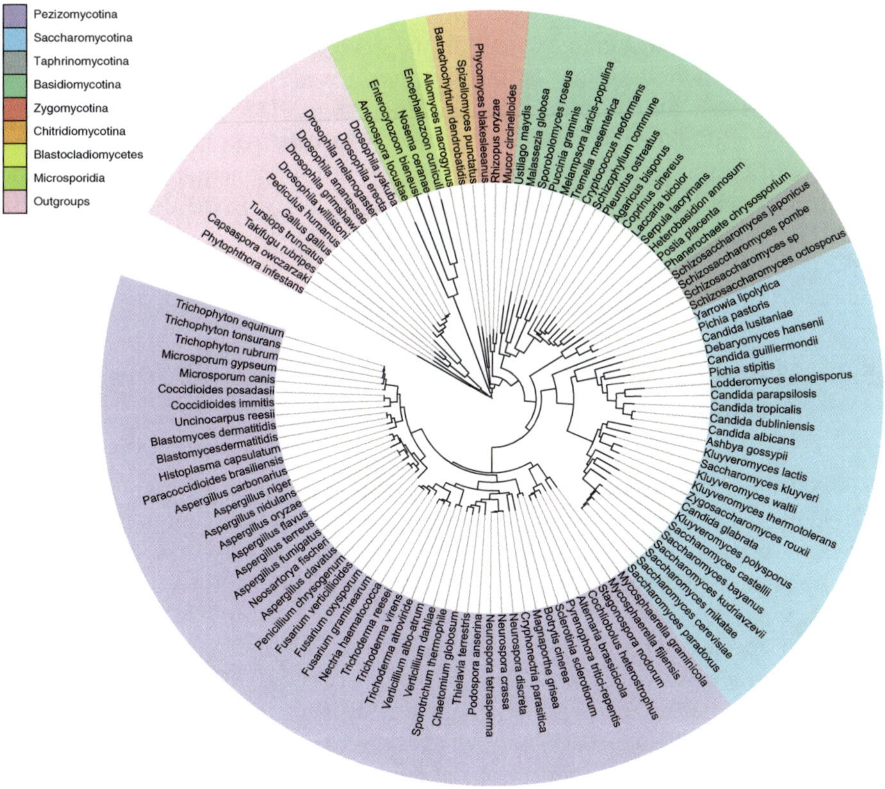

Fig. 15.1 Species tree representing the evolution of fungal species. In order to reconstruct the tree, 47 widespread proteins in 103 fungal species were concatenated and then the maximum likelihood tree was reconstructed. Representation was done using iTOL (Letunic and Bork 2007)

species and is based on the concatenation of 47 widespread proteins that complied with two conditions: they displayed a one-to-one orthology relationship and they were present in at least 90 species. Out-groups were chosen so that the number of proteins used was maximized. The resulting topology is largely similar to the ones published before. Still, some nodes present variability, such as the branching order of the three groups of Basidiomycota or the relative position of *C. glabrata* and *Saccharomyces castellii* in reference to the *Saccharomyces* group.

Traditional methods such as the bootstrap measure (Felsenstein 1985) or the approximate likelihood ratio rest (aLRT) (Anisimova and Gascuel 2006) have been used to determine the robustness of the nodes in a tree. Unfortunately, trees derived from gene concatenation tend to have high support values independently of the tree robustness. New methods are therefore needed to assess the robustness of nodes in the species tree. The phylome support, as implemented in treeKO (http://treeko.cgenomics.org), is one such method. A phylome is the complete collection of phylogenetic trees based on each gene in a genome and this large collection of trees can be used as a tool to identify nodes that are not well supported by the information encoded in the genome. For each node in the tree, the phylome support algorithm groups species derived from the node into groups and then considers all the arrangements of the groups. Then, the support of each arrangement within the phylome is calculated. The more supported it is, the more often it will appear in the phylome. Nodes with a low phylome support are the ones that we will need to consider carefully while studying the evolution of the OXPHOS pathway.

15.3 Evolution of the Oxidative Phosphorylation Pathway in Fungi

The OXPHOS pathway shows an enormous evolutionary plasticity. Therefore, it is of great interest to see what changes have occurred and relate them to the different respiratory phenotypes in extant fungal species. Lavin et al. (2008) performed a first study by detecting homologs of OXPHOS components in 27 fully sequenced fungal species. We later extended this study to encompass a total of 60 completely sequenced species (Marcet-Houben et al. 2009). An important addition to our study was the reconstruction of the phylogenetic trees of all OXPHOS components across the 60 species under study. Such phylogenetic approach allowed us to reliably distinguish between orthology and paralogy relationships and to investigate the relative timing and taxonomic scope of the gene duplications that have affected OXPHOS components. Gene duplication is considered to be one of the main sources for functional diversification (Ohno 1970), and there are specific hypothesis that predict evolutionary constrains to such mode of evolution in multi-protein complexes (Papp et al. 2003). Thus, we wanted to investigate to which degree this process had affected a group of families mostly coding for large multi-protein complexes and, in principle, expected to retain their central functions. The details of this study can be found in (Marcet-Houben et al. 2009), but we will offer here a brief overview.

We started our search for homologs with a representative from each OXPHOS pathway family, preferably from a model species such as *S. cerevisiae*, *N. crassa*, or *C. albicans*. By searching against a database comprising 60 completely sequenced species, we formed initial groups of homologs that were subsequently curated manually to remove spurious hits or add homologs that had been missed due to annotation errors or methodological artifacts. Finally, these homologs were aligned and a Maximum Likelihood approach (Guindon and Gascuel 2003) was used to infer the phylogenies for each of these groups. By using a species-overlap algorithm (Huerta-Cepas et al. 2007), speciation and duplication events were mapped onto the tree to define orthology and paralogy relationships. This data was used to derive a phylogenetic profile for each family, comprising information about the number of gene copies present in each genome. The main findings of this study are

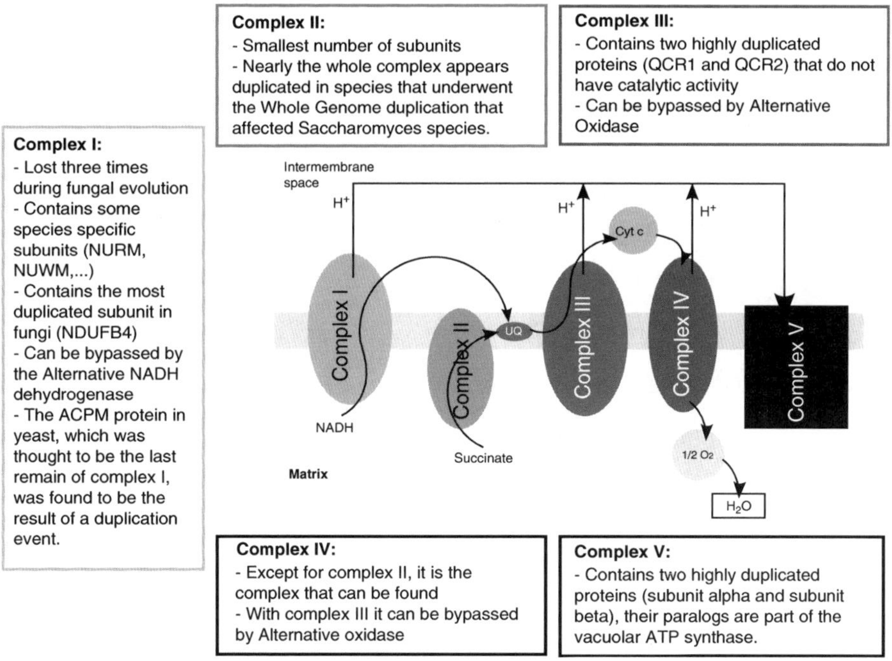

Fig. 15.2 Representation of the main oxidative phosphorylation pathway and the main findings for each of the five complexes that conform the pathway

summarized in Fig. 15.2. We will now center our attention to some of the details that came forth during the study.

15.3.1 Presence/Absence Pattern of the Fungal OXPHOS Genes

In Table 15.1, we show the percentage of fungal species that contain each subunit of the OXPHOS pathway. Darker colors represent widespread proteins while lighter colors represent genes that can only be found in some fungal groups. For instance, the lighter color found in all the proteins in complex I shows that the entire complex was lost in some fungal species. It is known that complex I was lost independently in three lineages, namely, in *Saccharomyces* species, in *Schizosaccharomyces* species, and in Microsporidia, which is confirmed by our analysis (Gabaldon et al. 2005). Thanks to the information provided by the fungal species tree, we can infer the relative timing (i.e., after which speciation events) of each of the three independent losses.

On the other hand, while almost all the proteins of the other complexes can be found in a high percentage of the fungal species, there are some proteins that have a limited taxonomic scope. For instance, NUVM and NUWM were at first considered

Table 15.1 Table of subunits that form the oxidative phosphorylation pathway. Cells are colored according to the presence rate of each subunit. The presence rate is calculated as the number of species that contain at least one copy of the subunit divided with the total number of species considered in the analysis. *Darker colors* represent higher presence rates and decrease in the following intervals: 1.0–0.9–0.75–0.5–0.33–0

Presence rates							
Complex I							
1.6.5.3	NAD1	NAD2	NAD3	NAD4	NAD4L	NAD5	NAD6
NDUFA1	NDUFA11	NDUFA12	NDUFA13	NDUFA2	NDUFA4	NDUFA5	NDUFA6
NDUFA8	NDUFA9	NDUFAB1	NDUFB3	NDUFB4	NDUFB7	NDUFB8	NDUFB9
NDUFS1	NDUFS2	NDUFS3	NDUFS4	NDUFS6	NDUFS7	NDUFS8	NDUFV1
NDUFV2	NI9M	NURM	NUVM	NUWM	NUXM	NUZM	–
Complex II				Alternative NADH dehydrogenase		Alternative oxidase	
SDHA	SDHB	SDHC	SDHD				
Complex III							
CytB	CytC	ISP	QCR1	QCR10	QCR2	QCR6	QCR7
QCR8	QCR9	–	–	–	–	–	–
Complex IV							
COX1	COX10	COX11	COX15	COX17	COX2	COX3	COX4
COX5	COX6	COX6A	COX6B	COX7	COX8	–	–
Complex V							
SUB 8	SUB A	SUB α	SUB B	SUB β	SUB C	SUB D	SUB δ
SUB ε	SUB F	SUB G	SUB γ	SUB H	SUB J	SUB K	SUB OSCP

to be specific for the OXPHOS pathway in *Yarrowia lipolytica* (Abdrakhmanova et al. 2004). Now, it has been seen that they are not so restricted in their taxonomic sampling. NUWM orthologs can be found in all the members of the Candida group while NUVM is extended to most Ascomycotina species.

There is a possibility that some of the proteins that have a restricted species distribution belong to the same protein family but that due to accelerated evolution, we are unable to identify them as such. We would then expect to have complementary species for each protein family. An interesting example of this possibility would be the NUWM and NURM families since the first one is found only in Saccharomycotina species that have complex I and the other one is only found in Pezizomycotina species.

In the work published by Cardol et al. (2005), NUVM was proposed to be orthologous to the mammalian NDUFB4. In light of our analysis this seems unlikely, as NDUFB4 has other orthologs in fungal species, and not even with the high level of duplicates found within this subunit were we able to find any NUVM homolog. In fact, performing a blast search using the human NDUFB4 as a starting point produced few, non-reliable, hits in the fungal database (e-values >1.0), showing that the fungal NDUFB4 and NUVM are either not homologous to the mammal NDUFB4 gene or that they have diverged too much to be able to safely trace their evolutionary history back. Either way, the two genes do not show any overlap and so we believe that they are two different subunits of complex I.

15.3.2 Duplication of OXPHOS Genes

Duplications are important events that can have a great impact in the evolution of a given species. According to our data, the OXPHOS pathway has not been exempt of such process despite the predicted low incidence of duplications occurring in a protein complex. In our analysis (Marcet-Houben et al. 2009), we found that more than 75% of the proteins that are part of the OXPHOS pathway contain, at least, one duplication event. Table 15.2 summarizes the average number of copies per species found for each protein in the OXPHOS pathway. These values were calculated as the total number of homologs found for each gene (orthologs and paralogs) divided by the number of species that contained at least one copy of the given gene.

The gene balance hypothesis theorizes that genes that are part of a complex have a lower chance of retaining both gene duplicates (Papp et al. 2003). The reason for that is found in the change of stoichiometric relationships that would occur if the expression level of only one of the genes in a complex was suddenly doubled. Contrary to these expectations, we find that most genes have an average number of copies per species superior to 1.

Seven genes have an average number of copies per species higher than 2: NDUFB4 (complex I), SDHA (complex II), QCR1 and QCR2 (complex III), subunits alpha and beta (complex V), and the alternative NADH dehydrogenase.

Table 15.2 Table of subunits that form the oxidative phosphorylation pathway. Cells are colored according to the average number of copies per species of each subunit. This value is calculated as the number of homologs of the subunit contained in each fungal species divided with the number of species that have at least one copy of the subunit. Darker colors represent higher average number of copies and go from 1.0 to 4.0 by increments of 1.0

Average number of copies							
Complex I							
1.6.5.3	NAD1	NAD2	NAD3	NAD4	NAD4L	NAD5	NAD6
NDUFA1	NDUFA11	NDUFA12	NDUFA13	NDUFA2	NDUFA4	NDUFA5	NDUFA6
NDUFA8	NDUFA9	NDUFAB1	NDUFB3	NDUFB4	NDUFB7	NDUFB8	NDUFB9
NDUFS1	NDUFS2	NDUFS3	NDUFS4	NDUFS6	NDUFS7	NDUFS8	NDUFV1
NDUFV2	NI9M	NURM	NUVM	NUWM	NUXM	NUZM	–
Complex II							
SDHA	SDHB	SDHC	SDHD	Alternative NADH dehydrogenase		Alternative oxidase	
Complex III							
CytB	CytC	ISP	QCR1	QCR10	QCR2	QCR6	QCR7
QCR8	QCR9	–	–	–	–	–	–
Complex IV							
COX1	COX10	COX11	COX15	COX17	COX2	COX3	COX4
COX5	COX6	COX6A	COX6B	COX7	COX8		
Complex V							
SUB 8	SUB A	SUB α	SUB B	SUB β	SUB C	SUB D	SUB δ
SUB ε	SUB F	SUB G	SUB γ	SUB H	SUB J	SUB K	SUB OSCP

This represents that 8% of the genes in the OXPHOS pathway have an average number of copies higher than 2. For comparison, we scanned the *C. albicans* phylome, namely, the collection of phylogenetic trees for each gene in the *C. albicans* genome, in order to find genes that had an average number of copies per species greater than 2. The *C. albicans* phylome was constructed using 83 fungal species and can be found in phylomeDB (http://www.phylomedb.org) (Huerta-Cepas et al. 2010). Out of 5,824 genes, 951 had an average number of copies per species greater than 2, representing about 16% of the genes, which doubles the amount of genes found in OXPHOS. On the other hand, we find that 76% of the genes in OXPHOS have at least one duplication event, while in the *C. albicans* phylome 77% of the genes have at least one duplicate. This indicates that while the genes that are part of the OXPHOS pathway have been duplicated as often as all the other genes in the *C. albicans* genome, they tend to conserve less copies than other genes in the genome.

15.4 Examples of Evolutionary Events That Have Shaped the OXPHOS Pathway in Fungi

15.4.1 QCR1 and QCR2

Complex III subunits are mainly nuclearly encoded. This complex usually has four redox centers that are involved in electron transfer. There are two core proteins facing the matrix (QCR1 and QCR2) that show homology to mitochondrial peptidases involved in processing newly imported proteins. However the complex III subunits in yeast are proteolitically inactive and are not involved in cytochrome c reductase activity. Even so, the two proteins are needed for the correct assembly of the complex. These two proteins are homologous to the mitochondrial processing peptideases (Mas1 and Mas2 in yeast). They evolved from ancient duplication events that happened before the divergence between humans, animals, and plants. After this first round of duplications, no other duplications have occurred in these complex III subunits except for a species-specific duplication resulting from the WGD in *Rhizopus oryzae* (Ma et al. 2009).

15.4.2 ACPM Protein in Complex I

The ACPM protein found in complex I was predicted to be the only remaining protein that could be found in *Saccharomyces* species after the loss of complex I. The protein was found as part of the synthesis of octanoic acid and this change of function served as explanation for this group of species to have retained it. We showed that the functional change occurred before the loss of complex I in

Saccharomyces species and that it was related to a duplication event that occurred in Saccharomycotina species. The ACPM protein identified in yeast was actually the paralog to the original complex I protein. This was seen due to the fact that *Candida* species still retain both copies of the protein and a phylogenetic tree clearly showed that a duplication event had happened followed by a loss of one of the copies in yeast.

Yarrowia lipolytica also contains two copies of the protein. It was seen that both copies were associated to complex I (Dobrynin et al. 2010). Deletion of ACPM1, which is orthologous to the complex I subunit in *N. crassa*, was viable but the organism was unable to assemble complex I, which clearly depicts an association between complex I and ACPM1. On the other hand, deletion of the second copy, the ortholog to the yeast ACPM protein, was not viable, hindering the identification of its function. It is possible that ACPM2 would then be associated to the synthesis of octanoic acid and that its association to complex I is only used as a mechanism to be recruited to the membrane. In any case, further experimental proofs are needed to establish the functional differences between both paralogous groups.

15.5 Concluding Remarks

The OXPHOS pathway in fungi has undergone numerous changes during the evolution of this diverse group of species. Events such as the loss of complex I in yeasts adapted to fermentative lifestyles, the massive loss of the whole pathway in parasitic Microsporidia, and the large number of duplications detected in all the complexes of the OXPHOS pathway, have shaped the respiratory mechanisms of fungi and allowed this kingdom to expand over numerous different environmental niches. Additionally, the presence of alternative pathways to the primary electron transport chain may have been a turning point in many evolutionary events such as the shift from nonpathogenic to pathogenic or the shift of aerobic to anaerobic lifestyles.

The large number of duplications detected in the OXPHOS pathway belies the notion that proteins that are part of complexes are less likely to retain both copies after a duplication event. We saw how the percentage of proteins in OXPHOS that have, at least, one duplication was the same as the one found in the *C. albicans* genome. The only difference observed was that OXPHOS proteins tend to have less duplicates than other *C. albicans* genes, so there may be some restrictions acting on the retention of too many duplicates of the complex.

Acknowledgment TG and MMH are founded through a grant of the Spanish Ministry of Science (BFV2009-09168).

References

Abdrakhmanova A, Zickermann V, Bostina M, Radermacher M, Schagger H, Kerscher S, Brandt U (2004) Subunit composition of mitochondrial complex I from the yeast *Yarrowia lipolytica*. Biochim Biophys Acta 1658(1–2):148–156. doi:10.1016/j.bbabio.2004.04.019; S0005272804001446 [pii]

Anisimova M, Gascuel O (2006) Approximate likelihood-ratio test for branches: a fast, accurate, and powerful alternative. Syst Biol 55(4):539–552. doi:T808388N86673K61 [pii]; 10.1080/10635150600755453

Burri L, Williams BA, Bursac D, Lithgow T, Keeling PJ (2006) Microsporidian mitosomes retain elements of the general mitochondrial targeting system. Proc Natl Acad Sci USA 103 (43):15916–15920. doi:0604109103 [pii]; 10.1073/pnas.0604109103

Cardol P, Gonzalez-Halphen D, Reyes-Prieto A, Baurain D, Matagne RF, Remacle C (2005) The mitochondrial oxidative phosphorylation proteome of *Chlamydomonas reinhardtii* deduced from the Genome Sequencing Project. Plant Physiol 137(2):447–459. doi:137/2/447 [pii]; 10.1104/pp. 104.054148

Delsuc F, Brinkmann H, Philippe H (2005) Phylogenomics and the reconstruction of the tree of life. Nat Rev Genet 6(5):361–375. doi:nrg1603 [pii]; 10.1038/nrg1603

Dobrynin K, Abdrakhmanova A, Richers S, Hunte C, Kerscher S, Brandt U (2010) Characterization of two different acyl carrier proteins in complex I from *Yarrowia lipolytica*. Biochim Biophys Acta 1797(2):152–159. doi:S0005-2728(09)00263-1 [pii]; 10.1016/j.bbabio.2009.09.007

Duarte M, Peters M, Schulte U, Videira A (2003) The internal alternative NADH dehydrogenase of *Neurospora crassa* mitochondria. Biochem J 371(Pt 3):1005–1011. doi:10.1042/BJ20021374; BJ20021374 [pii]

Felsenstein J (1985) Confidence limits on phylogenies: an approach using the bootstrap. Evolution 39(4):783–791

Fitzpatrick DA, Logue ME, Stajich JE, Butler G (2006) A fungal phylogeny based on 42 complete genomes derived from supertree and combined gene analysis. BMC Evol Biol 6:99. doi:1471-2148-6-99 [pii]; 10.1186/1471-2148-6-99

Gabaldon T, Huynen MA (2003) Reconstruction of the proto-mitochondrial metabolism. Science 301(5633):609. doi:10.1126/science.1085463; 301/5633/609 [pii]

Gabaldon T, Huynen MA (2004) Shaping the mitochondrial proteome. Biochim Biophys Acta 1659(2–3):212–220. doi:S0005-2728(04)00248-8 [pii]; 10.1016/j.bbabio.2004.07.011

Gabaldon T, Rainey D, Huynen MA (2005) Tracing the evolution of a large protein complex in the eukaryotes, NADH:ubiquinone oxidoreductase (complex I). J Mol Biol 348(4):857–870. doi: S0022-2836(05)00237-8 [pii]; 10.1016/j.jmb.2005.02.067

Guindon S, Gascuel O (2003) A simple, fast, and accurate algorithm to estimate large phylogenies by maximum likelihood. Syst Biol 52(5):696–704. doi:54QHX07WB5K5XCX4 [pii]

Hackstein JH, Tjaden J, Huynen M (2006) Mitochondria, hydrogenosomes and mitosomes: products of evolutionary tinkering! Curr Genet 50(4):225–245. doi:10.1007/s00294-006-0088-8

Hawksworth DL (1991) The fungal dimension of biodiversity: magnitude, significance, and conservation. Mycol Res 95(4):641–655

Helmerhorst EJ, Murphy MP, Troxler RF, Oppenheim FG (2002) Characterization of the mitochondrial respiratory pathways in *Candida albicans*. Biochim Biophys Acta 1556(1):73–80. doi:S0005272802003080 [pii]

Huerta-Cepas J, Dopazo H, Dopazo J, Gabaldon T (2007) The human phylome. Genome Biol 8(6): R109. doi:gb-2007-8-6-r109 [pii]; 10.1186/gb-2007-8-6-r109

Huerta-Cepas J, Capella-Gutierrez S, Pryszcz LP, Denisov I, Kormes D, Marcet-Houben M, Gabaldon T (2010) PhylomeDB v3.0: an expanding repository of genome-wide collections of trees, alignments and phylogeny-based orthology and paralogy predictions. Nucleic Acids Res. doi:doi:gkq1109 [pii]; 10.1093/nar/gkq1109

Joseph-Horne T, Hollomon DW, Wood PM (2001) Fungal respiration: a fusion of standard and alternative components. Biochim Biophys Acta 1504(2–3):179–195. doi:S0005272800002516 [pii]

Keeling PJ, Corradi N, Morrison HG, Haag KL, Ebert D, Weiss LM, Akiyoshi DE, Tzipori S (2010) The reduced genome of the parasitic microsporidian *Enterocytozoon bieneusi* lacks genes for core carbon metabolism. Genome Biol Evol 2:304–309. doi:evq022 [pii]; 10.1093/gbe/evq022

Kerscher SJ (2000) Diversity and origin of alternative NADH:ubiquinone oxidoreductases. Biochim Biophys Acta 1459(2–3):274–283. doi:S0005-2728(00)00162-6 [pii]

Lavin JL, Oguiza JA, Ramirez L, Pisabarro AG (2008) Comparative genomics of the oxidative phosphorylation system in fungi. Fungal Genet Biol 45(9):1248–1256. doi:S1087-1845(08)00108-4 [pii]; 10.1016/j.fgb.2008.06.005

Letunic I, Bork P (2007) Interactive Tree Of Life (iTOL): an online tool for phylogenetic tree display and annotation. Bioinformatics 23(1):127–128. doi:btl529 [pii]; 10.1093/bioinformatics/btl529

Li Q, Bai Z, O'Donnell A, Harvey LM, Hoskisson PA, McNeil B (2010) Oxidative stress in fungal fermentation processes: the roles of alternative respiration. Biotechnol Lett. doi:10.1007/s10529-010-0471-x

Ma LJ, Ibrahim AS, Skory C, Grabherr MG, Burger G, Butler M, Elias M, Idnurm A, Lang BF, Sone T, Abe A, Calvo SE, Corrochano LM, Engels R, Fu J, Hansberg W, Kim JM, Kodira CD, Koehrsen MJ, Liu B, Miranda-Saavedra D, O'Leary S, Ortiz-Castellanos L, Poulter R, Rodriguez-Romero J, Ruiz-Herrera J, Shen YQ, Zeng Q, Galagan J, Birren BW, Cuomo CA, Wickes BL (2009) Genomic analysis of the basal lineage fungus *Rhizopus oryzae* reveals a whole-genome duplication. PLoS Genet 5(7):e1000549. doi:10.1371/journal.pgen.1000549

Magnani T, Soriani FM, Martins Vde P, Policarpo AC, Sorgi CA, Faccioli LH, Curti C, Uyemura SA (2008) Silencing of mitochondrial alternative oxidase gene of *Aspergillus fumigatus* enhances reactive oxygen species production and killing of the fungus by macrophages. J Bioenerg Biomembr 40(6):631–636. doi:10.1007/s10863-008-9191-5

Marcet-Houben M, Gabaldon T (2009) The tree versus the forest: the fungal tree of life and the topological diversity within the yeast phylome. PLoS ONE 4(2):e4357. doi:10.1371/journal.pone.0004357

Marcet-Houben M, Marceddu G, Gabaldon T (2009) Phylogenomics of the oxidative phosphorylation in fungi reveals extensive gene duplication followed by functional divergence. BMC Evol Biol 9:295. doi:1471-2148-9-295 [pii]; 10.1186/1471-2148-9-295

Marella M, Seo BB, Yagi T, Matsuno-Yagi A (2009) Parkinson's disease and mitochondrial complex I: a perspective on the Ndi1 therapy. J Bioenerg Biomembr 41(6):493–497. doi:10.1007/s10863-009-9249-z

Mueller G, Schmit J (2007) Fungal biodiversity: what do we know? What can we predict? Biodivers Conserv 16:1–5

Ohno S (1970) Evolution by gene duplication. Springer, New York

Papp B, Pal C, Hurst LD (2003) Dosage sensitivity and the evolution of gene families in yeast. Nature 424(6945):194–197. doi:10.1038/nature01771; nature01771 [pii]

Schmit J, Mueller G (2007) An estimate of the lower limit of global fungal diversity. Biodivers Conserv 16(1):99–111

Snel B, Huynen MA, Dutilh BE (2005) Genome trees and the nature of genome evolution. Annu Rev Microbiol 59:191–209. doi:10.1146/annurev.micro.59.030804.121233

Suvase SA, Annapure US, Singhal RS (2010) Gellan gum as an immobilization matrix for the production of cyclosporin A. J Microbiol Biotechnol 20(7):1086–1091. doi:JMB020-07-05 [pii]

Takaya N (2009) Response to hypoxia, reduction of electron acceptors, and subsequent survival by filamentous fungi. Biosci Biotechnol Biochem 73(1):1–8. doi:JST.JSTAGE/bbb/80487 [pii]

Visser W, Scheffers WA, Batenburg-van der Vegte WH, van Dijken JP (1990) Oxygen requirements of yeasts. Appl Environ Microbiol 56(12):3785–3792

Voncken F, Boxma B, Tjaden J, Akhmanova A, Huynen M, Verbeek F, Tielens AG, Haferkamp I, Neuhaus HE, Vogels G, Veenhuis M, Hackstein JH (2002) Multiple origins of hydrogenosomes: functional and phylogenetic evidence from the ADP/ATP carrier of the anaerobic chytrid *Neocallimastix* sp. Mol Microbiol 44(6):1441–1454. doi:2959 [pii]

Wang H, Xu Z, Gao L, Hao B (2009) A fungal phylogeny based on 82 complete genomes using the composition vector method. BMC Evol Biol 9:195. doi:1471-2148-9-195 [pii]; 10.1186/1471-2148-9-195

Williams BA, Elliot C, Burri L, Kido Y, Kita K, Moore AL, Keeling PJ (2010) A broad distribution of the alternative oxidase in microsporidian parasites. PLoS Pathog 6(2):e1000761. doi:10.1371/journal.ppat.1000761

Wittig I, Carrozzo R, Santorelli FM, Schagger H (2006) Supercomplexes and subcomplexes of mitochondrial oxidative phosphorylation. Biochim Biophys Acta 1757(9–10):1066–1072. doi: S0005-2728(06)00130-7 [pii]; 10.1016/j.bbabio.2006.05.006

Wolf YI, Rogozin IB, Grishin NV, Tatusov RL, Koonin EV (2001) Genome trees constructed using five different approaches suggest new major bacterial clades. BMC Evol Biol 1:8

Yagi T, Seo BB, Nakamaru-Ogiso E, Marella M, Barber-Singh J, Yamashita T, Matsuno-Yagi A (2006) Possibility of transkingdom gene therapy for complex I diseases. Biochim Biophys Acta 1757(5–6):708–714. doi:S0005-2728(06)00024-7 [pii]; 10.1016/j.bbabio.2006.01.011

Chapter 16
Genome Structure and Gene Expression Variation in Plant Mitochondria, Particularly in the Genus *Silene*

Helena Storchova

Abstract The plant mt genomes are highly dynamic. Their evolution is driven by frequent rearrangements and gene transfers, whereas substitution rate is generally slow with several exceptions. The genus *Silene* (Caryophyllales) represents one of them and exhibits high mutation rate in mt DNA. The gynodioecious species (producing female and hermaphroditic individuals) of this genus show also a high polymorphism in mt DNA due to the balancing selection in favor of various mt genomes in the same population. Thus, *Silene* species possess plenty of mt markers, which facilitate the study of the impact of mt genome rearrangements on mt gene expression and function. They are also good models for the investigation of functional and evolutionary aspects of heteroplasmy, the situation when two or more organelles with distinct genomes co-occur in the same individual.

16.1 Introduction

Despite of the extensive investigation of plant mitochondrial (mt) genomes over the last decades, many questions remain to be solved. Understanding the relationship between the structure of mt genome and mt gene transcription is one of the topics of great importance. The species of a plant genus *Silene* have been extensively followed in population genetic and ecological studies. A recent progress in the analysis of their mt genomes enables to relate the evolution of mt genome to the evolution of populations and species (McCauley and Olson 2008). The genus *Silene* has become an emerging model for the study of evolution and function of the plant mt genome.

H. Storchova
Institute of Experimental Botany v.v.i., Academy of Sciences of the Czech Republic, Rozvojová 223, 165 00 Prague 6, Lysolaje, Czech Republic
e-mail: storchova@ueb.cas.cz

P. Pontarotti (ed.), *Evolutionary Biology – Concepts, Biodiversity, Macroevolution and Genome Evolution*, DOI 10.1007/978-3-642-20763-1_16,

16.1.1 Characteristics of Plant Mitochondrial Genomes

The structure of plant mt genomes is fluidic and varies substantially across angiosperms. Rearrangement, duplications, insertions, and deletions of small or large genomic regions are the prominent processes underlying the evolution of plant mt genomes, in addition to single nucleotide substitutions. The basic features have been documented by the fast growing number of completely sequenced plant mt genomes (e.g. *Arabidopsis thaliana* – Unseld et al. 1997; *Beta vulgaris* – Kubo et al. 2000; rice – Tian et al. 2006; maize – Allen et al. 2007; *Vitis vinifera* – Goremykin et al. 2009). Altogether, 23 completely sequence mt genomes of seed plants were reported by September 2010, but the speed of sequencing will accelerate owing to a broad application of the next-generation sequencing.

Plant mt genomes are much larger than their animal and fungal counterparts. Their size range spans from 200 kb to the astonishing 2,900 kb in *Cucumis melo* (Ward et al. 1981), whereas it reaches 15–20 kb in animals (Boore 1999) and 17–100 kb in fungi (Cummings et al. 1990; Lee and Young 2009). The large size of plant mt genomes is not associated with higher gene content than in animals or fungi, but it is rather due to the large portion of non-coding intergenic DNA of unknown origin. The described number of protein-coding genes in seed plant mitochondria varies between 41 (*Cycas taitungensis* – Chaw et al. 2008) and 25 (*Silene latifolia* – Sloan et al. 2010a). Ongoing gene loss or transfer of the genes between mitochondria, nucleus, and chloroplast has been observed across a plant kingdom (reviewed in Adams and Palmer 2003). Plant mt genes use the universal genetic code for translation, whereas animal mt genes are translated according to a modified genetic code. The identity of mt and nuclear genetic codes in plants facilitates the transfer of functional genes from mitochondria to the nucleus (Adams et al. 2000).

Despite of the common habit to present mitochondrial genomic maps as the circles, the actual morphological structure of plant mitochondrial genome is much more complex. Direct microscopic observations revealed linear molecules of various sizes, branched molecules, and small subgenomic circles, whereas large circular molecules consistent with genomic maps were rather rare (Oldenburg and Bendich 1996; Manchekar et al. 2009). Some of the complex molecular structures (Y- and H-shaped molecules), as well as the frequent single-stranded regions may be the intermediates of recombination and/or replication (Manchekar et al. 2006).

16.1.2 Three Categories of Recombination in Plant mt Genomes

Recombination is a major force shaping the dynamic structure of the plant mt genomes (Mackenzie 2007). Recombination events occurring in plant mitochondria could be classified to three categories (reviewed in Marechal and Brisson 2010). First of them involves frequent recombinations mediated by long direct or inverted repeats (>1 kb), found in the majority of completely sequenced angiosperm mt genomes. Their sequence and number vary. For example, two inverted repeats (7.3 kb) were revealed in mt genome of *Cucurbita* (Alverson et al. 2010), three copies of 6.2 kb repeat unit were described in *Beta vulgaris* (Kubo et al. 2000). Some mt genomes (e.g., in maize) contain several sets of unrelated large repeats (Allen et al. 2007). Homologous recombination in those repeats generates rearranged isoforms in case of inverted repeats or subgenomic circular molecules in case of direct repeats. Owing to the reversible character and high frequency of recombination, various (sub)genomic molecules coexist in a stoichiometric equilibrium (Palmer and Shields 1984).

The second class of recombination events is mediated by short repeats, ranging approximately from about 50 bp to 1 kb. This kind of recombination is much less frequent than previous one and often results in asymmetric products (Abdelnoor et al. 2003; Shedge et al. 2007). It may cause a sudden change of copy numbers of particular (sub)genomic molecule in the course of single generation, which is termed substoichiometric shifting (Small et al. 1987, 1989). The preexisting rare molecules (sublimons) are amplified by this process, or they are repeatedly created by constant recombination (Woloszynska and Trojanowski 2009).

The frequency of asymmetric recombination across short repeats is controlled by the nuclear genes *MSH1* and *RECA3* in *Arabidopsis thaliana* (Abdelnoor et al. 2003; Shedge et al. 2007). The *msh1 recA3* double mutants are heavily impaired in growth and exhibit high level of reorganization of mt and chloroplast genomes (Arrieta-Montiel et al. 2009; Shedge et al. 2010). *MSH1* probably modifies initial stages of recombination, inhibiting DNA exchange and favoring the process of gene conversion, which maintains sequnce identity of short repeats across mt genomes (Shedge et al. 2007; Arrieta-Montiel 2009). The *MSH1* and *RECA3* genes and their homologs are therefore responsible for maintaining genomic stability in plant organelles, together with the members of *OSB* family (organellar single-stranded DNA-binding proteins) (Zaegel et al. 2006). Surprisingly, the only animals possessing a *MSH1* homolog are corals (Culligan et al. 2000), showing a sessile lifestyle similar to plants.

Finally, rare illegitimate recombination in the regions of microhomology (around 10 bp) was observed (Moyekens et al. 1995; Feng et al. 2009). It may be the result of microhomology-mediated break-induced replication, which is an error-prone process. Single-stranded DNA-binding proteins from Whirly family support accurate repair of double-stranded DNA breaks (Cappadocia et al. 2010) and prevent illegitimate recombination in organellar DNA.

16.1.3 Cytoplasmic Male Sterility and a Cytonuclear Conflict

Recombination mediated by short repeats or by microhomology regions sometimes generates chimeric open reading frames (ORF) with mozaic structure. Some chimeric genes are expressed and play an important role in the reproduction system of numerous plant species. They are responsible for the cytoplasmic male sterility (CMS). Novel proteins encoded by CMS genes interfere with anther development and block pollen production (Hanson and Bentolila 2004). CMS genes are composed of various portions of mt and chloroplast genes or of unknown ORFs (Dewey et al. 1986; Balk and Leaver 2001).

CMS lineages are widely exploited in agriculture to produce hybrid seed and their structure and function have been thoroughly investigated in crops; however, much less effort has been focused on determining the molecular action of male sterility in natural populations of wild plants. CMS is associated with a reproduction system termed gynodioecy, which is characterized by a co-occurrence of hermaphrodites and functional females (lacking viable pollen) in the same population. It represents the second most widespread breeding system in angiosperms (Richards 1997), being present in about 7% of angiosperm species.

The progeny of male sterile plants contains variable portion of male fertile (hermaphroditic) individuals, despite of prevailing maternal inheritance of mt DNA. This variation is caused by nuclear fertility restorer (Rf) genes which interact with the CMS genes to create a gender phenotype. Each *Rf* locus is assumed to have both restorer alleles (*Rf*) and non-restoration alleles (*rf*). *Rf* alleles are most commonly dominant to *rf*; thus only one copy is necessary to restore male fertility. Accumulating evidence (reviewed in Bentolila et al. 2002; Schmitz-Linneweber and Small 2008) suggests that *Rf* genes encode pentatricopeptide (PPR) proteins. PPR proteins are known to be localized in organelles and regulate all stages of gene expression. It is thought that these proteins are sequence-specific RNA-binding proteins capable to direct effector enzymes to organellar mRNA (reviewed in Andres et al. 2007). Although the exact structure and action of only few *Rf* genes is currently known, it is likely that these PPR proteins may affect size, stability, or translatability of CMS transcripts, and restore functional development of anthers (Wang et al. 2006; Gillman et al. 2007; Barr and Fishman 2010).

As CMS genes reside in mt genomes, they are inherited predominantly via maternal transmission. In contrast, nuclear genes are transmitted by both pollen and seeds. The contrasting transmission mode may lead to the cytonuclear conflict between mt and nuclear genomes, which mutually compete for resources. Higher production of seeds in females, partially due to the allocation of nutrients into ovules, was reported in a number of gynodioecious species (Gouyon and Couvet 1987).

16.1.4 mt Genome Reorganization Influences Transcription of mt Genes

The generation of chimeric genes responsible for CMS by DNA reorganization is not the only impact of genomic rearrangements on mt function. The association of mt genes or particular exons with different flanking regions may affect transcription initiation, processing of transcript termini, or splicing (Forner et al. 2005, 2008).

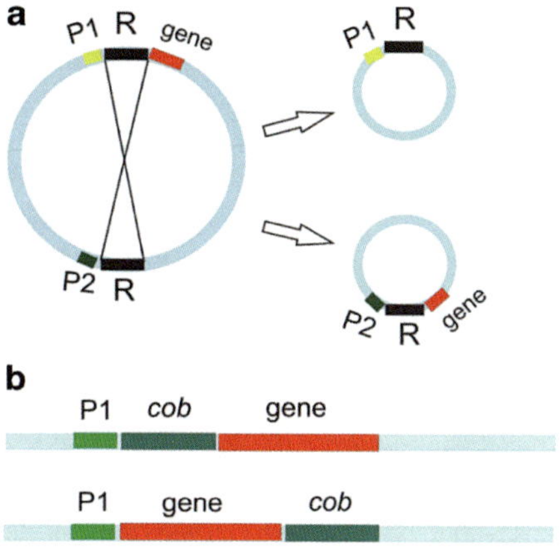

Fig. 16.1 The impact of mt genome rearrangements on transcription of mt genes. The recombination across large repeats places the gene under the control of one of two promoters, which may differ in strength (**a**). Rare recombinations mediated by short repeats or microhomology regions transfer the gene to the vicinity of the essential gene (e.g., *cob*), with which they are co-transcribed (**b**)

All three categories of recombination (see Sect. 16.1.2) taking place in plant mt DNA may influence mt gene expression. For example, one of two copies of large direct repeat present in maize mt genome lies between the gene encoding subunit II of cytochrome c oxidase (*cox2*) and its promoter (Lupold et al. 1999). Another promoter is located upstream of the second copy. Recombination between the two copies places the *cox2* gene alternatively under the control of the first or second promoter. Because the strength of both promoters varies, the degree of *cox2* expression within the same individual depends on the promoter-gene combination, which is more frequent (Fig. 16.1a).

Recombination across short repeats (50 bp–1 kb), controlled by the *MSH1* and *RECA3* genes, can also move the genes to the vicinity of different promoters and/or processing regulating elements. Such reorganization is responsible for the variation in transcript profiles observed among the ecotypes of *A. thaliana* (Forner et al. 2008), or among the lineages of sugar beet (Kubo et al. 1999). Temporary attenuation of *MSH1* suppression of asymmetric recombination via short repeats may lead to the differences in mt genomes and transcriptomes reported among accessions of *A. thaliana* (Arrieta-Montiel et al. 2009) and other species (Janska et al. 1998).

Rare recombination mediated by short microhomology sequences also contributes to the genomic rearrangements exhibiting the impact on the transcript size (Forner et al. 2005). Changes in genome configuration may lead to the association of two or more genes into the same transcription unit (Forner et al. 2007). Co-transcription facilitates expression of some CMS genes, which are placed upstream (Hedgcoth et al. 2002) or downstream (Kim et al. 2007) of important housekeeping genes (Fig. 16.1b). Transcriptional "hitchhiking" with the essential genes may bedevil downregulation of CMS genes by nuclear factors and thus favor these genes in the cytonuclear conflict (Hanson and Bentolila 2004).

The process opposite to the fusion of transcriptional units – the fragmentation of genes into separate pieces belonging to different transcriptional units also takes place in plant mt genome. Gene fission was described in the group II introns present in the genes coding for NADH dehydrogenase *nad1* (Chapdelaine and Bonen

1991), *nad2* (Binder et al. 1992), and *nad5* (Knoop et al. 1991). Putting the remote pieces of a fragmented intron together to generate a functional mRNA requires mutual recognition of the separated intron halves. A conserved secondary structure of group II introns plays a key role in this process termed *trans*-spicing (reviewed in Bonen 2008). As *cis*-spliced introns homologous to their *trans*-spliced counterparts were found in bryophytes and ferns (Malek and Knoop 1998), *trans*-splicing is considered to be evolutionary derived (Malek et al. 1997). The transition from *cis*-spliced to *trans*-spliced introns occurred several times during flowering plants evolutionary history (Qiu and Palmer 2004), but it is still a very rare event in comparison with DNA rearrangements responsible for within-species mt genomic variation described above (Forner et al. 2008; Arrieta-Montiel et al. 2009). Group II introns behave like mobile elements in bacteria (reviewed in Toro et al. 2007) owing to self-encoded reverse transcriptase/endonuclease activities. However, their ability to transpose is limited in plant mitochondria as documented by their occurrence in three *nad* genes only. Only one mt gene with maturase activity *matR* located in the fourth intron of *nad1* gene was reported in flowering plants (Qiu and Palmer 2004). Fragmentations of group II introns are caused by recombination events across repeats of various sizes. In addition, selection stabilizing a current mode of splicing may play an important role (Bonen 2008).

16.1.5 General Features of Plant mt Gene Expression

The viability and function of rearranged genomic structures is facilitated by the existence of multiple promoters with quite diverse sequence in plant mt genomes. Transcription of the transferred gene can be ensured by a so far silent promoter located nearby. For example, the consensus motif YRTA (Y = pyrimidine, R = purine) found in numerous plant mt promoters (Kühn et al. 2005) was discovered in only 2 of 12 promoters characterized in rice (Zhang and Liu 2006). Moreover, transcription initiation sites are rather scattered along the specific area, then concentrated in a single nucleotide. Transcription termination mechanism is not known in plant mitochondria, which suggests a relaxed control of transcription. This view is in agreement with the existence of large transcripts of unknown function derived from intergenic regions (Giege et al. 2000). The abundance of mt transcripts is rather regulated at the postranscriptional level by the control of their stability and longevity (Leino et al. 2005).

The large family of PPR proteins involves many members responsible for maturation, splicing, or translation of plant mt RNA (see Sect. 16.1.3). All these proteins important in mt RNA metabolism are encoded by the nucleus, as well as single-subunit phage-type RNA polymerases performing the transcription (Hedtke et al. 2002; Kuhn et al. 2005).

Additional RNA polymerase, encoded by a linear plasmid, exists in plant mitochondria. It was reported from *Beta maritima* (Saumitou-Laprade et al. 1991) and *Daucus carota* (Robison and Wolyn 2005). Whether it is capable to transcribe mt genes on the main chromosome is not currently known (Handa 2008). Linear plasmids, found in plant mitochondria, contain terminal inverted repeats and the proteins covalently linked to their 5′ ends (Handa et al. 2002), similarly to, e.g., phage phi-29 DNA termini (Mellado et al. 1980). The impact of the plasmids on mt function and evolution is not known. They may participate in mt DNA rearrangements (Newton et al. 1996) and be integrated to the main mt chromosome (Robison and Wolyn 2005).

16.1.6 The Variation in Substitution Rate Among Plant mt Genes

The capacity of plant mt genome for rapid changes in copy number and gene order, insertions and deletions, transposition and intergenomic transfer, including horizontal gene transfer across species (Bergthorsson et al. 2004) is enormous; it is sometimes termed fluidity of mt genome. In opposition to high genomic fluidity, the rate of nucleotide substitution in plant mt genes is slower than in chloroplast and nuclear genomes (Wolfe et al. 1987). An exception to this rule is a recent finding of elevated substitution rates in some lineages (*Pelargonium* – Palmer et al. 2000, *Plantago* – Cho et al. 2004, *Silene* – Mower et al. 2007). As no increase in sequence divergence of chloroplast and nuclear genes was observed in these lineages, rate acceleration was most likely attributed to the activity of DNA replication and repair enzymes in mitochondria causing high mutation rate (Mower et al. 2007; Sloan and Taylor 2010).

16.1.7 RNA Editing

The impact of substitution rate on mRNA coding capacity is modified by the process of RNA editing, which converts C to U (or less frequently U to C) (Yu and Schuster 1995). Editing affects protein-coding genes in plant organelles. It modifies non-synonymous codon sites more often than synonymous sites (Gray 2003) and contributes to the protein conservation across species (Covello and Gray 1989). In *Pelargonium*, highly elevated substitution rate was associated with dramatic loss of editing sites, which suggests that the two processes are interconnected (Parkinson et al. 2005). Another plant lineage with accelerated substitution rate – the genus *Silene* (Mower et al. 2007) provides the unique opportunity to study not only editing and mutation rate but more generally the relationship between mt genome evolution and breeding system, speciation or ecological requirements.

16.2 Structure of the mt Genome of a Plant Genus *Silene*

The plant genus *Silene* attracted attention of famous scientists of nineteenth century. Charles Darwin investigated *Silene* flower morphology and gender (Darwin 1877), Johann Gregor Mendel worked with dioecious *Silene latifolia* (Weiling 1991), blooming in his monastery backyard till recent days. *Silene* has become an emerging plant model to study evolution of sex chromosome, interaction between the plant and its fungal parasites, pollination, invasiveness, and the history of migration. *Silene* also maintains lot of variation in breeding system within the genus, making it a valubale comparative system (reviewed in Bernasconi et al. 2009).

16.2.1 mt Substitution Rate in Silene

High mt substitution rate described in *Silene vulgaris* (Mower et al. 2007) was exceeded by substitution rates found in *Silene noctiflora* (Mower et al. 2007) and *Silene conica* (Sloan et al. 2009). The rates estimated in the latter two species approach the levels measured in *Pelargonium* and *Plantago* (Mower et al. 2007). Similarly to *Pelargonium* and *Plantago* lineages, absolute mt synonymous substitution rate varies across two orders of magnitude among *Silene* species and is not accompanied by increased chloroplast or nuclear substitution rates (Sloan et al. 2009). In addition to among-species variation, rate variation among lineages within *Silene s*pecies was also observed (Barr et al. 2007; Sloan et al. 2008). Substitution rate in synonymous sites is not uniform across all the genes of the same mt genome, but varies substantially (Barr et al. 2007). The *atp9* gene evolved more than 40 times faster than *nad9* (Sloan et al. 2009). The reason for rate acceleration in some genes is not clear. Sloan et al. (2009) suggest that an increased mutation rate is responsible for rate divergence rather than horizontal gene transfer, balancing selection in favor of ancient gene variants, or transfer to the nucleus. Coexistence of multiple gene copies in the same individual (heteroplasmy), which was well documented in *S. vulgaris* (McCauley et al. 2005; Pearl et al. 2009), may also contribute to the high substitution rate in particular genes.

16.2.2 Polymorphism in mt Genes and Breeding System of Silene

Within-species polymorphism of mt genes is very high in some *Silene* species (Houliston and Olson 2006; Barr et al. 2007). As increased substitution rate may produce different gene variants, the question arises, whether within-species polymorphism correlates with the level of substitution rate or with some other phenomenon. Touzet and Delph (2009) demonstrated that it was the breeding system which was predictive of polymorphism. Gynodioecious species representing separated clades of the genus (*S. vulgaris*, *S. nutans* and *S. acaulis*) exhibited much more polymorphism in the *cob* (*cytochrome b*) and *cox1* genes than dioecious or hermaphroditic species of *Silene*, including *S. noctiflora* with very high substitution rate. The authors concluded that numerous haplotypes of the gynodioecious species are maintained over long evolutionary timescale owing to the balancing selection. This scenario was consistent with a previous study of *cob* polymorphism in *S. acaulis* (Stadler and Delph 2002).

16.2.3 Heteroplasmy and Paternal Transmission of Mitochondria in S. vulgaris

Earlier studies documented heteroplasmy, or within-individual diversity in mt genomes, under rather artificial conditions such as in plants regenerated from tissue cultures (Vitart et al. 1992) or nucleus–cytoplasm hybrids (Hattori et al. 2002). More recently, organellar heteroplasmy in both chloroplasts and mitochondria was recognized as a common phenomenon in natural populations of flowering plants (reviewed in Woloszynska 2010).

Gynodioecious species of *Silene* are very nice models to investigate the consequences of mt heteroplasmy owing to their high within-species polymorphism in mt markers (Touzet and Delph 2009). Multiple bands in Southern-RFLPs corresponding to the mt genes accompanied by various flanking regions and present in the same individual of *S. vulgaris* or *S. acaulis* were observed by Olson and McCauley (2000), Storchova and Olson (2004), Klass and Olson (2006). These studies also documented non-complete linkage disequilibrium between chloroplast and mitochondrial markers, which suggested deviation from strictly maternal transmission of these organelles.

Rare paternal transmission of mt genome in *S. vulgaris*, which resulted in heteroplasmy was described by McCauley et al. (2005). Quantification of *atp1* (encoding ATP synthase subunit 1) variants occurring in the same individual was achieved by qPCR (Welch et al. 2006). The rate of paternal transmission estimated by Pearl et al. (2009) in 18 natural populations of *S. vulgaris* reached 8%, albeit in most cases, paternal mitochondria represented just a minor portion of mt population of a heteroplasmic offspring individual. The rate of paternal leakage varied among populations and seemed to be dependent on the pollen donor (Bentley et al. 2010).

16.2.4 Heteroplasmy and Recombination in S. vulgaris

Heteroplasmy leads to the situation when recombination between two divergent mt genomes may occur. The observation of mt genotypes which might have originated due to the recombination was published by Stadler and Delph (2002) in *S. acaulis*, and by Houliston and Olson (2006) and McCauley and Ellis (2008) in *S. vulgaris*.

Thus, heteroplasmy facilitates intermolecular recombination between mt genomes. On the other hand, heteroplasmy may arise due to the asymmetric recombination across short repeats (Shedge et al. 2007), which creates new genomic configuration. Then, the portion of newly generated molecules may suddenly change in the course of substoichiometric shift (Small et al. 1987). The relationship between heteroplasmy and recombination is therefore reciprocal, heteroplasmy facilitates recombination, and recombination may lead to the heteroplasmy.

Whether heteroplasmy in *Silene* also originates owing to other processes than rare paternal transmission is not currently known. However, Elansary et al. (2010) observed within-individual and within-sibship variation in Southern-RFLPs, explicable by substoichiometric shifting. Complete sequencing of mt genomes in *Silene* is necessary to understand the extent of mt DNA rearrangements and its impact on genome structure, mt gene expression, and phenotype.

16.2.5 Complete mt Genome of **S. latifolia**

The first complete sequence of mt genome of *Silene* has been published recently (Sloan et al. 2010a), but additional complete genome sequences are expected to appear soon. High polymorphism in *S. vulgaris* and other gynodioecious species suggests that numerous divergent mt genomes exist in each species.

The mt genome of *S. latifolia* (Fig. 16.2) is small (253 kb) and contains the fewest genes of any plant mt genome sequenced so far. Extensive loss of the genes associated with translation was observed. For example, *S. latifolia* mt genome encoded tRNAs capable to translate only 17 of 61 sense codons (Sloan et al. 2010). Six copies of a >1 kb repeat were found to be at or near "recombinational equilibrium" with evidence for all 36 possible pairings of single copy sequences flanking the repeats. No chimeric ORF, which could be associated with a putative gynodioecious ancestor of dioecious *S. latifolia* was found.

Extensive loss of editing sites was observed in *S. latifolia* mt genome. The extent of loss of editing sites seems to be correlated with substitution rate, as it reached the highest degree in *S. noctiflora*, which also exhibited very high substitution rate in mt genome (Sloan et al. 2010b). Gene conversion with reverse-transcribed mRNA (termed retroprocessing) acting at short regions covering one or two editing sites was considered to be responsible for accelerated loss of editing sites in rapidly evolving *Silene* mt genomes (Sloan et al. 2010b).

16.2.6 Heteroplasmy and Phenotype in **S. vulgaris**

The coexistence of two or more divergent mt genomes in the same individual may influence its phenotype. Genetic evidence suggested that multiple CMS genes existed in *S. vulgaris* (Charlesworth and Laporte 1998; Olson and McCauley

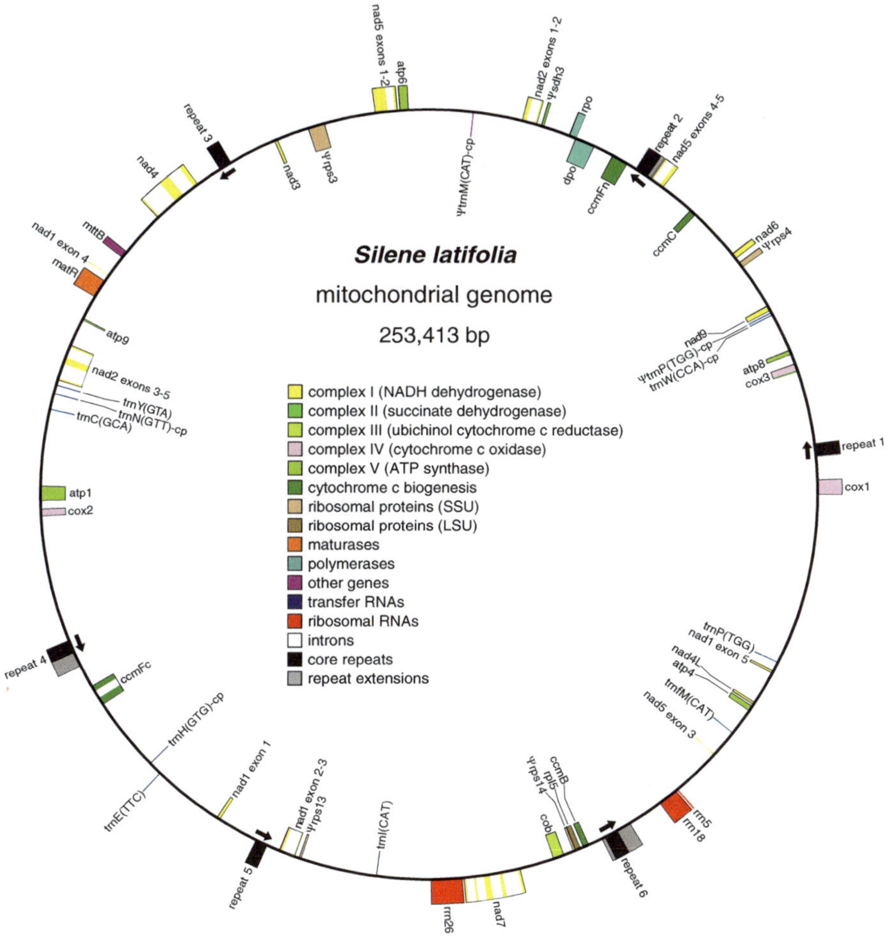

Fig. 16.2 The complete mt genome of *S. latifolia* belongs to the smallest plant mt genomes (From Sloan et al. 2010a)

2002) and *S. nutans* (Garraud et al. 2011). The presence of two different CMS genes restored by distinct restorers may influence the gender. Andersson (1999) observed female and hermaphroditic flowers on the same individual of *S. vulgaris*. The flowers produced offspring with different sex ratio after pollination by the same father. Andersson (1999) concluded that different CMS types prevailed in the branches producing flowers of different gender. The investigation of CMS heteroplasmy is hampered by the absence of sequence knowledge of CMS genes in *Silene*. The sequences of the first CMS candidate genes in *S. vulgaris* have been submitted (Storchova et al. under review), others will be identified soon after mt genomes of various haplotypes are completely sequenced.

Besides CMS determinants, heteroplasmy in the genes coding for essential enzymes involved in mt respiration may also influence plant phenotype. McCauley

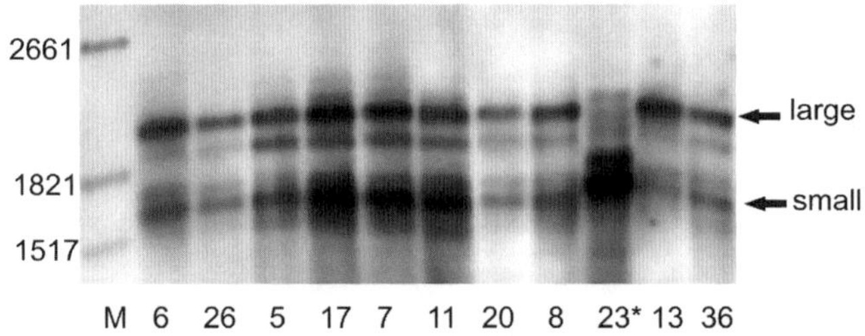

Fig. 16.3 The *atpI* transcription profile with two prominent bands detected by Northern was observed in a specific mt haplotype of *S. vulgaris*. The sibling individual (*23**) inherited mitochondria from a pollen donor and exhibited a different transcription profile (From Elansary et al. 2010)

et al. (2005) studied heteroplasmy in a coding region of the *atpI* gene. Information about the expression of particular gene variants will provide ideas on how respective copies participate in plant metabolism. Quantitative RT-PCR becomes a suitable tool for transcript level estimation, providing that the sequences of untranslated regions (UTRs) are available for the design of variant-specific primers.

The study of transcription patterns of mt genes in a natural population of *S. vulgaris* (Elansary et al. 2010) documented that high polymorphism in mt genome structure was accompanied by the high variation in transcript profiles of the *atpI* and *coxI* genes. For example, two bands corresponding to the transcripts derived from the *atpI* gene were found in Northerns performed with RNA of the specific mt haplotype (Fig. 16.3). Further studies showed that the transcription from an additional distant promoter was responsible for the larger one from the two transcripts. As various mt haplotypes differed in DNA regions located upstream of the *atpI* gene, genome configuration was predictive of the transcript pattern.

16.3 Conclusion

S. vulgaris and other gynodioecious species of *Silene* represent an ideal model for the investigation of transcription in plant mitochondria. The high natural polymorphism in gene flanking regions facilitates the study of *cis* elements important for transcription and/or translation. The vast variation in nuclear genes associated with worldwide distribution of some *Silene* species provides enough diversity to understand the role of nucleus-encoded proteins in mt gene expression. The only obstacle limiting the use of *Silene* as a model in research of plant mitochondria is a lack of reliable transformation protocol. However, there is a hope that it will be elaborated soon and *Silene* will become as useful and favored as *Arabidopsis*.

Acknowledgment I thank Daniel B Sloan from the University of Virginia, USA, for reading the manuscript and very helpful comments. Funding was graciously provided by the grants GA ČR number 521/09/0261 and MŠMT Kontakt ME09035.

References

Abdelnoor RV, Yule R, Elo A, Christensen AC, Meyer-Gauen G, Mackenzie SA (2003) Substoichiometric shifting in the plant mitochondrial genome is influenced by a gene homologous to *MutS*. Proc Natl Acad Sci USA 100:5968–5973

Adams KL, Palmer JD (2003) Evolution of mitochondrial gene content: gene loss and transfer to the nucleus. Mol Phylogenet Evol 29:380–395

Adams KL, Daley DO, Qiu YL, Whelan J, Palmer JD (2000) Repeated, recent and diverse transfers of a mitochondrial gene to the nucleus in flowering plants. Nature 408:354–357

Allen JO, Fauron CM, Minx P et al (2007) Comparisons among two fertile and three male sterile mitochondrial genomes of maize. Genetics 177:1173–1192

Alverson AJ, Wei XX, Rice DW, Stern DB, Barry K, Palmer JD (2010) Insights into the evolution of mitochondrial genome size from complete sequences of *Citrullus lanatus* and *Cucurbita pepo* (Cucurbitaceae). Mol Biol Evol 27:1436–1448

Andersson H (1999) Female and hermaphrodite flowers on a chimeric gynomonoecious *Silene vulgaris* plant produce offspring with different genders: a case of heteroplasmic sex determination? J Hered 90:563–565

Andres C, Lurin C, Small ID (2007) The multifarious roles of PPR proteins in plant mitochondrial gene expression. Physiol Plant 129:14–22

Arrieta-Montiel MP, Shedge V, Davila J, Christensen AC, Mackenzie SA (2009) Diversity of the *Arabidopsis* mitochondrial genome occurs via nuclear-controlled recombination activity. Genetics 183:1261–1268

Balk J, Leaver CJ (2001) The PET1-CMS mitochondrial mutation in sunflower is associated with premature programmed cell death and cytochrome c release. Plant Cell 13:1803–1818

Barr CM, Fishman L (2010) The nuclear component of a cytonuclear hybrid incompatibility in *Mimulus* maps to a cluster of pentatricopeptide repeat genes. Genetics 184:455, U198

Barr CM, Keller SR, Ingvarsson PI, Sloan DB, Taylor DR (2007) Variation in mutation rate and polymorphism among mitochondrial genes of *Silene vulgaris*. Mol Biol Evol 24:1783–1791

Bentley KE, Mandel JR, McCauley DE (2010) Paternal leakage and heteroplasmy of mitochondrial genomes in *Silene vulgaris*: evidence from experimental crosses. Genetics 185:961–968

Bentolila S, Alfonso AA, Hanson MR (2002) A pentatricopeptide repeat-containing gene restores fertility to cytoplasmic male-sterile plants. Proc Natl Acad Sci USA 99:10887–10892

Bergthorsson U, Richardson AO, Young GJ, Goertzen LR, Palmer JD (2004) Massive horizontal transfer of mitochondrial genes from diverse land plant donors to the basal angiosperm *Amborella*. Proc Natl Acad Sci USA 101:17747–17752

Bernasconi G, Antonovics J, Biere A, Charlesworth D, Delph LF, Filatov D, Giraud T, Hood ME, Marais GAB, McCauley DE, Pannell JR, Shykoff JA, Vyskot B, Wolfe L, Wimer A (2009) *Silene* as a model system in ecology and evolution. Heredity 103:5–14

Binder S, Marchfelder A, Brennicke A, Wissinger B (1992) RNA editing in transsplicing intron sequences of *nad2* messenger-RNAs in *Oenothera* mitochondria. J Biol Chem 267:7615–7623

Bonen L (2008) *Cis*- and *trans*-splicing of group II introns in plant mitochondria. Mitochondrion 8:26–34

Boore JL (1999) Animal mitochondrial genomes. Nucleic Acids Res 27:1767–1780

Cappadocia L, Marechal A, Parent JS, Lepage E, Sygusch J, Brisson N (2010) Crystal structures of DNA-Whirly complexes and their role in *Arabidopsis* organelle genome repair. Plant Cell 22:1849–1867

Chapdelaine Y, Bonen L (1991) The wheat mitochondrial gene for subunit I of the NADH dehydrogenase complex: a transsplicing model for this gene-in-pieces. Cell 65:465–472

Charlesworth D, Laporte V (1998) The male-sterility polymorphism of *Silene vulgaris*: analysis of genetic data from two populations and comparison with *Thymus vulgaris*. Genetics 150:1267–1282

Chaw SM, Shih ACC, Wang D, Wu YW, Liu SM, Chou TY (2008) The mitochondrial genome of the gymnosperm *Cycas taitungensis* contains a novel family of short interspersed elements, Bpu sequences, and abundant RNA editing sites. Mol Biol Evol 25:603–615

Cho Y, Mower JP, Qiu YL, Palmer JD (2004) Mitochondrial substitution rates are extraordinarily elevated and variable in a genus of flowering plants. Proc Natl Acad Sci USA 101: 17741–17746

Covello PS, Gray MW (1989) RNA editing in plant mitochondria. Nature 341:662–666

Culligan KM, Meyer-Gauen G, Lyons-Weiler J, Hays JB (2000) Evolutionary origin, diversification and specialization of eukaryotic MutS homolog mismatch repair proteins. Nucleic Acids Res 28:463–471

Cummings DJ, McNally KL, Domenico JM, Matsuura ET (1990) The complete DNA sequence of the mitochondrial genome of *Podospora anserina*. Curr Genet 17:375–402

Darwin CR (1877) The different forms of flowers on plants of the same species. Murray, London

Dewey RE, Levings CS, Timothy DH (1986) Novel recombinations in the maize mitochondrial genome produce a unique transcriptional unit in the Texas male-sterile cytoplasm. Cell 44:439–449

Elansary HO, Müller K, Olson MS, Storchova H (2010) Transcription profiles of mitochondrial genes correlate with mitochondrial DNA haplotypes in a natural population of *Silene vulgaris*. BMC Plant Biol 10:11

Feng X, Kaur AP, Mackenizie SA, Dweikat IM (2009) Substoichiometric shifting in the fertility reversion of cytoplasmic male sterile pearl millet. Theor Appl Genet 118:1361–1370

Forner J, Weber B, Wietholter C, Meyer RC, Binder S (2005) Distant sequences determine 5′ end formation of *cox3* transcripts in *Arabidopsis thaliana* ecotype C24. Nucleic Acids Res 33:4673–4682

Forner J, Weber B, Thuss S, Wildum S, Binder S (2007) Mapping of mitochondrial mRNA termini in *Arabidopsis thaliana*: t-elements contribute to 5′ and 3′ end formation. Nucleic Acids Res 35:3676–3692

Forner J, Hölzle A, Jonietz C, Thuss S, Schwarzländer M, Weber B, Meyer RC, Binder S (2008) Mitochondrial mRNA polymorphisms in different *Arabidopsis* accessions. Plant Physiol 148:1106–1116

Garraud C, Brachi B, Dufay M, Touzet P, Shykoff JA (2011) Genetic determination of male sterility in gynodioecious *Silene nutans*. Heredity 106:757–764. doi:10.1038/hdy.2010.116

Giege P, Hoffmann M, Binder S, Brennicke A (2000) RNA degradation buffers asymmetries of trancription in *Arabidopsis* mitochondria. EMBO Rep 1:164–170

Gillman JD, Bentolila S, Hanson MR (2007) The petunia restorer of fertility protein is part of a large mitochondrial complex that interacts with transcripts of the CMS-associated locus. Plant J 49:217–227

Goremykin VV, Salamini F, Velasco R, Viola R (2009) Mitochondrial DNA of *Vitis vinifera* and the issue of rampant horizontal gene transfer. Mol Biol Evol 26:99–110

Gouyon PH, Couvet D (1987) A conflict between two sexes, females and hermaphrodites. In: Stearns SC (ed) The evolution of sex and its consequences. Birkhauser, Basel, pp 245–261

Gray MW (2003) Diversity and evolution of mitochondrial RNA editing systems. IUBMB Life 55:227–233

Handa H (2008) Linear plasmids in plant mitochondria: peaceful coexistences or malicious invasions? Mitochondrion 8:15–25

Handa H, Itani K, Sato H (2002) Structural features and expression analysis of a linear mitochondrial plasmid in rapeseed (*Brassica napus L.*). Mol Genet Genomics 267:797–805

Hanson MR, Bentolila S (2004) Interactions of mitochondrial and nuclear genes that affect male gametophyte development. Plant Cell 16:S154–S169

Hattori N, Kitigawa K, Takumi S, Nakamura C (2002) Mitochondrial DNA heteroplasmy in wheat, *Aegilops* and their nucleus-cytoplasm hybrids. Genetics 160:1619–1630

Hedgcoth C, El-Shehawi AM, Wei P, Clarkson M, Tamalis D (2002) A chimeric open redaing frame associated with cytoplasmic male sterility in alloplasmic wheat with *Triticum timopheevi* mitochondria is present in several *Triticum* and *Aegilops* species, barley, and rye. Curr Genet 41:357–365

Hedtke B, Legen J, Weihe A, Herrmann RG, Borner T (2002) Six active phage-type RNA polymerase genes in *Nicotiana tabacum*. Plant J 30:625–637

Houliston GJ, Olson MS (2006) Nonneutral evolution of organelle genes in *Silene vulgaris*. Genetics 174:1983–1994

Ingvarsson PK, Taylor DR (2002) Genealogical evidence for epidemics of selfish genes. Proc Natl Acad Sci USA 99:11265–11269

Janska H, Sarria R, Woloszynska M, Arrieta-Montiel M, Mackenzie SA (1998) Stoichiometric shifts in the common bean mitochondrial genome leading to male sterility and spontaneous reversion to fertility. Plant Cell 10:1163–1180

Kim DH, Kang JG, Kim BD (2007) Isolation and characterization of the cytoplasmic male sterility-associated orf456 gene of chilli pepper (*Capsicum annuum* L.). Plant Mol Biol 63:519–532

Klass AL, Olson MS (2006) Spatial distributions of cytoplasmic types and sex expression in Alaskan populations of *Silene acaulis*. Int J Plant Sci 167:179–189

Knoop V, Schuster W, Wissinger B, Brennicke A (1991) Transsplicing integrates an exon of 22 nucleotides into the *nad5* messenger-RNA in higher-plant mitochondria. EMBO J 10:3483–3493

Kubo T, Nishizawa S, Mikami T (1999) Alterations in organization and transcription of the mitochondrial genome of cytoplasmic male sterile sugar beet (*Beta vulgaris L.*). Mol Gen Genet 262:283–290

Kubo T, Nishizawa S, Sugawara A, Itchoda N, Estiati A, Mikami T (2000) The complete nucleotide sequence of the mitochondrial genome of sugar beet (*Beta vulgaris L.*) reveals a novel gene for tRNA(Cys)(GCA). Nucleic Acids Res 28:2571–2576

Kühn K, Weihe A, Börner T (2005) Multiple promoters are a common feature of mitochondrial genes in Arabidopsis. Nucleic Acids Res 33:337–346

Lee J, Young JPW (2009) The mitochondrial genome sequence of the arbuscular mycorrhizal fungus *Glomus intraradices* isolate 494 and implications for the phylogenetic placement of Glomus. New Phytol 183:200–211

Leino M, Landgren M, Glimelius K (2005) Alloplasmic effects on mitochondrial transcriptional activity and RNA turnover result in accumulated transcripts of *Arabidopsis* orfs in cytoplasmic male-sterile *Brassica napus*. Plant J 42:469–480

Lupold DS, Caoile AGFS, Stern DB (1999) The maize mitochondrial *cox2* gene has five promoters in two genomic regions, including a complex promoter consisting of seven overlapping units. J Biol Chem 274:3897–3903

Mackenzie SA (2007) The unique biology of mitochondrial genome instability in plants. In: Logan D (ed) Plant mitochondria. Blackwell Publishing, Oxford, pp 36–46

Malek O, Knoop V (1998) Trans-splicing group II introns in plant mitochondria: the complete set of cis-arranged homologs in ferns, fern allies, and a hornwort. RNA 4:1599–1609

Malek O, Brennicke A, Knoop V (1997) Evolution of trans-splicing plant mitochondrial introns in pre-Permian times. Proc Natl Acad Sci USA 94:553–558

Manchekar M, Scissum-Gunn KD, Song DQ, Khazi F, McLean SL, Nielsen BL (2006) DNA recombination activity in soybean mitochondria. J Mol Biol 356:288–299

Manchekar M, Scissum-Gunn KD, Hammett LA, Backert S, Nielsen BL (2009) Mitochondrial DNA recombination in *Brassica campestris*. Plant Sci 177:629–635

Marechal A, Brisson N (2010) Recombination and the maintenance of plant organelle genome stability. New Phytol 186:299–317

McCauley DE, Ellis JR (2008) Recombination and linkage disequilibrium among mitochondrial genes in structured populations of the gynodioecious plant *Silene vulgaris*. Evolution 62:823–832

McCauley DE, Olson MS (2008) Do recent findings in plant mitochondrial molecular and population genetics have implications for the study of gynodioecy and cytonuclear conflict? Evolution 62:1013–1025

McCauley DE, Olson MS, Emery SN, Taylor DR (2000) Population structure influences sex ratio evolution in gynodioecious plant. Am Nat 155:814–819

McCauley DE, Bailey MF, Sherman NA, Darnell MZ (2005) Evidence for paternal transmission and heteroplasmy in the mitochondrial genome of *Silene vulgaris*, a gynodioecious plant. Heredity 95:50–58

Mellado RP, Penalva MA, Inciarte MR, Salas M (1980) The protein covalently linked to the 5′ termini of the DNA of *Bacillus subtilis* phage phi-29 is involved in the initiation of DNA-replication. Virology 104:84–96
Mower JP, Touzet P, Gummow JS, Delph LF, Palmer JD (2007) Extensive variation in synonymous substitution rates in mitochondrial genes of seed plants. BMC Evol Biol 7:135
Moyekens CA, Mackenzie SA, Shoemaker RC (1995) Mitochondrial genome diversity in soybean: repeats and rearrangements. Plant Mol Biol 29:245–254
Newton KJ, Mariano JM, Gibson CM, Kuzmin E, Gabay-Laughnan S (1996) Involvement of S2 episomal sequences in the generation of NCS4 deletion mutation in maize mitochondria. Dev Genet 19:277–286
Oldenburg DJ, Bendich J (1996) Size and structure of replicating mitochondrial DNA in cultured tobacco cells. Plant Cell 8:447–461
Olson MS, McCauley DE (2000) Linkage disequilibrium and phylogenetic congruence between chloroplast and mitochondrial haplotypes in *Silene vulgaris*. ProcR Soc Lond B 267: 1801–1808
Olson MS, McCauley DE (2002) Mitochondrial DNA diversity, population structure, and gender association in the gynodioecious plant *Silene vulgaris*. Evolution 56:253–262
Palmer JD, Shields CR (1984) Tripartite structure of the *Brassica-campestris* mitochondrial genome. Nature 307:437–440
Palmer JD, Adams KL, Cho Y, Parkinson CL, Qiu YL, Song K (2000) Dynamic evolution of plant mitochondrial genomes: mobile genes and introns and highly variable mutation rates. Proc Natl Acad Sci 97:6960–6966
Parkinson CL, Mower JP, Qiu YL, Shirk AJ, Song K, Young ND, dePamphilis CW, Palmer JD (2005) Multiple major increases and decreases in mitochondrial substitution rates in the plant family Geraniaceae. BMC Evol Biol 5:73
Pearl SA, Welch ME, McCauley DE (2009) Mitochondrial heteroplasmy and paternal leakage in natural populations of *Silene vulgaris*, a gynodioecious plant. Mol Biol Evol 26:537–545
Qiu YL, Palmer JD (2004) Many independent origins of trans splicing of a plant mitochondrial group II intron. J Mol Evol 59:80–89
Richards AJ (1997) Plant breeding systems. Chapman and Hall, London
Robison MM, Wolyn DJ (2005) A mitochondrial plasmid and plasmid-like RNA and DNA polymerases encoded within the mitochondrial genome of carrot (*Daucus carota L.*). Curr Genet 47:57–66
Saumitou-Laprade P, Pannenbecker G, Boutin-Stadler V, Michaelis G, Vernet P (1991) Plastid DNA diversity in natural populations of *Beta maritima* showing additional variation in sexual phenotype and mitochondrial DNA. Theor Appl Genet 81:533–536
Schmitz-Linneweber C, Small I (2008) Pentatricopeptide repeat proteins: a socket set for organelle gene expression. Trends Plant Sci 13:663–670
Shedge V, Arrieta-Montiel M, Christensen AC, Mackenzie SA (2007) Plant mitochondrial recombination surveillance requires unusual *RecA* and *MutS* homologs. Plant Cell 19:1–14
Shedge V, Davile J, Arrieta-Montiel MP, Mohammed S, Mackenzie SA (2010) Extensive rearrangement of the *Arabidopsis* mitochondrial genome elicits cellular conditions for thermotolerance. Plant Physiol 152:1960–1970
Sloan DB, Taylor DR (2010) Testing for selection on synonymous sites in plant mitochondrial DNA: the role of codon bias and RNA editing. J Mol Evol 70:479–491
Sloan DB, Barr CM, Olson MS, Keller SR, Taylor DR (2008) Evolutionary rate variation at multiple levels of biological organization in plant mitochondrial DNA. Mol Biol Evol 25:243–246
Sloan DB, Oxelman B, Rautenberg A, Taylor DR (2009) Phylogenetic analysis of mitochondrial substitution rate variation in the angiosperm tribe Sileneae. BMC Evol Biol 9:260
Sloan DB, Alverson AJ, Storchova H, Palmer JD, Taylor DR (2010a) Extensive loss of translational genes in the structurally dynamic mitochondrial genome of the angiosperm *Silene latifolia*. BMC Evol Biol 10:274

Sloan DB, MacQueen AH, Alverson AJ, Palmer JD, Taylor DR (2010b) Extensive loss of RNA editing sites in rapidly evolving *Silene* mitochondrial genomes: selection vs. retroprocessing as the driving force. Genetics 185:1369–1380

Small I, Isaac P, Leaver C (1987) Stoichiometric differences in DNA molecules containing the *atpA* gene suggest mechanisms for the generation of mitochondrial genome diversity in maize. EMBO J 6:865–869

Small I, Suffolk R, Leaver CJ (1989) Evolution of plant mitochondrial genomes via substoichiometric intermedates. Cell 58:69–76

Stadler T, Delph LF (2002) Ancient mitochondrial haplotypes and evidence for intragenic recombination in a gynodioecious plant. Proc Natl Acad Sci USA 99:11730–11735

Storchova H, Olson MS (2004) Comparison between mitochondrial and chloroplast DNA variation in the native range of *Silene vulgaris*. Mol Ecol 13:2909–2919

Storchova H, Müller K, Lau S, Olson MS Mosaic origins of a complex chimeric mitochondrial gene in *Silene vulgaris*. Under review

Tian XJ, Zheng J, Hu SN, Yu J (2006) The rice mitochondrial genomes and their variations. Plant Physiol 140:401–410

Toro N, Jimenez-Zurdo JI, Garcia-Rodriguez FM (2007) Bacterial group II introns: not just splicing. FEMS Microbiol Rev 31:342–358

Touzet P, Delph LF (2009) The effect of breeding system on polymorphism in mitochondrial genes of *Silene*. Genetics 181:631–644

Unseld M, Marienfeld JR, Brandt B, Brennicke A (1997) The mitochondrial genome of *Arabidopsis thaliana* contains 57 genes in 366,924 nucleotides. Nat Genet 15:57–61

Vitart V, DePaeper R, Mathieu C, Chetrit P, Vedel F (1992) Amplification of substoichiometric recombinant mitochondrial DNA sequences in a nuclear, male sterile mutant regenerated from protoplast culture in *Nicotiana sylvestris*. Mol Gen Genet 233:193–200

Wang Z, Zou Y, Li X, Zhang Q, Chen L, Wu H, Su D, Chen Y, Guo J, Luo D, Zhong Y, Liu YG (2006) Cytoplasmic male sterility of rice with Boro II cytoplasm is caused by a cytotoxic peptide and is restored by two related PPR motif genes via distinct modes of mRNA silencing. Plant Cell 18:676–687

Ward BL, Anderson RS, Bendich AJ (1981) The mitochondrial genome is large and variable in a family of plants (Cucurbitaceae). Cell 25:793–803

Weiling F (1991) Historical study: Johann Gregor Mendel 1822–1884. Am J Med Genet 40:1–25

Welch ME, Darnell MZ, McCauley DE (2006) Variable populations within variable populations: quantifying mitochondrial heteroplasmy in natural populations of the gynodioecious plant *Silene vulgaris*. Genetics 174:829–837

Wolfe KH, Li WH, Sharp PM (1987) Rates of nucleotide substitution vary greatly among plant mitochondrial, chloroplast, and nuclear DNAs. Proc Natl Acad Sci USA 84:9054–9058

Woloszynska M (2010) Heteroplasmy and stoichiometric complexity of plant mitochondrial genomes – though this be madness, yet there's method in't. J Exp Bot 61:657–671

Woloszynska M, Trojanowski D (2009) Counting mtDNA molecules in *Phaseolus vulgaris*: sublimons are constantly produced by recombination via short repeats and undergo rigorous selection during substoichiometric shifting. Plant Mol Biol 70:511–521

Yu W, Schuster W (1995) Evidence for a site specific cytidine deamination reaction involved in C to U RNA editing of plant mitochondria. J Biol Chem 270:18227–18233

Zaegel V, Guermann B, Le Ret M, Andres C, Meyer D, Erhardt M, Canaday J, Gualberto JM, Imbault P (2006) The plant-specific ssDNA binding protein OSB1 is involved in the stoichiometric transmission of mitochondrial DNA in *Arabidopsis*. Plant Cell 18:3548–3563

Zhang QY, Liu YG (2006) Rice mitochondrial genes are transcribed by multiple promoters that are highly diverged. J Integr Plant Biol 48:1473–1477

Chapter 17
Evolutionary Dynamics and Genomic Impact of Prokaryote Transposable Elements

Nicolas Cerveau, Sébastien Leclercq, Didier Bouchon, and Richard Cordaux

Abstract Transposable elements (TEs) are one of the major forces that drive prokaryote genome evolution. Analyses of TE evolutionary dynamics revealed extensive variability in TE density between prokaryote genomes, even closely related ones. To explain this variability, a model of recurrent invasion/proliferation/extinction cycles has been proposed. In this chapter, we examine different parameters that influence these cycles in two of the simplest TE classes: insertion sequences and group II introns. In particular, we discuss TE transposition efficiency (mechanisms and regulation), ability to transfer horizontally (through plasmids and phages), and impact on genome evolution (gene activation/inactivation and structural variation). Finally, we describe TE dynamics in bacterial endosymbionts, especially in *Wolbachia*, to illustrate the importance of host population size in prokaryote TE evolution.

17.1 Introduction

Mobile genetic elements are one of the major forces that drive genome evolution in all living organisms. Among mobile genetic elements, transposable elements (TEs) can be defined as elements able to move from one genomic location to another. While TEs sometimes represent a large fraction of eukaryote genomes (up to 80% in maize), they generally do not account for more than a few percent of prokaryote genomes (Siguier et al. 2006). In this chapter, we focus on two of the simplest prokaryotic TEs, namely, insertion sequences (IS) and group II introns.

IS elements range in size from 0.8 to 2.5 kb and encode a transposase (Tpase) protein allowing mobility (Chandler and Mahillon 2002) (Fig. 17.1). IS are

N. Cerveau and S. Leclercq are co-first authors of the chapter.

N. Cerveau • S. Leclercq • D. Bouchon • R. Cordaux
Université de Poitiers, UMR CNRS 6556 Ecologie Evolution Symbiose, 40 Avenue du Recteur Pineau, 86022 Poitiers, France
e-mail: richard.cordaux@univ-poitiers.fr

P. Pontarotti (ed.), *Evolutionary Biology – Concepts, Biodiversity, Macroevolution and Genome Evolution*, DOI 10.1007/978-3-642-20763-1_17,

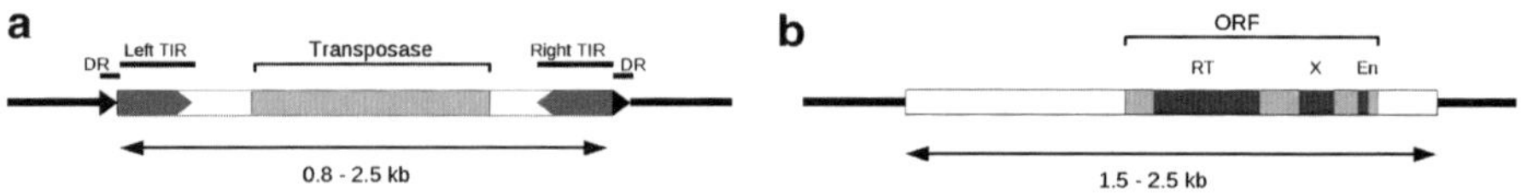

Fig. 17.1 Schematic representation of an insertion sequence (**a**) and a group II intron (**b**). *DR* direct repeat, *TIR* terminal inverted repeat, *ORF* open reading frame, *RT* reverse transcriptase, *X* maturase, *EN* endonuclease (lacking in several group II introns). Flanking sequences are shown in *black*. Drawings are not to scale

typically bounded by terminal inverted repeats (TIRs) ranging in size from 10 to 40 bp, which are recognized and bound by the Tpase during the transposition process. Most IS elements create 2–8 bp long direct repeats when inserting in a new genomic location. They are divided in around 20 families (Chandler and Mahillon 2002). Bacterial group II introns are 1.5–2.5 kb long elements, which generally encode a multi-domain protein promoting self-splicing of the element and reintegration into another genomic location (Fig. 17.1). They are distributed in nine major classes (Lambowitz and Zimmerly 2010). Contrary to IS elements which use a DNA intermediate during their transposition, group II introns use an RNA intermediate with a typical 6-domain secondary structure.

The recent sequencing of hundreds of genomes revealed that IS and group II introns are not uniformly distributed among prokaryotes (Touchon and Rocha 2007; Leclercq et al. 2011). This trend holds true even at the strain-level scale (Sawyer et al. 1987; Tourasse and Kolsto 2008; Leclercq et al. 2011; Qiu et al. 2010). The basis of this variability depends on the underlying TE evolutionary dynamics. Here, we summarize several aspects of TE dynamics, including mechanisms of transpositional activity and ability to horizontally transfer, which are essential for TE invasion and propagation. Next, we focus on TE genomic impact and on the selective pressures, which determine TE maintenance or loss in prokaryotic genomes.

17.2 Mobility of Prokaryote Transposable Elements

17.2.1 IS Transposition

17.2.1.1 Transposition Mechanisms

Excellent reviews on IS transposition mechanisms have already been published (Chandler and Mahillon 2002; Curcio and Derbyshire 2003). Briefly, IS transposition typically starts with the transcription of the Tpase gene. Most IS elements are autonomous as they carry their own promoter. After translation, the resulting Tpase binds to the TIRs of the target IS element, and cleaves both DNA strands at the 5′ and 3′ ends of the element to physically excise the IS element from the donor

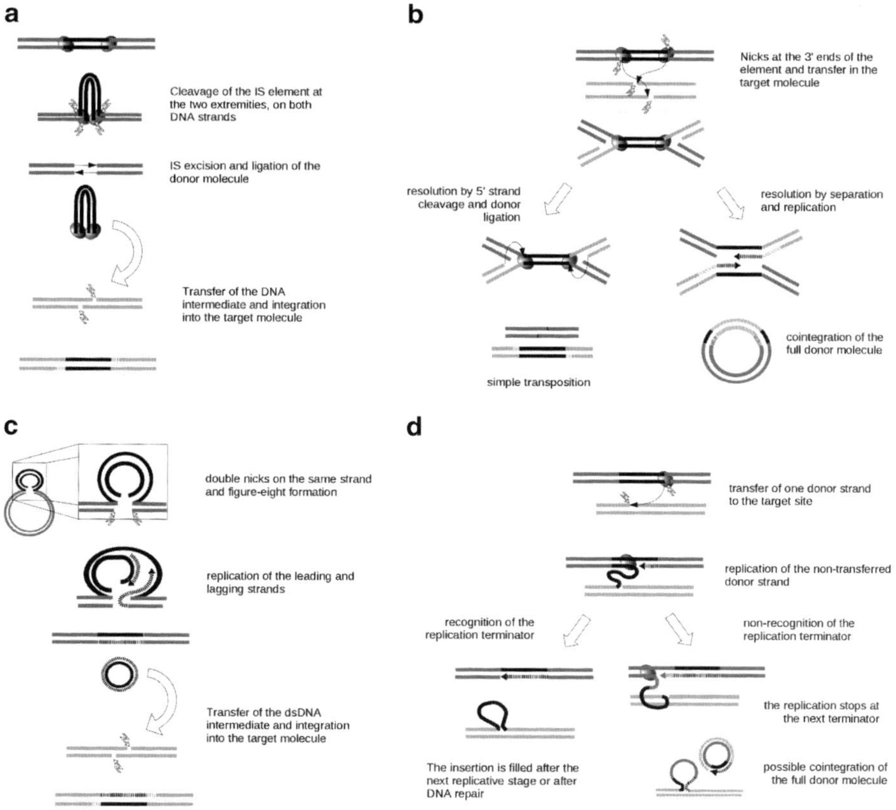

Fig. 17.2 Major IS transposition mechanisms: classical excision-based transposition (**a**), cointegrative transposition (**b**), figure-eight transposition (**c**), and rolling-circle transposition (**d**). IS DNA is represented in *black*, the donor molecule in *dark gray*, and the target molecule in *light gray*. Newly synthetized DNA is hatched. *Shaded ovals* represent Tpase proteins, and *sparks* represent cleavage events

location (Fig. 17.2a). The linear excised IS element forms a synaptic complex with the Tpase, which targets a new genomic location and promotes element integration. This pathway normally does not increase IS copy number in the genome, and is referred to as non-replicative transposition. However, if transposition occurs during replication, the IS element can excise from one newly replicated molecule and insert in the other (or at a position not yet replicated), thus leading to duplication of the element in the target molecule.

Other IS elements, such as those of the IS1 and IS6 families, use an alternative pathway called cointegrative transposition, in which a single DNA strand at the 5′ and 3′ ends of the element is cleaved. The element is thus not excised when it is integrated into the new genomic location, resulting in a covalent association between the donor and the target locations (Fig. 17.2b). The second strand at the 5′ and 3′ ends can then be cleaved, which leads to the simple transfer of the element from the donor position to the target position. The covalent association can also be resolved by IS strand separation and replication. In this case, the IS element is duplicated at the new genomic location and not excised from the donor location. This replicative process also leads to the integration of the whole donor molecule within the target molecule in a case of an intermolecular transfer, and to genomic inversion in the case of an intramolecular transfer.

Another replicative pathway, termed “figure-eight” transposition, was identified for IS911 elements and uses a circularized DNA intermediate (Duval-Valentin et al. 2004). Tpase binding creates specific molecular figure-eight structures, which are resolved by replication of the IS element using the host replication machinery (Fig. 17.2c). The circularized replicated IS DNA sequence can be inserted into a new genomic location. This transposition pathway is probably used by members of the IS3 family, such as IS2 and IS3, which produce circularized DNA intermediates (Lewis and Grindley 1997; Ohtsubo et al. 2004). Contrary to cointegrative transposition, figure-eight transposition just duplicates the IS element without integrating the donor molecule in the target molecule.

Rolling-circle transposition is another mechanism used by elements such as IS91, in which a single-stranded IS molecule is transferred to the target molecule, and concurrently replicated in the donor molecule (Garcillan-Barcia et al. 2002) (Fig. 17.2d). Replication stops at a replication terminator located at the IS termini, leading to the duplication of the IS element. However, the whole cointegration of the donor molecule sometimes happens when the host replication system does not recognize the replication terminator and performs several rounds of replication.

The four major transposition mechanisms discussed above are the most documented pathways, but other more atypical pathways have been described or probably remain to be discovered.

17.2.1.2 Control of IS Transposition

IS elements use several strategies to regulate their transposition, presumably to limit their negative effect on the host genome, as detailed in Nagy and Chandler (2004). For example, the production of a fully active Tpase may be conditional on a ribosome frameshift during translation (Escoubas et al. 1991; Vögele et al. 1991; Lewis and Grindley 1997). This reduces transposition activity by half in the IS3 family (Vögele et al. 1991) and by more than 99% for IS1 and IS2 elements (Escoubas et al. 1991; Lewis and Grindley 1997). In other cases, transposition may be regulated through impinging transcription, i.e., sequestering of the Tpase translation initiation site in a secondary structure of the mRNA, generally induced by inverted repeat sequences (Beuzon et al. 1999). This hinders the ribosomal complex to initiate Tpase translation.

Other elements, such as IS10 and IS50, carry Dam sites that may be methylated, leading to transcriptional inactivation (Roberts et al. 1985; Tomcsanyi and Berg 1989). During genome replication, methylated sites become hemimethylated, leading to the reactivation of the IS element. This suggests that such IS elements increase their transposition rate during the replication phase of their host genome (Roberts et al. 1985; Dodson and Berg 1989). Moreover, as explained above, these IS elements use a non-replicative transposition pathway and they may benefit from DNA replication to promote their duplication. Thus, replication-induced activity could be an efficient evolutionary strategy for proliferation of these elements. Another example of intrinsic regulation is given by the recently described IS608 elements. They are excised only as single-stranded DNA. Thus, they can transpose only when the DNA molecule is opened, i.e., at replication forks (Ton-Hoang et al. 2010) or during repair after DNA fragmentation (Pasternak et al. 2010).

Transposition may also be limited by the availability of free insertion sites. Although most IS families are able to insert at nonspecific positions, some families show nonrandom insertion patterns. Insertion generally targets 2–5 bp DNA sequences (Chandler and Mahillon 2002). Target insertion sites are sometimes much more restrictive, as for elements of the IS30 family, which precisely insert in a ~25-bp long palindromic sequence that resembles their own TIRs (Olasz et al. 1998; Kiss et al. 2007). Lack of this specific motif in a genome is a strong limitation for insertion, as exemplified by *Salmonella typhimurium*, which naturally lacks the IS30 target site and shows a very low IS30 integration rate (Casadesus et al. 1999). When the relevant insertion site was experimentally added, IS30 transposition greatly increased. This demonstrates that the site specificity of IS insertions is a key factor for IS invasion and proliferation in bacterial genomes.

17.2.2 Group II Intron Mobility

Contrary to IS elements, group II introns transpose via an RNA-intermediate, which necessarily leads to element duplication. They do not carry transcription promoters, so they must be inserted in a transcribed region to be active. Mobility starts with the transcription of the region containing the intron. The intron-encoded protein (IEP) produced from the intron mRNA binds to the intron ribozyme to form a ribonucleoprotein (RNP) complex and catalyzes intron self-splicing. The remaining mRNA is religated and can then be normally translated, while the RNP complex targets a new genomic location to insert the intron mRNA via reverse-splicing/reverse-transcription mechanisms, as reviewed in Toro et al. (2007); Lambowitz and Zimmerly (2010). Many group II introns insert directly within double-stranded DNA using the endonuclease activity of their IEP to open one DNA strand. However, some group II introns lack the endonuclease domain and are thought to insert in single-stranded DNA at replication forks.

Group II intron mobility is often called retrohoming, as introns were primarily observed to target intronless alleles of orthologous genes, defined as homing sites. Intron insertions are highly site-specific and require a conserved region of approximatively 30 bp. This very stringent insertion capacity presumably relies on the need of the intron ribozyme to bind to specific DNA motifs to be correctly spliced. As intron survival depends on its splicing ability (when inserted within genes and to promote mobility), it is more relevant to target genomic locations that may promote intron transcription and activity (Mohr et al. 2010). Insertions at ectopic sites, i.e., at genomic locations with limited similarity with the typical insertion site motif were also observed (Cousineau et al. 2000; Martinez-Abarca and Toro 2000). These events occur with a much lower frequency but they are believed to contribute to group II intron proliferation, by ultimately diversifying insertion sites (Leclercq et al. 2011; Mohr et al. 2010). Finally, some group II intron families preferentially target structural regions rather than nucleotide motifs, such as class C introns, which insert downstream of transcriptional terminators (Robart et al. 2007) and the Avi.GroEL group II intron and relatives, which insert at or near initiation/stop codons (Michel et al. 2007).

Regulation of group II intron mobility is poorly documented, except site specificity and the need to be inserted in a transcribed region to be active. It was recently observed, though, that environmental conditions, such as temperature, may affect mobility of natural group II introns (Mohr et al. 2010).

17.2.3 Horizontal Transfers

TE survival and evolutionary success in bacteria is intimately linked to their ability to spread through horizontal transfers (HT). HT can be unambiguously detected when two divergent bacteria share identical or almost identical TEs. Evidence for HT within bacterial genera has been reported for both IS elements (Lawrence et al. 1992; Bisercic and Ochman 1993; Wagner and de la Chaux 2008) and group II introns (Dai and Zimmerly 2002; Fernandez-Lopez et al. 2005; Tourasse and Kolsto 2008; Leclercq et al. 2011). A typical example is provided by ISWpi1 elements in *Wolbachia*. Average nucleotide divergence between copies in 22 different strains is only of 0.22%, while the average divergence between highly conserved housekeeping genes for the same strains is ~3.7% (Cordaux et al. 2008).

More ancient HT can also be detected by comparing presence/absence patterns of IS families or intron classes in a set of prokaryote genomes and the phylogenetic relationships of these prokaryotes. Using this method, it was inferred that no more than 30 detectable HT are needed to explain the distribution of the 20 most abundant IS elements in 450 fully sequenced bacterial genomes (Wagner and de la Chaux 2008). HT events were unequally distributed among nine IS families, with seven HT inferred for the IS1 family, while other IS families displayed only one HT.

IS elements are commonly viewed as vectors for genetic exchanges between bacterial strains because they can form composite transposons that may carry virulence, resistance, or metabolic genes (Toussaint and Merlin 2002). However, IS elements and composite transposons, just like group II introns, lack the genetic material enabling HT between bacterial cells, and consequently, they are unable to perform HT by themselves (Toussaint and Merlin 2002). They need to shuttle via larger mobile elements, such as plasmids and bacteriophages, to be horizontally transferred (Frost et al. 2005).

17.2.3.1 Plasmid-Mediated Transfers

TE shuttling via plasmids has long been proposed, as IS elements and group II introns are recurrently detected in plasmid sequences of prokaryote species (Hall et al. 1989; Ng et al. 1998; Sundin 2007), in which they sometimes represent more than half of the detected open reading frames (ORF), as in *Shigella* plasmids (Venkatesan et al. 2001).

The first step for a plasmid-mediated HT is the transposition of the TE from a chromosomal location to a plasmid location. This was demonstrated in vitro and in vivo with mobility assays for IS (Schwartz et al. 1988; Wilde et al. 2003) and group II introns (Martinez-Abarca and Toro 2000; Ichiyanagi et al. 2003). Chromosome-to-plasmid transposition also occurs in natural populations, as exemplified in *Bacillus subtilis*. Many strains of *B. subtilis* carry multiple ISBsu2 copies in their chromosome and some strains also carry two plasmids only differentiated by an ISBu2 insertion (Poluektova et al. 2002).

After HT of a TE-containing plasmid from one bacterial cell to another by conjugation (Frost et al. 2005), the TE must move from the plasmid to the bacterial chromosome. This transfer can be achieved through several ways. Direct transfer through transposition is the most obvious possibility, and it is used in IS and group II intron mobility assays in vitro and in vivo (Vögele et al. 1991; Olasz et al. 1998; Cousineau et al. 2000). TEs can also be transferred to the host chromosome indirectly via the integration of genomic islands, also known as conjugative transposons, which are DNA regions containing virulence or adaptive genes and occasionally TEs (Mullany et al. 1996; Burrus et al. 2002). Finally, integrative plasmids are able to fully integrate into host chromosomes (Burrus et al. 2002), leading to concomitant integration of plasmid-borne TEs.

17.2.3.2 Phage-Mediated Transfers

HT through bacteriophages is also commonly assumed for TEs but far less documented than HT through plasmids. IS elements and group II introns are frequently found in prophage sequences, i.e., silent phages integrated in the host genome, but infrequently in active bacteriophage sequences. For example, several IS and group II intron copies are found in WO prophages of *Wolbachia* genomes (Leclercq et al. 2011; unpublished results). However, no intron and only one IS element is inserted in the genome of the active WO phage sequenced by Tanaka et al. (2009). When present, IS in active phages are found in only one or two copies, and they often are defective (Lobocka et al. 2004; Creuzburg et al. 2005). One exception is the C neurotoxin-converting phage of *Clostridium botulinum*, which contains 12 IS elements belonging to 7 families (Sakaguchi et al. 2005). To our knowledge, no intact group II intron has been reported to date in an active bacteriophage sequence.

17.3 Genomic Impact of Prokaryote Transposable Elements

IS elements and group II introns may impact genomic instability and variation in various ways, which can be classified in two major categories: (1) insertional mutagenesis, which refers to genomic consequences that directly follow insertions at novel genomic sites, and (2) structural variation, which refers to genomic consequences that take place at a post-insertional stage and are coupled with cell mechanisms like recombination. These two major types of genomic impact mediated by TEs are well illustrated by studies of experimental evolution. For example, analyses of two *Escherichia coli* populations after 10,000 generations of growth on glucose minimal medium led to the identification of several IS-associated mutations (Schneider et al. 2000), such as four IS150 copies disrupting genes and two additional IS150 copies, which had recombined, resulting in the inversion of the intervening sequence.

17.3.1 Insertional Mutagenesis

17.3.1.1 Coding Sequence Disruption

Gene inactivation by TE insertion, particularly IS, is very frequent in bacterial genomes. A study performed on *E. coli* demonstrated that the mutational spectrum of a reporter gene is largely linked to IS insertions (60% of the mutants) that disrupt the ORF (Rodriguez et al. 1992). Beyond studies using experimental systems, many cases of specific genes interrupted by IS have been reported to naturally occur. For example, the study of an unusual case of nonmobile and nonpathogenic strain of *Rickettsia peacockii* led to the discovery that two genes are disrupted by IS insertions (Simser et al. 2005). One disrupted gene is involved in actin-tail polymerization and the second gene is suspected to be involved in cell adhesion and bacterial virulence. These observations provided new insight into pathogenesis mechanisms in these bacteria.

While TE insertion in a coding gene is often thought to inactivate the gene through introduction of a premature stop codon resulting in truncated, nonfunctional proteins, this is not always the case. For example, IS insertion in the gene coding the ribosomal protein S1 of *E. coli* did not preclude protein production, although it lacked the last of six imperfect repeats (Skorski et al. 2007). Growth of the *E. coli* strain expressing the smaller S1 protein was lower compared to the wild type. However, growth delay is not due to the absence of the final repeat. Indeed, experimental introduction of an early stop codon that suppressed the final repeat in the wild-type S1 gene had no impact on bacterial growth. Instead, it was the presence of the IS in the S1 gene that probably created an unnatural 3′ end, which favored exonuclease degradation and induced growth delay.

The increasing availability of bacterial genomes provides the opportunity to more systematically track IS-mediated gene disruptions. For example, analysis of *Anabaena* sp. strain PCC 7120 genome identified 145 IS divided into several families (Wolk et al. 2010). More than 20% of these IS are inserted in putative protein-coding genes. By contrast, a systematic analysis of the pseudogenes in the *Sodalis glossinidius* genome revealed that only 18 out of 1,051 pseudogenes were associated with IS insertions, suggesting that IS elements are not a significant source of gene disruption in this species (Belda et al. 2010).

In comparison with IS elements, there are far fewer examples of gene inactivation mediated by group II introns. This is at least partly attributable to their splicing ability, which restores normal ORFs at the mRNA level. Furthermore, group II introns are less prevalent than IS elements in bacterial genomes, so less likely to disrupt genes (Touchon and Rocha 2007; Leclercq et al. 2011). One case of gene disrupted by a group IIC intron has been reported in *Geobacillus stearothermophilus* (Moretz and Lampson 2010). Group IIC introns generally target transcription terminators and thus avoid the disruption of host genes. Consistently, all but one of the 20 intron copies identified in *G. stearothermophilus* strain 10 are inserted in transcription terminators. The remaining copy is inserted in the rRNA methylase gene. Experiments were performed to detect splicing of the intron copy without success (Moretz and Lampson 2010). Thus, the authors suspected that this intron is able to splice in vivo to avoid blocking of methylase synthesis, which is an essential gene for the bacterial host.

Wolbachia is one of the few bacterial organisms for which information on the mutagenic potential of both IS and group II introns is currently available. Three of the four completely sequenced *Wolbachia* genomes harbor genes interrupted by IS and intron insertions: up to 45 genes are disrupted by IS in the *Wolbachia* *w*Mel, *w*Ri, and *w*Pel genomes (Wu et al. 2004; Klasson et al. 2008; Klasson et al. 2009), and 12 of 18 introns detected in these *Wolbachia* genomes are inserted in conserved genes (Leclercq et al. 2011). By contrast, the genome devoid of IS- and intron-disrupted genes is also the only one that lacks introns and potentially functional IS copies (Foster et al. 2005; Cordaux 2009; Leclercq et al. 2011).

TE insertions in genes being mainly deleterious, they are widely used as experimental tools to identify gene function or biosynthesis pathways (Reznikoff 2008). For example, the last unknown gene of the histidine biosynthesis pathway of *Corynebacterium glutamicum* was identified by random IS6100 insertional mutagenesis (Mormann et al. 2006). IS6100 had initially been described in a *C. glutamicum* plasmid and experimentally shown to be potentially active in this species using a transposition assay (Tauch et al. 2002). IS6100 was subsequently used to create a transposon mutant library of *C. glutamicum*. One mutant exhibited a histidine-auxotrophic phenotype. Analysis of the IS6100-disrupted gene in the mutant revealed that it encoded an L-histidinol-phosphate phosphatase. This example nicely illustrates the usefulness of IS elements as experimental tools.

17.3.1.2 Impact on Gene Expression

In the previous section, we discussed consequences of IS and group II intron insertions in the coding sequences of bacterial genomes. In this section, we focus on TE insertions in non-coding regulatory regions.

Some IS elements carry transcriptional promoters (Chandler and Mahillon 2002; Nagy and Chandler 2004). Thus, IS insertions can radically change expression levels of neighboring genes. For example, the internal promoter of IS3 has been shown to activate *argE* transcription in *E. coli* (Charlier et al. 1982). More recently, it was shown that glycerol use is modified by an IS insertion in *E. coli* (Zhang and Saier 2009). Expression of essential proteins for glycerol use (encoded by the *glpFK* operon) is normally activated by the cyclic AMP receptor protein encoded by the *crp* gene. However, expression of the *glpFK* operon was found to be activated by an IS5 insertion in mutant strains lacking *crp* (Zhang and Saier 2009). IS5 partial truncation experiments further demonstrated that only a short sequence is fully responsible for the activation of the *glpFK* operon. Thus, IS insertions followed by degradation can lead to the creation of new bacterial promoters.

IS insertions cannot only activate gene and operon expression, but they can also increase expression levels of already expressed genes. For example, the virulence level of group B *Streptococcus*, which mainly causes neonatal sepsis and meningitis, is linked to an IS1548 insertion in the *scpB-lmb* intergenic region (Al Safadi et al. 2010). This leads to overexpression of the *lmb* gene, which encodes laminin, a surface protein that probably plays a crucial role in binding and invasion of different host surfaces. Consequently, laminin-binding ability is increased and its density on cell surface rises, which results in the induction of neonatal meningitis. In another example, *ampC* gene transcription and β-lactamase protein production were increased by 20-fold following an IS2 insertion in the *ampC* promoter of *E. coli* (Jaurin and Normark 1983). In this case, the increase was not due to the internal IS2 promoter, but rather to a cryptic −35 box-like sequence, which became activated following IS2 insertion in a configuration restoring an optimal distance between this cryptic −35 box and the endogenous −10 box of the *ampC* gene.

IS insertions can also interact with gene expression by inactivating their repression. For example, the expression of *SrpABC*, a gene encoding multidrug efflux pump, was derepressed by IS insertions in *Pseudomonas putida*. It was shown that the vast majority of *P. putida* strains are able to resist to 1% toluene shock, resulting from IS*S12* insertion-mediated inactivation of *SrpS*, which is a *SrpABC* repressor (Wery et al. 2001). In another example, extended incubation of a nonmotile *E. coli* strain led to discrimination of two motile subpopulations harboring an IS5 insertion at one of two different sites in the *flhD* operon promoter region, which is the master operon of the flagellar regulon (Barker et al. 2004). The IS5 insertions did not alter the transcriptional start site of the operon. In addition, they cannot activate *flhd* operon transcription because they are inserted in opposite orientation relative to the operon. Thus, the authors suggested a disturbance of transcriptional repression due to IS5 insertions (Barker et al. 2004).

Other effects on gene expression involve transcriptional attenuation. Transcriptional terminators were recently identified in IS elements (Naville and Gautheret 2010). Two types of terminators were identified: (1) terminators located upstream of the Tpase gene, which could limit IS proliferation, and (2) terminators located in IS-borne sequences and immediately upstream of cellular genes. Many IS-related terminators are conserved, suggesting that they may have an important impact on genome evolution.

17.3.2 Impact on Genomic Structural Variation

17.3.2.1 Genome Size Variation

IS elements generally represent less than ~3% of prokaryote genomes (Siguier et al. 2006). However, IS elements sometimes cover more than 10% of the genome, as in *Sulfolobus solfataricus*, *Orientia tsutsugamushi*, and *Wolbachia* (Brügger et al. 2004; Nakayama et al. 2008; unpublished results). Overall, genome size is positively correlated with IS number in prokaryote genomes (Touchon and Rocha 2007). This correlation is also suggested in *Wolbachia* genomes (Fig. 17.3). This may not be so surprising because these bacterial endosymbionts have reduced genomes in the Mb size range and high IS densities (48–118 copies/Mb, unpublished results), as compared to other bacterial genomes in which IS density is usually closer to ~3.5 IS copies/Mb on average (Touchon and Rocha 2007).

17.3.2.2 Genomic Rearrangements

The presence of multiple TE copies in a genome may induce a variety of changes on chromosomal structure. TEs are generally of recent origin in bacterial genomes

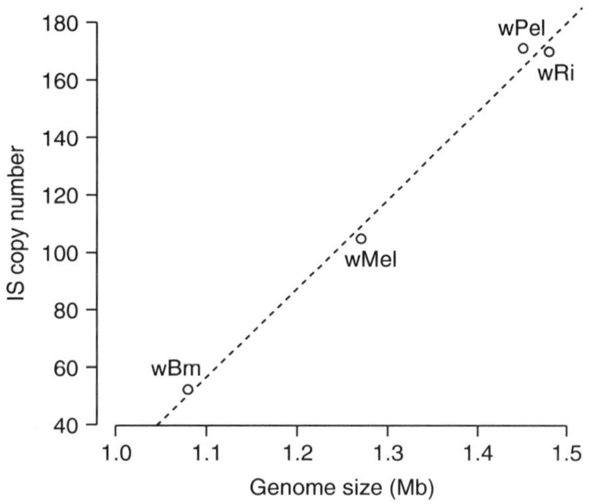

Fig. 17.3 Scatter plot of IS copy number according to genome size in the four completely sequenced *Wolbachia* genomes (*empty circles*). The dotted line indicates the correlation between genome size and IS copy number

(Wagner 2006), and their high sequence similarity makes them ideal substrates for ectopic (nonhomologous) recombination events. In addition, sequence homogeneity between copies within genomes may be maintained by gene conversion, as described for IS elements in *Wolbachia w*Bm (Cordaux 2009).

Sequence inversions are one of the potential genome rearrangements mediated by recombination between TE sequences. Inversions may result from recombination between the two TIRs of a single IS copy (Ling and Cordaux 2010), but they are more easily detected as recombination events between IS copies. For example, the *sgaA* mutant in *E. coli* has been shown to result from a rearrangement between two IS5 copies located ~60 kb apart from each other (Zinser et al. 2003). The IS5-mediated genomic inversion in the *sgaA* mutant relocated the *ybeJ* operon under the control of the promoter of another gene. As a result, the *sgaA* mutant was able to grow with aspartate and grow faster with glutamate, asparagine, and proline as carbon sources compared to the wild-type strain.

Deletions can also be mediated by IS elements. Growth of *E. coli* on glucose minimal medium during 2,000 generations resulted in rapid and parallel losses of D-ribose catabolism function (Cooper et al. 2001). The mutation frequency between the wild type and the mutant was $\sim 5.1 \times 10^{-5}$ per cell generation. PCR investigation of the *rbs* operon in 11 strains demonstrated that a fragment of variable size of the operon was deleted in all strains. Interestingly, an IS150 insertion constituted a boundary shared by all deletions. The authors suggested that the deletion process involved a second IS150 insertion at various sites in all strains, followed by recombination events between the IS150 pairs.

In the last decade, the number of available genome sequences dramatically increased. Comparative analyses have allowed quantifying IS-mediated rearrangements. For example, the comparison of *Rickettsia peacockii* and *R. rickettsii* genomes highlighted a lack of synteny, which is associated with the recent presence of 42 ISRpe1 copies in the *R. peacockii* genome (while ISRpe1 is lacking in *R. rickettsii*) (Felsheim et al. 2009). It turns out that 84% of syntenic block rearrangements are associated with ISRpe1 and 71% of genomic deletions are flanked by at least one ISRpe1 copy. These IS-mediated rearrangements might have deeply impacted *R. peacockii* lifestyle, as *R. rickettsii* is virulent whereas *R. peacockii* is nonpathogenic (Felsheim et al. 2009). Similar patterns of chromosomal rearrangements linked to IS elements have been reported in other genomes, such as *Bordetella* (Parkhill et al. 2003) and *Shigella* (Yang et al. 2005). *Wolbachia* genomes do not infringe the rule. Comparison between the two closely related strains *w*Mel and *w*Ri identified 35 gene-order breakpoints, half of which are flanked by IS elements (Klasson et al. 2009). Comparison of *Wolbachia* genomes also provided evidence for the implication of group II introns in genomic rearrangements, notably inversions (Leclercq et al. 2011). This is in contrast with the traditional view that group II introns are rarely involved in recombination events (Tourasse and Kolsto 2008). Thus, TEs may have an important impact on prokaryote chromosomal architecture.

17.4 Transposable Element Evolutionary Dynamics in Prokaryote Genomes

In the previous sections, we described mechanisms of TE mobility and impact on bacterial genome integrity. However, these properties cannot fully explain TE abundance in bacteria without considering population effects. Theoretical models of TE dynamics have been developed, involving positive or negative selection, horizontal transfers, and/or genetic drift (Sawyer et al. 1987; Basten and Moody 1991; Bichsel et al. 2010). How do these parameters influence TE spread and maintenance in bacteria?

17.4.1 Positive Selection

The most obvious way for a TE to be maintained in a genome is to be beneficial enough to the host to be positively selected. Several cases of gene activation following IS insertions have been shown to increase host fitness. For instance, the *fucAO* operon in *E. coli* became constitutively expressed after an IS5 insertion, allowing growth on propanediol substrates (Chen et al. 1989). Also, IS-mediated disruption of gene repressors can increase gene expression in a beneficial way, as in the case of the multidrug efflux pump gene repressor disrupted by IS*Ppu21* in *Pseudomonas putida*, which resulted in a 17,000-fold higher resistance of the mutant to a solvent stress compared to the wild type (Sun and Dennis 2009). Positive selection of IS elements was also observed in studies of experimental evolution of *E. coli* populations, in which IS insertions interrupted host genes and conferred selective advantages in minimal-glucose environment or under conditions of freezing/thawing cycles (Cooper et al. 2001; Sleight et al. 2008).

In a more general context, TEs are often considered as promoters of adaptability, via the various genomic rearrangements they may induce (Naas et al. 1994; Yang et al. 2005; Felsheim et al. 2009). A nice example is an inversion of a genomic fragment located between two IS5 copies in *E. coli* that induced a change in an operon expression pattern (Zinser et al. 2003) (see Sect. 17.3.2.2 above). This led to faster growth on diverse substrates, but reduced fitness during starvation. TE proliferation might thus be positively selected in bacteria living in highly unstable environments. However, the exact beneficial effect of TE-mediated recombination and resulting genomic variability on adaptability is still a matter of great debate (Dale and Moran 2006; Rocha 2008). Indeed, most bacteria for which high TE-mediated recombination levels were reported have small population sizes, thus allowing the fixation of slightly deleterious recombination events because of genetic drift and not necessarily because they are beneficial (see Sect. 17.4.4 below).

17.4.2 Negative Selection

Despite several cases of positively selected IS insertions, TE effect on bacterial genomes is thought to be mainly deleterious. Indeed, our vision of selection underestimates TE deleteriousness due to several biases. First, one of the major drawbacks of experimental evolution studies is that they proceed in controlled environments. Consequently, IS-interrupted genes may not be considered as essential although they could be in natural environments. Observed fitness increases may thus reflect a deregulation only sustainable in the experimental environment. A second and more important bias is that we cannot detect lethal insertions in genes. Similarly, sublethal transposition events are quickly discarded by natural selection and they are difficult to observe in natural prokaryote genomes. As protein-coding genes generally cover at least 80% of bacterial genomes (Lawrence et al. 2001), deleterious insertions in these genomic regions are expected to occur frequently. Consistently, cases of TE deleterious activity are regularly reported. For example, IS carry transcriptional terminators that negatively influence downstream gene expression (Naville and Gautheret 2010), and they can also induce mRNA destabilization (Skorski et al. 2007). Interestingly, some IS elements and group II introns have been found to specifically insert in other TEs, which may be a way to attenuate their deleteriousness in other genomic regions (Copertino and Hallick 1991; Olasz et al. 1998; Fernandez-Lopez et al. 2005). Overall, negative selection is an important evolutionary force in TE dynamics.

17.4.3 The Influence of Genetic Drift

In addition to TE intrinsic properties and deleteriousness, host effective population size is another critical factor in TE dynamics because it determines the efficiency of natural selection. Most bacterial species are assumed to have huge population sizes, at least large enough to rapidly discard most deleterious mutations. In this case, the spread of TE copies within genomes is strongly limited by negative selection, leading to the relatively reduced TE abundance generally observed in free-living bacteria (Moran and Plague 2004; Touchon and Rocha 2007). However, a reduction in effective population size leads to decreased efficiency of selection on slightly deleterious mutations, as well as increased fixation probability because of enhanced genetic drift. This can ultimately lead to an increase in TE copies. There is ample empirical evidence for this population genetics prediction, e.g., *Pyrococcus*, a thermophilic archaeon with a fragmented habitat (Escobar-Paramo et al. 2005), and recent pathogenic bacteria of human populations, cultured plants, and domesticated animals (Mira et al. 2006).

17.4.4 Cycles of Invasions and Extinctions

Deleteriousness raises the question of TE persistence in prokaryote genomes. Analyses of IS elements among related strains of given species usually show very high similarity between copies within strains, and striking differences in copy number between closely related strains (Sawyer et al. 1987; Lawrence et al. 1992; Parkhill et al. 2003). This suggests that IS elements are not stably maintained in bacterial genomes, but they rather undergo periodic phases of invasion, spread, and extinction (Wagner 2006). Interestingly, our data on *Wolbachia* IS elements provide direct evidence for such long-term dynamics (unpublished results).

The extinction–recolonization hypothesis is based on a balance between HT rate, transposition rate, and strength of natural selection. HT is a major driving force of bacterial genome evolution (Gogarten and Townsend 2005). Indeed, it has been estimated that up to 17% of bacterial genome sequences were acquired by recent HT (Ochman et al. 2000). Specifically, the HT rate depends on TE ability to invade vectors such as plasmids or phages (see Sect. 17.2.3 above). However, genes acquired by HT, and particularly TEs, are eliminated more rapidly than core genes (Fuxelius et al. 2008). Transposition rates are directly linked to transposition mechanisms (replicative vs. non-replicative), transposition regulation, and insertion site availability (see Sect. 17.2.1 above). Thus, the issue of selection is critical in this model. Hence, some authors consider that at least some TE copies must be beneficial to the host for elements to be maintained (Blot 1994; Schneider and Lenski 2004). Yet, recent models suggest that TE invasion and persistence in bacterial genomes does not necessarily require positive selection (Bichsel et al. 2010).

Interestingly, the extinction–recolonization hypothesis seems to also apply to group II intron dynamics (Tourasse and Kolsto 2008; Leclercq et al. 2011), despite their splicing ability that should minimize their deleteriousness. In fact, the deleterious effect of group II introns could arise from less-than-100% splicing efficiency (Chillon et al. 2011), potential loss of splicing activity after inactivating mutations, or because of their recombinational power (Leclercq et al. 2011).

17.4.5 Transposable Elements in Bacterial Endosymbionts

Bacterial endosymbionts nicely illustrate the different forces that play on TE dynamics. Indeed, the first step of endosymbiosis, i.e., the shift from a free-living lifestyle to intracellularity, is generally linked to a sharp increase in TE density (Moran and Plague 2004). Such TE proliferations reflect drastic reductions in effective population size and ensuing relaxed selection and enhanced genetic drift. Effective population size reduction is initially caused by the founding effect during which a bacterial population becomes intracellular, and subsequently accentuated by the recurrent bottlenecks undergone at each cellular transmission

(Mira and Moran 2002; Moran and Plague 2004; Wernegreen 2005). Subsequently, TEs are slowly degraded, but recolonization is prevented by the cellular confinement, which limits HT. Finally, long-term endosymbionts are completely depleted in TEs, as illustrated by the genomes of mutualistic endosymbionts such as *Buchnera aphidicola* (Shigenobu et al. 2000). TE loss in long-term endosymbionts can be attributed to genomic reduction, which corresponds to a sharp decrease in genome size, by loosing all nonessential genes, including mobile genetic elements (Wernegreen 2002).

Interestingly, *Wolbachia* endosymbionts harbor many characteristics of long-term endosymbionts (i.e., small genome, host dependence). Yet, their genomes are littered with IS elements, group II introns and phages, a good part of which is of recent origin (Cordaux 2008; Cordaux et al. 2008; Kent and Bordenstein 2010; Leclercq et al. 2011; unpublished results). Such TE dynamics is analogous to that observed in recently host-associated bacteria. It is probably linked to *Wolbachia* ability to switch hosts, leading to coinfections and breaking intracellular confinement (Cordaux et al. 2001). This may facilitate DNA exchange and import of new TEs between strains (Bordenstein and Reznikoff 2005).

17.5 Conclusion

A complex interplay between activity and HT, genomic impact and various selective pressures shapes TE dynamics in prokaryote genomes. These observations suggest a model in which TEs are recurrently acquired and rapidly lost. By contrast, eukaryote genomes often contain more ancient TE sequences, thereby offering a perspective on TE dynamics at broader evolutionary timescales (Kapitonov and Jurka 2003; Han et al. 2005; Cordaux and Batzer 2009). The fast-growing number of available bacterial genomes will certainly provide the opportunity to further our understanding of prokaryote TE dynamics at broader evolutionary timescales.

References

Al Safadi R, Amor S, Hery-Arnaud G, Spellerberg B, Lanotte P, Mereghetti L, Gannier F, Quentin R, Rosenau A (2010) Enhanced expression of lmb gene encoding laminin-binding protein in *Streptococcus agalactiae* strains harboring IS1548 in scpB-lmb intergenic region. PLoS ONE 5:e10794

Barker CS, Pruss BM, Matsumura P (2004) Increased motility of *Escherichia coli* by insertion sequence element integration into the regulatory region of the flhD operon. J Bacteriol 186:7529–7537

Basten CJ, Moody ME (1991) A branching-process model for the evolution of transposable elements incorporating selection. J Math Biol 29:743–761

Belda E, As M, Bentley S, Silva FJ (2010) Mobile genetic element proliferation and gene inactivation impact over the genome structure and metabolic capabilities of *Sodalis glossinidius*, the secondary endosymbiont of tsetse flies. BMC Genomics 11:449

Beuzon CR, Marquas S, Casadesus J (1999) Repression of IS200 transposase synthesis by RNA secondary structures. Nucleic Acids Res 27:3690–3695

Bichsel M, Barbour AD, Wagner A (2010) The early phase of a bacterial insertion sequence infection. Theor Popul Biol 78:278–288

Bisercic M, Ochman H (1993) The ancestry of insertion sequences common to *Escherichia coli* and *Salmonella typhimurium*. J Bacteriol 175:7863–7868

Blot M (1994) Transposable elements and adaptation of host bacteria. Genetica 93:5–12

Bordenstein SR, Reznikoff WS (2005) Mobile DNA in obligate intracellular bacteria. Nat Rev Microbiol 3:688–699

Brügger K, Torarinsson E, Redder P, Chen L, Garrett RA (2004) Shuffling of *Sulfolobus* genomes by autonomous and non-autonomous mobile elements. Biochem Soc Trans 32:179–183

Burrus V, Pavlovic G, Decaris B, Guedon G (2002) Conjugative transposons: the tip of the iceberg. Mol Microbiol 46:601–610

Casadesus J, Naas T, Garzon A, Arini A, Torreblanca J, Arber W (1999) Lack of hotspot targets: a constraint for IS30 transposition in *Salmonella*. Gene 238:231–239

Chandler M, Mahillon J (2002) Insertion sequences revisited. In: Craig NL et al (eds) Mobile DNA II. ASM Press, Washington, DC, pp 305–366

Charlier D, Piette J, Glansdorff N (1982) IS3 can function as a mobile promoter in *E. coli*. Nucleic Acids Res 10:5935–5948

Chen YM, Lu Z, Lin EC (1989) Constitutive activation of the fucAO operon and silencing of the divergently transcribed fucPIK operon by an IS5 element in *Escherichia coli* mutants selected for growth on L-1,2-propanediol. J Bacteriol 171:6097–6105

Chillon I, Martinez-Abarca F, Toro N (2011) Splicing of the *Sinorhizobium meliloti* RmInt1 group II intron provides evidence of retroelement behavior. Nucleic Acids Res 39:1095–1104

Cooper VS, Schneider D, Blot M, Lenski RE (2001) Mechanisms causing rapid and parallel losses of ribose catabolism in evolving populations of *Escherichia coli* B. J Bacteriol 183:2834–2841

Copertino DW, Hallick RB (1991) Group II twintron: an intron within an intron in a chloroplast cytochrome b-559 gene. EMBO J 10:433–442

Cordaux R (2008) ISWpi1 from *Wolbachia pipientis* defines a novel group of insertion sequences within the IS5 family. Gene 409:20–27

Cordaux R (2009) Gene conversion maintains nonfunctional transposable elements in an obligate mutualistic endosymbiont. Mol Biol Evol 26:1679–1682

Cordaux R, Batzer MA (2009) The impact of retrotransposons on human genome evolution. Nat Rev Genet 10:691–703

Cordaux R, Michel-Salzat A, Bouchon D (2001) *Wolbachia* infections in crustaceans: novel hosts and potential routes for horizontal transmission. J Evol Biol 14:237–243

Cordaux R, Pichon S, Ling A, Perez P, Delaunay C, Vavre F, Bouchon D, Greve P (2008) Intense transpositional activity of insertion sequences in an ancient obligate endosymbiont. Mol Biol Evol 25:1889–1896

Cousineau B, Lawrence S, Smith D, Belfort M (2000) Retrotransposition of a bacterial group II intron. Nature 404:1018–1021

Creuzburg K, Jr R, Kuhle V, Herold S, Hensel M, Schmidt H (2005) The Shiga toxin 1-converting bacteriophage BP-4795 encodes an NleA-like type III effector protein. J Bacteriol 187:8494–8498

Curcio MJ, Derbyshire KM (2003) The outs and ins of transposition: from mu to kangaroo. Nat Rev Mol Cell Biol 4:865–877

Dai L, Zimmerly S (2002) The dispersal of five group II introns among natural populations of *Escherichia coli*. RNA 8:1294–1307

Dale C, Moran NA (2006) Molecular interactions between bacterial symbionts and their hosts. Cell 126:453–465

Dodson KW, Berg DE (1989) Factors affecting transposition activity of IS50 and Tn5 ends. Gene 76:207–213

Duval-Valentin G, Marty-Cointin B, Chandler M (2004) Requirement of IS911 replication before integration defines a new bacterial transposition pathway. EMBO J 23:3897–3906

Escobar-Paramo P, Ghosh S, DiRuggiero J (2005) Evidence for genetic drift in the diversification of a geographically isolated population of the hyperthermophilic archaeon *Pyrococcus*. Mol Biol Evol 22:2297–2303

Escoubas JM, Prere MF, Fayet O, Salvignol I, Galas D, Zerbib D, Chandler M (1991) Translational control of transposition activity of the bacterial insertion sequence IS1. EMBO J 10:705–712

Felsheim RF, Kurtti TJ, Munderloh UG (2009) Genome sequence of the endosymbiont *Rickettsia peacockii* and comparison with virulent *Rickettsia rickettsii*: identification of virulence factors. PLoS ONE 4:e8361

Fernandez-Lopez M, Munoz-Adelantado E, Gillis M, Willems A, Toro N (2005) Dispersal and evolution of the *Sinorhizobium meliloti* group II RmInt1 intron in bacteria that interact with plants. Mol Biol Evol 22:1518–1528

Foster J, Ganatra M, Kamal I, Ware J, Makarova K, Ivanova N, Bhattacharyya A, Kapatral V, Kumar S, Posfai J, Vincze T, Ingram J, Moran L, Lapidus A, Omelchenko M, Kyrpides N, Ghedin E, Wang S, Goltsman E, Joukov V, Ostrovskaya O, Tsukerman K, Mazur M, Comb D, Koonin E, Slatko B (2005) The *Wolbachia* genome of *Brugia malayi*: endosymbiont evolution within a human pathogenic nematode. PLoS Biol 3:e121

Frost LS, Leplae R, Summers AO, Toussaint A (2005) Mobile genetic elements: the agents of open source evolution. Nat Rev Microbiol 3:722–732

Fuxelius H-H, Darby AC, Cho N-H, Andersson SGE (2008) Visualization of pseudogenes in intracellular bacteria reveals the different tracks to gene destruction. Genome Biol 9:R42

Garcillan-Barcia MP, Bernales I, Mendiola MV, De la Cruz F (2002) IS91 Rolling-circle transposition. In: Craig NL et al (eds) Mobile DNA II. ASM Press, Washington, DC, pp 891–904

Gogarten JP, Townsend JP (2005) Horizontal gene transfer, genome innovation and evolution. Nat Rev Microbiol 3:679–687

Hall BG, Parker LL, Betts PW, DuBose RF, Sawyer SA, Hartl DL (1989) IS103, a new insertion element in *Escherichia coli*: characterization and distribution in natural populations. Genetics 121:423–431

Han K, Xing J, Wang H, Hedges DJ, Garber RK, Cordaux R, Batzer MA (2005) Under the genomic radar: the stealth model of Alu amplification. Genome Res 15:655–664

Ichiyanagi K, Beauregard A, Belfort M (2003) A bacterial group II intron favors retrotransposition into plasmid targets. Proc Natl Acad Sci USA 100:15742–15747

Jaurin B, Normark S (1983) Insertion of IS2 creates a novel ampC promoter in *Escherichia coli*. Cell 32:809–816

Kapitonov VV, Jurka J (2003) Molecular paleontology of transposable elements in the *Drosophila melanogaster* genome. Proc Natl Acad Sci USA 100:6569–6574

Kent BN, Bordenstein SR (2010) Phage WO of *Wolbachia*: lambda of the endosymbiont world. Trends Microbiol 18:173–181

Kiss J, Nagy Z, Toth G, Kiss G, Jakab J, Chandler M, Olasz F (2007) Transposition and target specificity of the typical IS30 family element IS1655 from *Neisseria meningitidis*. Mol Microbiol 63:1731–1747

Klasson L, Walker T, Sebaihia M, Sanders MJ, Quail MA, Lord A, Sanders S, Earl J, O'Neill SL, Thomson N, Sinkins SP, Parkhill J (2008) Genome evolution of *Wolbachia* strain wPip from the *Culex pipiens* group. Mol Biol Evol 25:1877–1887

Klasson L, Westberg J, Sapountzis P, Naslund K, Lutnaes Y, Darby AC, Veneti Z, Chen L, Braig HR, Garrett R, Bourtzis K, Andersson SGE (2009) The mosaic genome structure of the *Wolbachia* wRi strain infecting *Drosophila simulans*. Proc Natl Acad Sci USA 106:5725–5730

Lambowitz AM, Zimmerly S (2010) Group II introns: mobile ribozymes that invade DNA. Cold Spring Harb Perspect Biol. doi:10.1101/cshperspect.a003616

Lawrence JG, Ochman H, Hartl DL (1992) The evolution of insertion sequences within enteric bacteria. Genetics 131:9–20

Lawrence JG, Hendrix RW, Casjens S (2001) Where are the pseudogenes in bacterial genomes? Trends Microbiol 9:535–540

Leclercq S, Giraud I, Cordaux R (2011) Remarkable abundance and evolution of mobile group II introns in *Wolbachia* bacterial endosymbionts. Mol Biol Evol 28:685–697

Lewis LA, Grindley ND (1997) Two abundant intramolecular transposition products, resulting from reactions initiated at a single end, suggest that IS2 transposes by an unconventional pathway. Mol Microbiol 25:517–529

Ling A, Cordaux R (2010) Insertion sequence inversions mediated by ectopic recombination between terminal inverted repeats. PLoS ONE 5:e15654

Lobocka MB, Rose DJ, Plunkett G, Rusin M, Samojedny A, Lehnherr H, Yarmolinsky MB, Blattner FR (2004) Genome of bacteriophage P1. J Bacteriol 186:7032–7068

Martinez-Abarca F, Toro N (2000) RecA-independent ectopic transposition in vivo of a bacterial group II intron. Nucleic Acids Res 28:4397–4402

Michel F, Costa M, Doucet AJ, Ferat J-L (2007) Specialized lineages of bacterial group II introns. Biochimie 89:542–553

Mira A, Moran NA (2002) Estimating population size and transmission bottlenecks in maternally transmitted endosymbiotic bacteria. Microb Ecol 44:137–143

Mira A, Pushker R, Rodriguez-Valera F (2006) The neolithic revolution of bacterial genomes. Trends Microbiol 14:200–206

Mohr G, Ghanem E, Lambowitz AM (2010) Mechanisms used for genomic proliferation by thermophilic group II introns. PLoS Biol 8:e1000391

Moran NA, Plague GR (2004) Genomic changes following host restriction in bacteria. Curr Opin Genet Dev 14:627–633

Moretz SE, Lampson BC (2010) A group IIC-type intron interrupts the rRNA methylase gene of *Geobacillus stearothermophilus* strain 10. J Bacteriol 192:5245–5248

Mormann S, Lomker A, Ruckert C, Gaigalat L, Tauch A, Puhler A, Kalinowski J (2006) Random mutagenesis in *Corynebacterium glutamicum* ATCC 13032 using an IS6100-based transposon vector identified the last unknown gene in the histidine biosynthesis pathway. BMC Genomics 7:205

Mullany P, Pallen M, Wilks M, Stephen JR, Tabaqchali S (1996) A group II intron in a conjugative transposon from the gram-positive bacterium, *Clostridium difficile*. Gene 174:145–150

Naas T, Blot M, Fitch WM, Arber W (1994) Insertion sequence-related genetic variation in resting *Escherichia coli* K-12. Genetics 136:721–730

Nagy Z, Chandler M (2004) Regulation of transposition in bacteria. Res Microbiol 155:387–398

Nakayama K, Yamashita A, Kurokawa K, Morimoto T, Ogawa M, Fukuhara M, Urakami H, Ohnishi M, Uchiyama I, Ogura Y, Ooka T, Oshima K, Tamura A, Hattori M, Hayashi T (2008) The Whole-genome sequencing of the obligate intracellular bacterium *Orientia tsutsugamushi* revealed massive gene amplification during reductive genome evolution. DNA Res 15:185–199

Naville M, Gautheret D (2010) Premature terminator analysis sheds light on a hidden world of bacterial transcriptional attenuation. Genome Biol 11:R97

Ng WV, Ciufo SA, Smith TM, Bumgarner RE, Baskin D, Faust J, Hall B, Loretz C, Seto J, Slagel J, Hood L, DasSarma S (1998) Snapshot of a large dynamic replicon in a halophilic archaeon: megaplasmid or minichromosome? Genome Res 8:1131–1141

Ochman H, Lawrence JG, Groisman EA (2000) Lateral gene transfer and the nature of bacterial innovation. Nature 405:299–304

Ohtsubo E, Minematsu H, Tsuchida K, Ohtsubo H, Sekine Y (2004) Intermediate molecules generated by transposase in the pathways of transposition of bacterial insertion element IS3. Adv Biophys 38:125–139

Olasz F, Kiss J, Konig P, Buzas Z, Stalder R, Arber W (1998) Target specificity of insertion element IS30. Mol Microbiol 28:691–704

Parkhill J, Sebaihia M, Preston A, Murphy LD, Thomson N, Harris DE, Holden MTG, Churcher CM, Bentley SD, Mungall KL, Cerdeno-Tarraga AM, Temple L, James K, Harris B, Quail MA, Achtman M, Atkin R, Baker S, Basham D, Bason N, Cherevach I, Chillingworth T, Collins M, Cronin A, Davis P, Doggett J, Feltwell T, Goble A, Hamlin N, Hauser H, Holroyd S, Jagels K, Leather S, Moule S, Norberczak H, O'Neil S, Ormond D, Price C, Rabbinowitsch E, Rutter S, Sanders M, Saunders D, Seeger K, Sharp S, Simmonds M, Skelton J, Squares R, Squares S, Stevens K, Unwin L, Whitehead S, Barrell BG, Maskell DJ (2003) Comparative analysis of the genome sequences of *Bordetella pertussis*, *Bordetella parapertussis* and *Bordetella bronchiseptica*. Nat Genet 35:32–40

Pasternak C, Ton-Hoang B, Coste G, Bailone A, Chandler M, Sommer S (2010) Irradiation-induced *Deinococcus radiodurans* genome fragmentation triggers transposition of a single resident insertion sequence. PLoS Genet 6:e1000799

Poluektova EU, Holsappel S, Gagarina EI, Bron S, Prozorov AA (2002) The ISBsu2 mobile element is present in a plasmid of a soil strain and in the chromosomes of several other strains of *Bacillus subtilis*. Genetika 38:1719–1722

Qiu N, He J, Wang Y, Cheng G, Li M, Sun M, Yu Z (2010) Prevalence and diversity of insertion sequences in the genome of *Bacillus thuringiensis* YBT-1520 and comparison with other *Bacillus cereus* group members. FEMS Microbiol Lett 310:9–16

Reznikoff WS (2008) Transposon Tn5. Annu Rev Genet 42:269–286

Robart AR, Seo W, Zimmerly S (2007) Insertion of group II intron retroelements after intrinsic transcriptional terminators. Proc Natl Acad Sci USA 104:6620–6625

Roberts D, Hoopes BC, McClure WR, Kleckner N (1985) IS10 transposition is regulated by DNA adenine methylation. Cell 43:117–130

Rocha EPC (2008) The organization of the bacterial genome. Annu Rev Genet 42:211–233

Rodriguez H, Snow ET, Bhat U, Loechler EL (1992) An *Escherichia coli* plasmid-based, mutational system in which supF mutants are selectable: insertion elements dominate the spontaneous spectra. Mutat Res 270:219–231

Sakaguchi Y, Hayashi T, Kurokawa K, Nakayama K, Oshima K, Fujinaga Y, Ohnishi M, Ohtsubo E, Hattori M, Oguma K (2005) The genome sequence of *Clostridium botulinum* type C neurotoxin-converting phage and the molecular mechanisms of unstable lysogeny. Proc Natl Acad Sci USA 102:17472–17477

Sawyer SA, Dykhuizen DE, DuBose RF, Green L, Mutangadura-Mhlanga T, Wolczyk DF, Hartl DL (1987) Distribution and abundance of insertion sequences among natural isolates of *Escherichia coli*. Genetics 115:51–63

Schneider D, Lenski RE (2004) Dynamics of insertion sequence elements during experimental evolution of bacteria. Res Microbiol 155:319–327

Schneider D, Duperchy E, Coursange E, Lenski RE, Blot M (2000) Long-term experimental evolution in *Escherichia coli*. IX. Characterization of insertion sequence-mediated mutations and rearrangements. Genetics 156:477–488

Schwartz E, Herberger C, Rak B (1988) Second-element turn-on of gene expression in an IS1 insertion mutant. Mol Gen Genet 211:282–289

Shigenobu S, Watanabe H, Hattori M, Sakaki Y, Ishikawa H (2000) Genome sequence of the endocellular bacterial symbiont of aphids *Buchnera* sp. APS. Nature 407:81–86

Siguier P, Filee J, Chandler M (2006) Insertion sequences in prokaryotic genomes. Curr Opin Microbiol 9:526–531

Simser JA, Rahman MS, Dreher-Lesnick SM, Azad AF (2005) A novel and naturally occurring transposon, ISRpe1 in the *Rickettsia peacockii* genome disrupting the rickA gene involved in actin-based motility. Mol Microbiol 58:71–79

Skorski P, Proux F, Cheraiti C, Dreyfus M, Hermann-Le Denmat S (2007) The deleterious effect of an insertion sequence removing the last twenty percent of the essential *Escherichia coli* rpsA gene is due to mRNA destabilization, not protein truncation. J Bacteriol 189:6205–6212

Sleight SC, Orlic C, Schneider D, Lenski RE (2008) Genetic basis of evolutionary adaptation by *Escherichia coli* to stressful cycles of freezing, thawing and growth. Genetics 180: 431–443
Sun X, Dennis JJ (2009) A novel insertion sequence derepresses efflux pump expression and preadapts *Pseudomonas putida* S12 for extreme solvent stress. J Bacteriol 191:6773–6777
Sundin GW (2007) Genomic insights into the contribution of phytopathogenic bacterial plasmids to the evolutionary history of their hosts. Annu Rev Phytopathol 45:129–151
Tanaka K, Furukawa S, Nikoh N, Sasaki T, Fukatsu T (2009) Complete WO phage sequences reveal their dynamic evolutionary trajectories and putative functional elements required for integration into the *Wolbachia* genome. Appl Environ Microbiol 75:5676–5686
Tauch A, Gotker S, Puhler A, Jr K, Thierbach G (2002) The 27.8-kb R-plasmid pTET3 from *Corynebacterium glutamicum* encodes the aminoglycoside adenyltransferase gene cassette aadA9 and the regulated tetracycline efflux system Tet 33 flanked by active copies of the widespread insertion sequence IS6100. Plasmid 48:117–129
Tomcsanyi T, Berg DE (1989) Transposition effect of adenine (Dam) methylation on activity of O end mutants of IS50. J Mol Biol 209:191–193
Ton-Hoang B, Pasternak C, Siguier P, Guynet C, Hickman AB, Dyda F, Sommer S, Chandler M (2010) Single-stranded DNA transposition is coupled to host replication. Cell 142:398–408
Toro N, Jimenez-Zurdo J, Garcia-Rodriguez FM (2007) Bacterial group II introns: not just splicing. FEMS Microbiol Rev 31:342–358
Touchon M, Rocha EPC (2007) Causes of insertion sequences abundance in prokaryotic genomes. Mol Biol Evol 24:969–981
Tourasse NJ, Kolsto A-B (2008) Survey of group I and group II introns in 29 sequenced genomes of the *Bacillus cereus* group: insights into their spread and evolution. Nucleic Acids Res 36:4529–4548
Toussaint A, Merlin C (2002) Mobile elements as a combination of functional modules. Plasmid 47:26–35
Venkatesan MM, Goldberg MB, Rose DJ, Grotbeck EJ, Burland V, Blattner FR (2001) Complete DNA sequence and analysis of the large virulence plasmid of *Shigella flexneri*. Infect Immun 69:3271–3285
Vögele K, Schwartz E, Welz C, Schiltz E, Rak B (1991) High-level ribosomal frameshifting directs the synthesis of IS150 gene products. Nucleic Acids Res 19:4377–4385
Wagner A (2006) Periodic extinctions of transposable elements in bacterial lineages: evidence from intragenomic variation in multiple genomes. Mol Biol Evol 23:723–733
Wagner A, de la Chaux N (2008) Distant horizontal gene transfer is rare for multiple families of prokaryotic insertion sequences. Mol Genet Genomics 280:397–408
Wernegreen JJ (2002) Genome evolution in bacterial endosymbionts of insects. Nat Rev Genet 3:850–861
Wernegreen JJ (2005) For better or worse: genomic consequences of intracellular mutualism and parasitism. Curr Opin Genet Dev 15:572–583
Wery J, Hidayat B, Kieboom J, de Bont JA (2001) An insertion sequence prepares *Pseudomonas putida* S12 for severe solvent stress. J Biol Chem 276:5700–5706
Wilde C, Escartin F, Kokeguchi S, Latour-Lambert P, Lectard A, Clement J-M (2003) Transposases are responsible for the target specificity of IS1397 and ISKpn1 for two different types of palindromic units (PUs). Nucleic Acids Res 31:4345–4353
Wolk CP, Lechno-Yossef S, Jager KM (2010) The insertion sequences of *Anabaena* sp. strain PCC 7120 and their effects on its open reading frames. J Bacteriol 192:5289–5303
Wu M, Sun LV, Vamathevan J, Riegler M, Deboy R, Brownlie JC, McGraw EA, Martin W, Esser C, Ahmadinejad N, Wiegand C, Madupu R, Beanan MJ, Brinkac LM, Daugherty SC, Durkin AS, Kolonay JF, Nelson WC, Mohamoud Y, Lee P, Berry K, Young MB, Utterback T, Weidman J, Nierman WC, Paulsen IT, Nelson KE, Tettelin H, O'Neill SL, Eisen JA (2004) Phylogenomics of the reproductive parasite *Wolbachia pipientis* wMel: a streamlined genome overrun by mobile genetic elements. PLoS Biol 2:E69

Yang F, Yang J, Zhang X, Chen L, Jiang Y, Yan Y, Tang X, Wang J, Xiong Z, Dong J, Xue Y, Zhu Y, Xu X, Sun L, Chen S, Nie H, Peng J, Xu J, Wang Y, Yuan Z, Wen Y, Yao Z, Shen Y, Qiang B, Hou Y, Yu J, Jin Q (2005) Genome dynamics and diversity of *Shigella* species, the etiologic agents of bacillary dysentery. Nucleic Acids Res 33:6445–6458

Zhang Z, Saier MH Jr (2009) A novel mechanism of transposon-mediated gene activation. PLoS Genet 5:e1000689

Zinser ER, Schneider D, Blot M, Kolter R (2003) Bacterial evolution through the selective loss of beneficial genes. Trade-offs in expression involving two loci. Genetics 164:1271–1277

Chapter 18
Transposable Elements in a Marginal Population of *Aegilops speltoides*: Temporal Fluctuations Provide New Insights into Genome Evolution of Wild Diploid Wheat

Alexander Belyayev and Olga Raskina

Abstract The Middle East is considered the primary center of cereal species variability that resulted from Pleistocene/Holocene climatic fluctuations. It is now on record that environmentally sensitive transposable elements (TE) may be among the most important internal sources for genotypic population change. Thus, we explore the temporal dynamics of several TEs in individual genotypes from a small, marginal population of *Aegilops speltoides*. The population is characterized by high heteromorphy and possesses a wide spectrum of chromosomal abnormalities. The dynamics of the TE complex was traced in three morphologically different genotypes and their progeny. It was discovered that: (1) various families of TEs vary tremendously in copy number between individuals from the same population and the selfed progenies; (2) the fluctuations in copy number are TE-family specific; (3) there is a great difference in TE copy number expansion or contraction between gametophytes and sporophytes; and (4) a small percentage of TEs that increase in copy number can actually insert at novel locations. We hypothesize that TE dynamics could promote or intensify morphological and karyotypical changes, some of which may be potentially important for the process of microevolution and allow species with plastic genomes to survive as new forms or even species in times of rapid climatic change.

18.1 Flora Dynamics and Climate Change

Speciation is the main event of evolution and the most intriguing enigma of biology. Despite the different opinions about the causes and underlying processes for speciation, evolutionary biologists agree that the climate fluctuations are the

---

A. Belyayev • O. Raskina
Laboratory of Plant Molecular Cytogenetics, Institute of Evolution, University of Haifa, Mt. Carmel, Haifa 31905, Israel
e-mail: belyayev@research.haifa.ac.il

P. Pontarotti (ed.), *Evolutionary Biology – Concepts, Biodiversity, Macroevolution and Genome Evolution*, DOI 10.1007/978-3-642-20763-1_18,

major factor causing the emergence of new species. In other words, the phyletic group keeps pace with the climatic and edaphic changes by means of a succession trend of species (Grant 1989).

We study speciation-related processes on species from the genus *Aegilops L* (*Poaceae*). Systematically, this genus takes an intermediate position between the genus *Triticum* and the genus *Agropyron* (Zhukovsky 1928). Among different *Aegilops* species, we focus our research on *Ae. speltoides*, a diploid cross-pollinated dimorphic grass species, a wild relative of the various wheat species, which belongs to section *Sitopsis*. The species was proposed to be the closest to the wild diploid progenitor of the G- and B-genomes of polyploid wheat (Sarkar and Stebbins 1956; Riley et al. 1958; Maestra and Naranjo 2000; Raskina et al. 2002; Feldman and Levy 2005), and is distributed in and around the Fertile Crescent. There are two subspecies, *ausheri* and *ligustica*, and phenotypical expression is controlled by several linked genes (Sears 1941). In nature, both subspecies exist in the mixed cross-pollinated populations. It was suggested that *ligustica* and *aucheri* should be considered not as subspecies but as two states of the dimorphic population. It was also noted that the dominance of one of the subtypes depended on environmental conditions. Such a dimorphic population state can be regarded as a speciation precursor (Zohary and Imber 1963).

It has repeatedly been stated that the Middle East is considered the primary center of *Triticum/Aegilops* species variability where local populations of wild progenitors of cultivated wheats exhibit significant genetic diversity (Nevo 1998), and up until the present time, this center has preserved its speciation potential (Zhukovsky 1928). The extant biotic composition of the region results from very complex chains of historical events, and there is no way to understand the recent distribution and processes in plant populations without looking carefully into the past (Tchernov 1988). Since the Tertiary period, when in the place of a modern Middle East a land bridge was formed across the Tethys Ocean between the parts of Laurasia and Gondwana, this region appears to be a "biogeographical crossroads" for movement and exchange of biotas. The beginning of the Quaternary period was marked by a sharp decrease of temperature and the formation of extensive ice sheets. Notably, the expansion of the Fennoscandian ice sheet to the South was maximal on the longitudes closest to longitudes of the Middle East. During the Pleistocene, thermophilic Tertiary Mediterranean flora was replaced by evolved mountainous and northern species, and plant zones significantly shifted to the South. At the beginning of the Holocene (from 11,000 years ago to the present), when the main trend of climate started to change in the opposite direction, and according to the Greenland ice cores the average temperature had risen by 7° in 50 years, another recession of Palearctic flora to the North took place. However, even though the Holocene climate was not stable: the temperate Boreal period gave way to a warm and humid Atlantic, which, in turn, was replaced by Sub-boreal and then by the relatively cold Sub-Atlantic (which includes the Little Ice Age) (Tchernov 1988; http://www.ipcc.ch) followed by modern global warming. Any change in climate, in turn, causes the movement of plant zones. Moreover, if cooling led to imminent death/retreat of southern flora, warming, and shifts in vegetation zones to the north does not mean compulsory extinction of all northern species in southern regions. As a result, the modern mosaic of the Eastern Mediterranean flora was formed.

Thus, over the past few million years, species repeatedly changed their areal, and plant populations on the periphery of the distribution area will be the first to suffer the impact of climate change. Certainly, this would have caused reactions of plant organisms and could lead not only to their extinction but possibly to the formation of new forms and species (Raskina et al. 2004b). The communication system of environment–organism is largely unexplored, but in recent years several possible mechanisms have been proposed (Martienssen 2008). One of these is the well-known expression of mobile elements (that may share up to 80% of repetitive DNA fraction of large cereal genomes) under environmental stress. It is now on record that transposable elements (TE) may be among the most important internal sources for genotypic population change as a result of their ability to create mutations, alter gene expression, and promote chromosomal aberrations (Kidwell and Lisch 1998, 2000; Kumar and Bennetzen 1999; Grandbastien et al. 2005). The main question is: whether and to what degree the TEs contribute to evolutionarily significant shifts in the genotypic structure of populations, especially in small stressed populations where microevolutionary processes are intensified?

18.2 Transposable Elements Activity in Marginal Population of *Ae. speltoides*

Extrapolating the Dobzhansky Central–Marginal Model (Da Cunha and Dobzhansky 1954) on TE fraction, and given the fact that TEs are sensitive to changes in the external environment, we hypothesize that in peripheral plant populations under abiotic stress, TEs may be activated and contribute to evolutionarily significant shifts in the genotypic structure of populations by producing an extended number of genomic variants for natural selection (Belyayev et al. 2010).

To find stressful populations in which significant microevolutionary events were expected, we analyzed 19 populations from different ecogeographical zones (Raskina et al. 2011). The criteria for stressful populations were: a position relative to the center of the species range; altitude; population size, local ecology; and the level of population destruction (mainly by human activity). After an analysis of populations of *Ae. speltoides* and closely related *Sitopsis* species by Inter Retrotransposons Amplified Polymorphism (IRAP) fingerprinting (Kalendar and Schulman 2006), was revealed that 17 populations were more or less similar, but two populations, namely, Technion 2 and Kishon showed an unusual distribution of TE. The Kishon population for a number of parameters fell under the definition of stressful. The population is small ($\approx$100 m^2) located at the sea level near Akko domain of desert plants and represents the southern extent of the species range.

Three original plants were selected from the Kishon population of *Ae. speltoides*. Each selected original genotype represents three groups of previously investigated plants (five to seven individual original spikes in each group), which have been clustered due to their morphological and cytogenetic similarity, e.g., spike morphology, B-chromosomes existence, appearance of additional 5S rDNA chromosomal clusters, and specific chromosomal rearrangements (Raskina et al. 2004a, b). The progeny from each genotype were obtained in three rounds of selfing. We simulated the situation in nature where, in marginal populations under critical external conditions, outcrossing plants and particularly *Ae. speltoides*, very often, transit to self-pollination. It is well known since the nineteenth century that the transition to self-pollination is a powerful stress–defense mechanism (Darwin 1859; Zohary and Imber 1963; Stebbins 1970). Theoretically, selfing genotypes may have an inheritance advantage over outcrossing genotypes because reproductive assurance can overweigh the disadvantages of inbreeding (Kelly 2005; Grant-Downton and Dickinson 2004), and it is a very important event because the change of mating system can be regarded as a precursor for reproductive isolation, i.e., speciation.

For the present research, we used total DNA of two types: one from young leaves, the other from spikes in the microsporogenesis stage. Thus, we expected to compare the dynamics of mobile elements in time in the vegetative and generative tissues. The copy numbers of several TE families were determined by qPCR for each original genotype and its offspring. Quantitative PCR data were verified by dot-blot analysis. The transpositional activities of TEs were inferred by IRAP retrotransposon display (Kalendar and Schulman 2006). We applied a set of non-TE markers (5S rDNA, *Spelt 1*, and *Spelt 52*) that could be followed both cytologically and by their copy number in order to separate TE activity from chromosomal aberrations and reorganization *per se*.

An analysis of spike size and morphology revealed significant intrapopulation morphological variability of *Ae. speltoides* (Fig. 18.1). We must emphasize that,

Fig. 18.1 Changes in spike morphology in three self-pollinated generations of three genotypes from the Kishon population of *Ae. speltoides*. Spike morphology of plants from the TS–84 population was used as the control (Belyayev et al. 2010)

except genotype 9, which is lethal in S2 generation, and despite pathological morphology, all plants were fertile. Moreover, in genotype 14, after several generations of pathological spikes, plants returned to normal morphology. Peculiarities of morphology guided the cytogenetical screening. We found a fan of chromosomal rearrangements. On the meiotic plate in Fig. 18.2c, the main chromosomal rearrangements can be seen: deletion, pericentric inversion on chromosome 2, and translocation.

When we traced the dynamics of TEs over three rounds of selfing, it was discovered that various families of TEs vary tremendously in copy number between individuals from the same population and the selfed progenies, and the fluctuations in copy number are TE-family specific (Fig. 18.3a). The TE copy numbers decreased or increased significantly, with a rise in one generation followed by a drop in the next and vice versa in an oscillatory fashion. The second remarkable feature is that in generative tissues, these oscillations displayed much higher amplitudes. We supposed that more than the differences between tissues, this phenomenon reflects differences between the alternating sporophyte and gametophyte generations. Regularities found are true for all explored elements that belong to Classes I and II. We compare temporal TE dynamics with dynamics of the inert tribe specific tandem repeats *Spelt 52* and 5S rDNA. These inert repeats exhibited absolutely different dynamics. These observations were supported by the principle component analysis (Fig. 18.3b–d). Thus, we propose that by modeling natural situations in small marginal plant population, we managed to stimulate intensive rearrangement of TE fraction. We also assume that increased variability in the TE complex may balance against inbreeding.

Can part of this activity be attributed to transpositional activity? To answer this question, we apply IRAP analysis for several TE in progeny of three genotypes. In IRAP, the insertion of a retrotransposon near another creates a new template for PCR amplification. The IRAP display data, in particular the appearance of unique novel bands in S2 and S3, are consistent with TE activation. Most of the unique bands were seen in G13. Of these, retrotransposons *WIS-2*, *Daniela* produced the greatest number. Cloning of the unique IRAP bands makes it possible to propose that a small percentage of TEs that increase in copy number are transpositionally active and can actually insert at novel locations (Belyayev et al. 2010). But still question remains: whether the detected bands considered transposition or do they may arise from ectopic exchanges? The result of ectopic exchange is the appearance of the structures of the two or more tandemly organized similar mobile elements. Indeed, such an organization is often observed in plants. But, in our case, sequencing of the regions of the TE insertion did not reveal the presence of related elements and, with reasonable certainty, we can assume that the case of transposition had been detected.

We might ask: what mechanisms induce wavelike TE copy number changes in succeeding generations? Most likely, mass amplification of the TE is caused by a combination of two factors: a permanent ecological shock and a genomic shock, as a consequence of inbreeding. Increases in TE copy numbers successfully enhance illegitimate recombination that is proposed as the main driving force behind

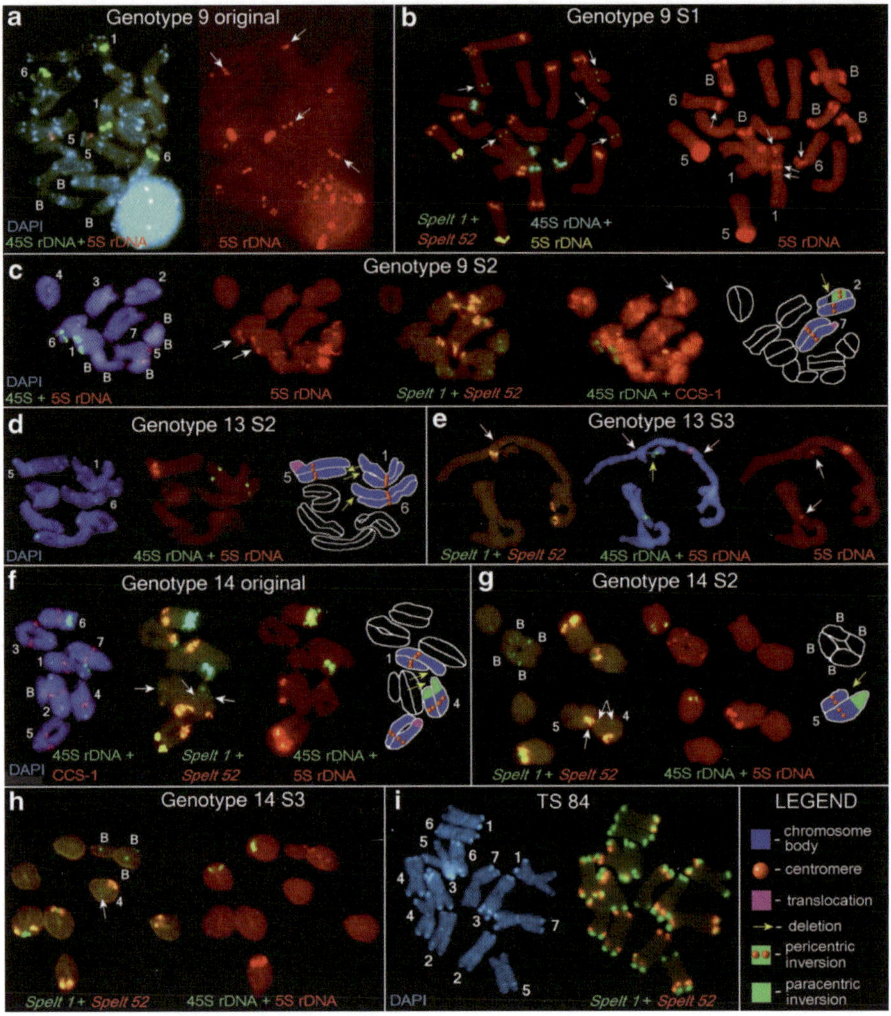

Fig. 18.2 In situ hybridization (FISH) and differential staining with DAPI on somatic and meiotic chromosomes of *Ae. speltoides*. (**a**) FISH with 5S rDNA (*red*), 45S rDNA (*green*), and differential staining with DAPI (*blue*) on somatic chromosomes of the original G9 plant (*left*). Chromosomes 1, 6 (*arrows*), and Bs carry additional 5S rDNA sites (*arrows*, *right*). (**b**) FISH with *Spelt 52* (*red*), *Spelt 1* (*green*, shown by *arrows* in B chromosomes), 5S rDNA (*yellow* pseudocolor), and 45S rDNA (*blue* pseudocolor) on the somatic chromosomes of the G9 S1 plant (*left*); FISH with 5S rDNA (*red*) on the same metaphase plate (*right*). Additional 5S rDNA sites on chromosomes 1 and 6 are shown by *arrows*. (**c**) From left to right: FISH with 5S rDNA (*red*) and 45S rDNA (*green*), and staining with DAPI on the meiotic chromosomes (late diakinesis stage) of the G9 S2 plant; second from the left: 5S rDNA probe alone, chromosomes 1, 6 (*two arrows*), and Bs carry additional 5S rDNA sites. Third from left: FISH with *Spelt 52* (*red*) and *Spelt 1* (*green*) on the same chromosomes. All B chromosomes carry intercalary *Spelt 1* clusters. Fourth from left: FISH with CCS-1 probe (*red*) and 45S rDNA (*green*). A pericentric inversion on chromosome 2 is shown by an *arrow*. Fifth from left: the scheme of the main chromosomal rearrangements

genome size decrease. Thus, one event (increase of TE amount) sooner or later induces the other (illegitimate recombination) as a response.

Taking these strands together, we can conclude that in small marginal population where the majority of plants transit to self-pollination: (1) various families of TEs vary tremendously in copy number between individuals from the same population and the selfed progenies; (2) the fluctuations in copy number are TE-family specific; (3) there is a great difference in TE copy number expansion or contraction between gametophytes and sporophytes; (4) a small percentage of TEs that increase in copy number can actually insert at novel locations and could serve as a bona fide mutagen.

What could be the consequences of TEs abnormal activity in stressed plant population? We propose that TE dynamics could promote or intensify morphological and karyotypical changes, some of which may be potentially important for the process of microevolution (Raskina et al. 2008), and allow species with plastic genomes to survive in times of rapid climatic change (Raskina et al. 2004b). The scenario could be as follows: in small marginal plant population under the influence of unusual ecology, transposable elements became active, especially in generative tissues. The mobilization of TE is known to be associated with high rates of karyotypic change, genetic variations (though limited in marginal populations), and epigenetic alterations (an important source for phenotypic variability). Some of

Fig. 18.2 (continued) (see *LEGEND*). (**d**) Differential staining with DAPI (*left*) and FISH (*middle*) with 5S rDNA (*red*) and 45S rDNA (*green*) on the meiotic chromosomes (diakinesis stage) of the G13 S2 plant. A scheme of the main chromosomal rearrangements is on the right (see *LEGEND*). (**e**) FISH with *Spelt 52* (*red*) and *Spelt 1* (*green*) on the diplotene stage of the G13 S3 plant (*left*). Paracentric inversion in the long arm of the chromosome 5 marked by a small *Spelt 52* cluster shown with an *arrow*. In the middle: FISH with 5S rDNA (*red*) and 45S rDNA (*green*) on the same chromosomes (stained with DAPI). Two points of heterologous synapses involving both short and long arms of chromosome 5 are shown by two *white arrows*; heterozygous deletion of the satellite of chromosome 6 is shown by the *yellow arrow*. On the right: additional 5S rDNA clusters on chromosomes 1 and 6 in the NOR regions (*arrowed*). (**f**) First from the left: differential staining with DAPI and FISH with the CCS-1 probe (*red*) and 45S rDNA (*green*) on the meiotic (late diakinesis stage) chromosomes of the original G14 plant. Second from the left: FISH with *Spelt 52* (*red*) and *Spelt 1* (*green*) on the same chromosomes. Sites of 45S rDNA are in *blue* pseudocolor, 5S rDNA are in *yellow* pseudocolor. The clusters of *Spelt 1* that mark a homozygous paracentric inversion on chromosome 4, and a *Spelt 1* cluster on the B chromosome are shown by *arrows*. Third from the left: FISH with 5S rDNA (*red*) and 45S rDNA (*green*) on the same chromosomes. Fourth from the left: a scheme of the main chromosomal rearrangements (see *LEGEND*). (**g**) FISH with *Spelt 52* (*red*) and *Spelt 1* (*green*) on the chromosomes at the metaphase I stage of the G14 S2 plant (*left*). The clusters of *Spelt 1* that mark a homozygous paracentric inversion on chromosome 4 and cluster of *Spelt 52* that marks a heterozygous inversion on chromosome 5 are shown by *arrows*. FISH with 5S rDNA (*red*) and 45S rDNA (*green*) is in the middle. B-chromosomes carry 5S rDNA clusters in both arms. The scheme of the main chromosomal rearrangements is shown on the right (see *LEGEND*). (**h**) FISH with *Spelt 52* (*red*) and *Spelt 1* (*green*) on the chromosomes at the metaphase I stage of the G14 S3 plant (*left*). Homozygous paracentric inversion on chromosome 4 is shown by *arrows*. FISH with 5S rDNA (*red*) and 45S rDNA (*green*) on the same chromosomes (*right*). (**i**) Somatic chromosomes of TS 84 differentially stained with DAPI (*left*). FISH with *Spelt 52* (*red*) and *Spelt 1* (*green*) on the same chromosomes (*right*) (Belyayev et al. 2010)

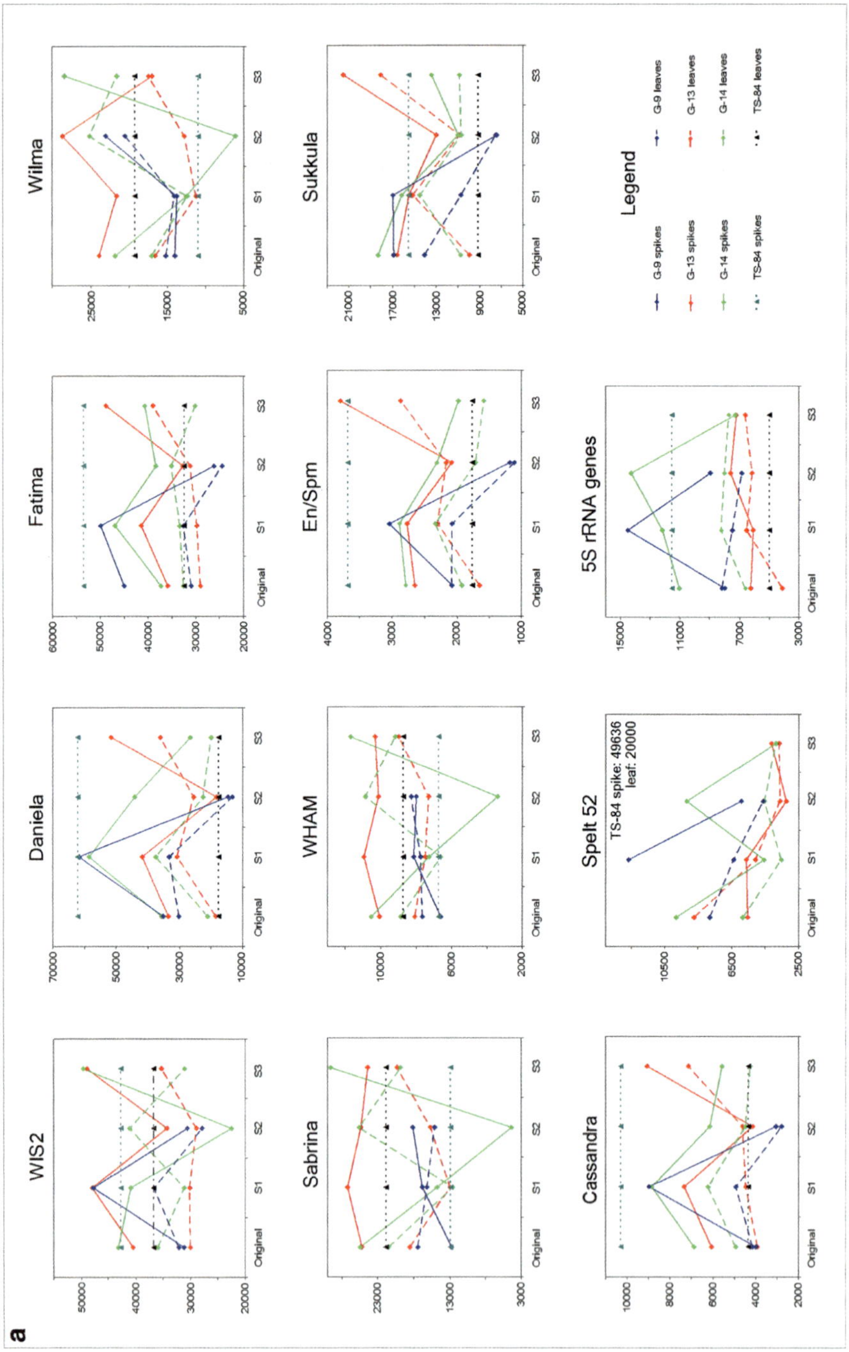

Fig. 18.3 Continued

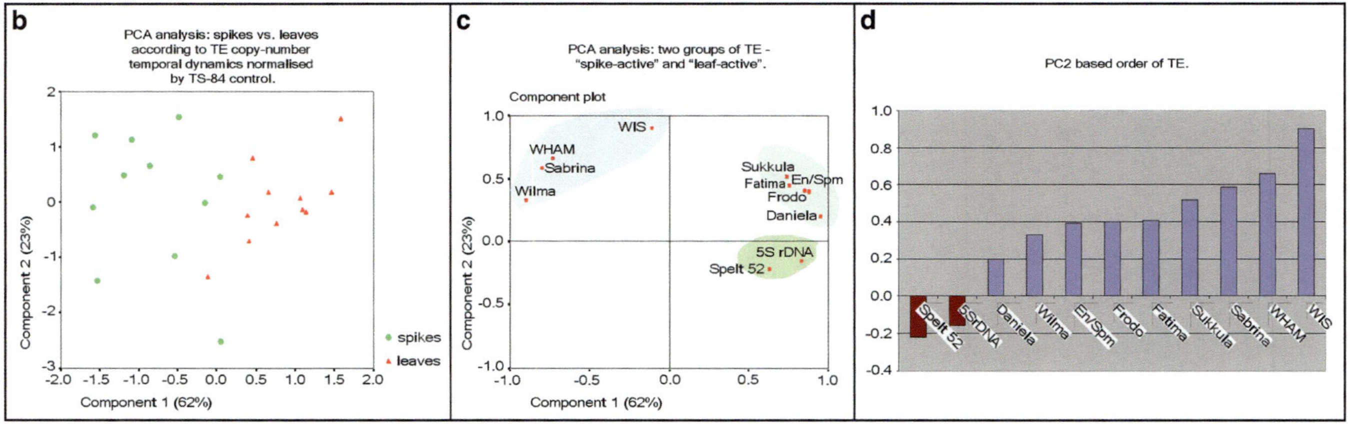

Fig. 18.3 (**a**) Dynamics of TE copy numbers in three self-pollinated generations of three genotypes from the Kishon population of *Ae. speltoides* (shown by *lines*). TE copy numbers of the TS-84 population were used as controls. TE copy numbers for available sibs in selfing generations are shown by separate *dots*. (**b**) PCA analysis: spikes vs. leaves for TE copy number changes over the generations, normalized by the TS-84 control. (**c**) PCA analysis: two groups of TE – spike-active and leaf-active. (**d**) PC2-based order of TE activation (Belyayev et al. 2010)

these mutations could be heritable. The combination of genetic/epigenetic alterations with karyotypic rearrangements allows some genotypes to survive as new forms or even species under intensive environmental pressure.

18.3 Conclusive Remarks

The value of wheat in the world's economy can hardly be overestimated. While crop yields, for the most part, increased recently, the genetic basis for most of the important food crops has been rapidly narrowing (Avery 1985). This is due to the global extension of modern pure breeding practices, which increase genetic homogeneity (Frankel and Soule 1981). The loss of genetic diversity of some of the world's crops has accelerated greatly in recent decades with many crops becoming increasingly susceptible to diseases, pests, and environmental stresses. Wild cereals are widely adaptive to all of these stressful factors. Genetic diversity in central and semi-isolated, and ecologically peripheral and marginal isolated populations include specific alleles and allele combinations predominating as coadapted blocks of genes that can adapt to diverse environments. This explains the interest in wild relatives of cultivated wheat, and in particular to species of genus *Aegilops*, which has been proposed by many researchers as the donor of B- and G-genomes of polyploid wheat (Sarkar and Stebbins 1956; Zohary and Feldman 1962; Kimber and Sears 1987; Feldman et al. 1995). The wild diploid species of *Aegilops*, enriched with genes of resistance to various diseases and drought, are widely used to create new varieties of wheat by chromosomal engineering (Feldman and Sears 1981). Exploration of the wild wheat natural variability and genetic diversity across a geographic range and timescale will allow the establishment of a novel view on the adaptation and speciation phenomena, highlight the evolution of wheat, and contribute to the development of new strategies in wheat improvement.

Acknowledgment We are most grateful to our collaborators Ruslan Kalendar, Alan Schulman, and Leonid Brodsky. This work was supported by the Israel Science Foundation under grant number 723/07.

References

Avery D (1985) US Farm dilemma: the global bad news is wrong. Science 230:408–412

Belyayev A, Kalendar R, Brodsky L, Nevo E, Schulman AH, Raskina O (2010) Transposable elements in a marginal plant population: temporal fluctuations provide new insights into genome evolution of wild diploid wheat. Mob DNA 1:6

Da Cunha AB, Dobzhansky T (1954) A further study of chromosomal polymorphism in *Drosophila willistoni* in its relation to the environment. Evolution 8:119–134

Darwin CD (1859) On the origin of species by means of natural selection, or preservation of favoured races in the struggle for life. John Murray, London

Feldman M, Levy AA (2005) Allopolyploidy – a shaping force in the evolution of wheat genomes. Cytogenet Genome Res 109:250–258

Feldman M, Sears ER (1981) The wild gene resources of wheat. Sci Am 244:102–112

Feldman M, Lipton FGH, Miller TE (1995) Wheats. *Triticum* spp. (Gramineae-Triticinae). In: Smartt J, Simmonds NW (eds) Evolution of crop plants. Longman Scientific and Technical Press, Harlow, pp 184–192

Frankel OH, Soule ME (1981) Conservation and evolution. Cambridge University Press, Cambridge

Grandbastien MA, Audeon C, Bonnivard E, Casacuberta JM, Chalhoub B, Costa AP, Le QH, Melayah D, Petit M, Poncet C, Tam SM, Van Sluys MA, Mhiri C (2005) Stress activation and genomic impact of Tnt1 retrotransposons in *Solanaceae*. Cytogenet Genome Res 110:229–241

Grant V (1989) The theory of speciation trends. Am Nat 133:604–612

Grant-Downton RT, Dickinson HG (2004) Plants, pairing and phenotypes – two's company? Trends Genet 20:188–195

Kalendar R, Schulman AH (2006) IRAP and REMAP for retrotransposon-based genotyping and fingerprinting. Nat Protoc 1:2478–2484

Kelly JK (2005) Family level inbreeding depression and the evolution of plant mating systems. New Phytol 165:55–62

Kidwell MG, Lisch DR (1998) Hybrid genetics. Transposons unbound. Nature 393:22–23

Kidwell MG, Lisch DR (2000) Transposable elements and host genome evolution. Trends Ecol Evol 15:95–99

Kimber G, Sears ER (1987) Evolution in the genus *Triticum* and the origin of cultivated wheat. In: Heyne EC (ed) Wheat and wheat improvement. American Society of Agronomy, Madison, pp 154–164

Kumar A, Bennetzen JL (1999) Plant retrotransposons. Annu Rev Genet 33:479–532

Maestra B, Naranjo T (2000) Genome evolution in Triticeae. Chromosomes Today 13:155–167

Martienssen R (2008) Great leap forward? Transposable elements, small interfering RNA and adaptive Lamarckian evolution. New Phytol 179:572–574

Nevo E (1998) Genetic diversity in wild cereals: regional and local studies and their bearing on conservation ex-situ and in-situ. Genet Resour Crop Evol 45:355–370

Raskina O, Belyayev A, Nevo E (2002) Repetitive DNAs of wild emmer wheat *Triticum dicoccoides* and their relation to S-genome species: molecular-cytogenetic analysis. Genome 45:391–401

Raskina O, Belyayev A, Nevo E (2004a) Activity of the *En/Spm*-like transposons in meiosis as a base for chromosome repatterning in a small, isolated, peripheral population of *Aegilops speltoides* Tausch. Chromosome Res 12:153–161

Raskina O, Belyayev A, Nevo E (2004b) Quantum speciation in Aegilops: molecular cytogenetic evidence from rDNA clusters variability in natural populations. Proc Natl Acad Sci USA 101:14818–14823

Raskina O, Barber J, Nevo E, Belyayev A (2008) Repetitive DNA and chromosomal rearrangements: speciation-related events in plant genomes. Cytogenet Genome Res 120:351–357

Raskina O, Brodsky L, Belyayev A (2011) Tandem repeats on an eco-geographical scale: outcomes from the genome of *Aegilops speltoides*. Chromosome Research 19: doi:10.1007/s10577-011-9220-9

Riley R, Unrau J, Chapman V (1958) Evidence on the origin of the B genome of wheat. J Hered 49:91–98

Sarkar P, Stebbins GL (1956) Morphological evidence concerning the origin of the B genome in wheat. Am J Bot 43:297–304

Sears E (1941) Amphidiploids in the seven-chromosome *Triticinae*. Univ Missouri Agric Exper Sta Res Bull 336:1–46

Stebbins GL (1970) Adaptive radiation of reproductive characteristics in angiosperms. 1. Pollination mechanisms. Annu Rev Ecol Syst 1:307–326

Tchernov E (1988) The biogeographical history of southern Levant. In: Yom-Tov Y, Tchernov E (eds) The zoogeography of Israel. The distribution and abundance at a zoogeographical crossroad. Kluwer, Dr. W. Junk Publishers, Dordrecht, pp 159–251

Zhukovsky PM (1928) A critical-systematical survey of the species of the genus *Aegilops* L. Proc Appl Bot Select Genet 18:417–609

Zohary D, Feldman M (1962) Hybridization between amphidiploids and the evolution of polyploids in the wheat (*Aegilops-Triticum*) group. Evolution 16:44–61

Zohary D, Imber D (1963) Genetic dimorphism in fruit types in *Aegilops speltoides*. Heredity 18:223–231

Chapter 19
Analysis of the Conservative Motifs in Promoters of miRNA Genes, Expressed in Different Tissues of Mammalians

Oleg V. Vishnevsky, Konstantin V. Gunbin, Andrey V. Bocharnikov, and Eugene V. Berezikov

Abstract Numerous miRNAs play an important role in translation regulation, modulating embryo development, stem cells proliferation, and tissue differentiation. Aberrant miRNA expression has been associated with diseases like cancer, microcephaly, and schizophrenia. It is too little known about regulation of miRNA expression. A computer approach was developed in order to reveal the significant oligonucleotide motifs in the regulatory regions of eukaryotic genes. The regulatory signals that are specific to the promoter regions of miRNA containing genes, which are expressed in different tissues of mammalians, were obtained and classified.

O.V. Vishnevsky
Institute of Cytology and Genetics, Siberian Branch of the Russian Academy of Sciences, 630090, Prospekt Lavrentyeva 10, Novosibirsk, Russia

Department of Natural Sciences, Novosibirsk State University, Novosibirsk, Russia
e-mail: oleg@bionet.nsc.ru

K.V. Gunbin
Institute of Cytology and Genetics, Siberian Branch of the Russian Academy of Sciences, 630090, Prospekt Lavrentyeva 10, Novosibirsk, Russia

A.V. Bocharnikov
Department of Natural Sciences, Novosibirsk State University, Novosibirsk, Russia

E.V. Berezikov
Institute of Cytology and Genetics, Siberian Branch of the Russian Academy of Sciences, 630090, Prospekt Lavrentyeva 10, Novosibirsk, Russia

Hubrecht Institute, RNAAS, Utrecht, the Netherlands

P. Pontarotti (ed.), *Evolutionary Biology – Concepts, Biodiversity, Macroevolution and Genome Evolution*, DOI 10.1007/978-3-642-20763-1_19,

19.1 Introduction

miRNAs are short (~22 bp) RNA sequences that bind to the 3′-untranslated region (3′-UTR) of the mRNAs of target human genes (Kim 2005; Bartel 2004; Pasquinelli et al. 2005). This binding event causes translation repression (Wightman et al. 1993) or stimulates rapid degradation of the transcript (Giraldez et al. 2006). Other types of regulation, such as translation activation (Filipowicz et al. 2008) and heterochromatin formation (Kim et al. 2008), have also been described. About 30% of all protein-coding human genes are targets for miRNAs (John et al. 2004). miRNAs are believed to particularly target genes of transcription factors, such as nuclear hormone receptors (John et al. 2004). One miRNA can target hundreds of downstream target mRNAs, while one mRNA can be targeted by multiple miRNAs. miRNAs control the expression of large number of genes (Lewis et al. 2003; Enright et al. 2003) and play an important role in the regulation of the main biological mechanisms that guide the development of organisms, stem cell proliferation, apoptosis, and the processes of tissue differentiation (Foshay and Gallicano 2009; Johnnidis et al. 2008; Shcherbata et al. 2006; Stadler and Ruohola-Baker 2008; Yi et al. 2008). Disturbances in the expression of miRNAs result in pathologies during development and serious diseases (Davis et al. 2008; Stark et al. 2008; Alvarez-Garcia and Miska 2005; Jiang et al. 2008). Recent studies have implicated miRNAs in numerous human diseases such as colorectal cancer, chronic lymphocytic leukemia, and Fragile X syndrome (Calin et al. 2002, 2004, 2005; Caudy et al. 2002; McManus 2003; Croce and Calin 2005).

The vast majority of miRNAs (Fig. 19.1) are transcribed by RNA polymerase II (Pol II) (Lee et al. 2004; Cai et al. 2004) as long primary transcripts (pri-miRNA) (Bartel 2004; Lee et al. 2002; Cullen 2004), that contain single or multiple hairpin-like structures (Cullen 2004; Cai et al. 2004; Lee et al. 2004; Altuvia et al. 2005). Approximately 50% of mammalian miRNA loci are found in close proximity to other miRNAs (Lee et al. 2002). Like other Pol II transcripts, pri-miRNAs are 5′- capped and 3′- polyadenylated (Cai et al. 2004; Lee et al. 2004; Bracht et al. 2004). pri-miRNAs are processed in the nucleus into approximately 70 nT hairpins (pre-miRNA) by Drosha, an RNAse-III like enzyme acting with DGCR8 and other cofactors (Lee et al. 2002, 2003). The pre-miRNAs are exported by Exportin 5 (EXP5) to the cytoplasm (Kim 2004). EXP5 is a member of the nuclear transport receptor family (Lund et al. 2004; Yi et al. 2005; Bohnsack et al. 2004). In the cytoplasm, a second RNAse-III like enzyme, Dicer cleaves the pre-miRNA into a mature double-stranded dsRNA duplex (Bernstein et al. 2001; Grishok et al. 2001; Hutvágner et al. 2001; Ketting et al. 2001; Knight and Bass 2001). Dicer is a highly conserved protein that is found in almost all eukaryotic organisms, including *Schizosaccharomyces pombe*, plants and animals. One strand of the 21–23 nucleotide dsRNA is packaged into the RNA-Inhibiting Silencing Complex (RISC) and guided to target mRNAs. Pairing between microRNAs and target mRNAs takes place in association with one or more Argonaute proteins, the major protein component of RISCs. Depending

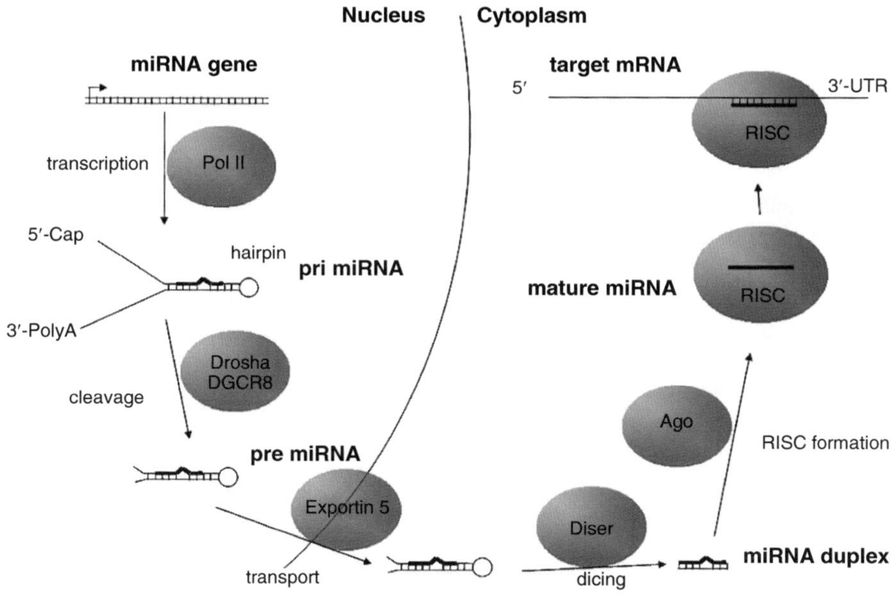

Fig. 19.1 Pathway of microRNA (miRNA) biogenesis and action

upon the degree of complementarity between the target mRNA and the miRNA, the mRNA is either subject to repression of translation or cleavage and degradation. It has been demonstrated that six nucleotides at positions 2–7 relative to the 5′ end of the miRNA (called seed) are crucial in base pairing with the target mRNA in animals, while the remaining portion of the 30-miRNA seems less important (Bartel 2004, 2009).

The transcription of most miRNA genes is mediated by RNA polymerase II, although a minor group of miRNAs that are associated with Alu repeats can be transcribed by Pol III (Borchert et al. 2006).

Approximately 50% of human miRNAs appear to be expressed from introns of protein-coding transcripts (Rodriguez et al. 2004). Some part of miRNAs was found in exons of protein-coding genes. Others are transcribed by Pol II as independent transcription units using its own promoter.

Expression analyses show that miRNAs are expressed in a tissue-specific manner at specific times (Lagos-Quintana et al. 2002; Sempere et al. 2004; Krichevsky et al. 2006; Landgraf et al. 2007). Brain development has been associated with highly dynamic and temporally regulated waves of miRNA expression, with specific groups of miRNA being expressed only at specific time points during embryonic development of the nervous system (Stadler and Ruohola-Baker 2008; Dogini et al. 2008; Miska et al. 2004; Wheeler et al. 2006).

A range of Pol II–associated transcription factors control miRNA gene transcription (Lee and Dutta 2009; Cao et al. 2006; Kosik 2006; Zeng 2009). For instance, myogenic transcription factors, such as MyoD1, induce the transcription of miR-1 and miR-133 during myogenesis (Chen et al. 2006; Kim et al. 2006; Rao et al. 2006). Also in the brain, transcription of miR-134, participating in the regulation of BDNF-stimulated synaptic plasticity (Schratt et al. 2006) is regulated by neuronal activity via factor Mef2 (Fiore et al. 2009; Pulipparacharuvil et al. 2008). Some miRNAs are under the control of tumor-suppressive or oncogenic transcription factors. The tumor suppressor p53 activates transcription of the miR-34 family of miRNAs (He et al. 2007), whereas the oncogenic protein MyC transactivates or represses a number of miRNAs that are involved in the cell cycle and apoptosis (He et al. 2005; Chang et al. 2008). The canonical TATA box motifs have been identified upstream of miRNA genes (Stormo 2000).

Epigenetic control also contributes to miRNA gene regulation (Lujambio et al. 2007; Saito et al. 2006). For instance, the miR-203 locus frequently undergoes DNA methylation in T-cell lymphoma but not in normal T lymphocytes (Bueno et al. 2008).

RNA editing is another possible way of regulating miRNA biogenesis. The alteration of adenines to inosines has been observed in miR-142 (Yang et al. 2006) and miR-151 (Kawahara et al. 2007). RNA editing can also change the target specificity of the miRNA (Kawahara et al. 2008).

The development of high-performance deep-sequencing techniques (Lu et al. 2005; Margulies et al. 2005) and in silico prediction methods (Lai et al. 2003; Nam et al. 2005; Li et al. 2006; Huang et al. 2007) has accelerated the discovery of miRNAs. Based on these technologies (Berezikov et al. 2005), the human genome is now believed to contain more than 1,000 miRNA genes. The diversity of the miRNA repertoire increases with the organism's complexity: humans and other mammals have about four times more annotated miRNA genes compared to insects and nematodes, suggesting the role of miRNAs in the development of this complexity.

Many of the animal miRNAs are phylogenetically conserved; ~55% of *Caenorhabditis elegans* miRNAs have homologues in humans. It means that miRNAs have had important roles throughout animal evolution (Ibáñez-Ventoso et al. 2008). At the same time, many miRNAs were conserved only between primates and some were even species-specific, suggesting the existence of recently evolved miRNA genes (Berezikov et al. 2006).

Although we know that miRNAs are expressed in tissue-specific manner, we know very little about peculiarities of regulation of miRNA expression. The aim of our research is to reveal and analyze the regulatory signals in promoter regions of miRNA genes, expressed in different tissues of mammalians.

19.2 Materials and Methods

The sets of the regulatory regions of miRNA genes expressed in the brain, lungs, and excretory system of 12 species of mammalians (*Homo sapiens*, *Pan troglodytes*, *Pongo pygmaeus*, *Gorilla gorilla*, *Macaca mulatta*, *Callithrix jacchus*, *Bos taurus*,

Equus caballus, *Canis familiaris*, *Mus musculus*, *Rattus norvegicus*, *Sus scrofa*) were generated. The [−1,000; +1] regions of miRNA genes relative to the transcription start site were analyzed.

For this, at the first stage, information about localization in the genome of human 1,572 miRNAs, which were predicted by experimental and computer approaches, was obtained from the Ensembl database (http://www.ensembl.org). Then the nearest transcription start sites in the 5′-region of miRNAs were located at a distance of not more than 3,000 nucleotides in relation to the beginning of the miRNA. This information was obtained from the FANTOM4 database (Kawaji et al. 2009). This database contains information about 5′-ends of more than 24 millions of human mRNAs and information about tissues, where these mRNAs were detected. Based on this information, the sets of promoters of human miRNA genes, expressed in brain, excretory system, and lungs in the [−1,000; +1] region relative to the transcription start site were generated. After this, homologous 1,000 bp genome regions were obtained for 11 species of mammals in Ensembl Compara database using Compara Perl API. This database contains the whole genomic alignments of different species and is widely used in methods of comparative analysis. Finally, the sets of promoters of miRNA genes of 12 species of mammalians, expressed in brain ("brain set"), excretory system ("excretory set"), and lungs ("lungs set") have been produced.

The 36 sets constructed strongly differ by size (Fig. 19.2). The biggest sets were obtained for primates. Non-primates have approximately two times less size of the

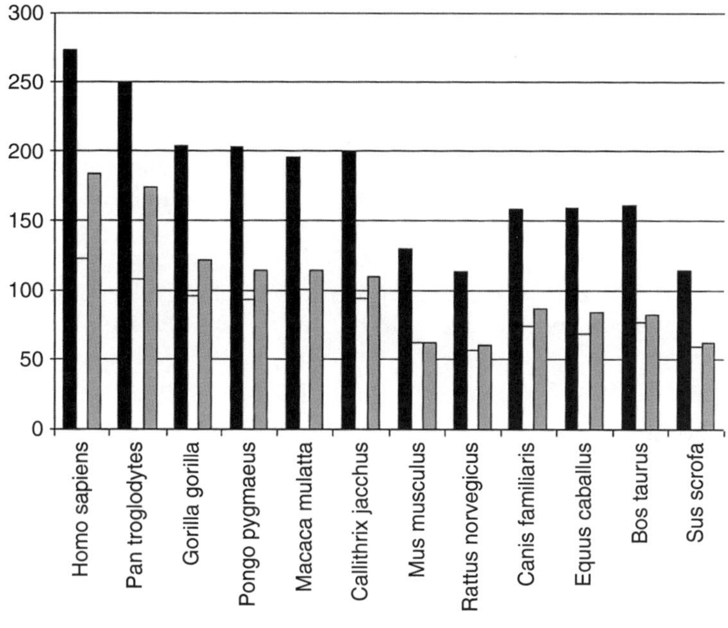

Fig. 19.2 The size of the brain set (*black bar*), excretory set (*white bar*), and lungs set (*gray bar*)

sets for all tissues. The sets of promoters of miRNA genes, expressed in brain, were the biggest. The size of the sets varies from 273 sequences for human brain set to 57 sequences for excretory set of rat.

The search for significant regulatory signals that are related to the structural–functional organization of promoters was performed in 36 sets of species- and tissue-specific sequences using the method (Vishnevsky and Kolchanov 2005) of revealing degenerate oligonucleotide motifs, i.e., short words of a fixed length written in the 15 single letter-based IUPAC code (A,T,G,C, R = G/A, Y = T/C, M = A/C, K = G/T, W = A/T, S = G/C, B = T/G/C, V = A/G/C, H = A/T/C, D = A/T/G, N = A/T/G/C). This approach is based on the analysis of oligonucleotide vocabularies of the promoters, finding and clustering of the similar oligonucleotides characterized by low Hamming distance and located in sequences of different promoters, with continuous iteration building of IUPAC-consensus for every class of the oligonucleotides. We consider the motif as significant, if it occurs in the high number of promoter sequences, located in the low number of random sequences and its probability $P(n, N)$ to occur by chance in the set of promoters is low.

The probability $P(n, N)$ is calculated as follows. Let us consider an oligonucleotide motif $M = m_1, m_2, \ldots, m_l$ of length l in the expanded 15 single letter-based IUPAC code. The probability of this motif to occur at a particular position in the DNA sequence S_k of length L is: $P(M) = \prod_{i=1}^{l} P_i$, where P_i is a frequency of a letter m_i assessed from the nucleotide content of S_k. The binomial probability $P(n, N)$ to observe the motif M in more or equal than n $(0 \leq n \leq N)$ sequences is: $P(n,N) = \sum_{i=n}^{N} C_N^i P^i (1-P)^{N-i}$, where $P(S_k) = 1 - e^{-(L-l+1)*P(M)}$.

This approach does not need the preliminary experimental information about transcription factor binding sites or multiple alignments to reveal significant signals in the analyzed set of the sequences.

The graphics accelerators (GPU) and chips with programmable logic (FPGA) were applied to increase the calculation speed. It allows us to increase the speed of calculation in 50 times for GPU and in 500 times for FPGA in comparison with single CPU.

To assess the similarities in the regulatory regions of genes based on oligonucleotide motifs, we developed a method (Vishnevsky and Kolchanov 2005) based on a comparison of the abundance and character of the distribution of motifs in sequences under study. As a measure of similarity between the jth promoter and the sequence studied, the value $P_j = -\sum_{k=1}^{L} \log p_k / L$ is used, where L is the length of the sequence analyzed and p_k is the product of frequencies of nucleotides, which are consistent with the motifs covering the kth position. The greater is P_j, the lower is the probability of chance occurrence of the motif set characteristic of the jth promoter in the sequence.

19.3 Results and Discussion

The number of the motifs, revealed using the ARGO program, strongly varied for different tissues in different species (Fig. 19.3). For example, it was found 363 motifs in the human excretory set. Against this, only 10 significant motifs were obtained from the lung promoters of mouse. These differences could be explained as by differences in size of the sets analyzed, as by different heterogeneity of the sets. In general, the most of the motifs were found in the sets of brain-specific promoters of primates.

Characteristics of the most significant oligonucleotide motifs found in the sets of human promoters are shown in Table 19.1. For example, the AGRRRGAA = (A)(G)(G/A)(G/A)(G/A)(G)(A)(A) motif is presented in 37% of human brain-specific promoters. It was found in 5% of random sequences, generated with the same mononucleotide content. The logarithm of binomial probability $P(n, N)$ to observe the motif by chance is −39. An analysis of the motifs demonstrates that some of them are rather complex words, but others look like polyA–polyT runs or TATA-like sequences.

The next step we estimated the tissue specificity of the motifs revealed. Figure 19.4 demonstrates that along with the motifs, overrepresented in one set of promoters, there are motifs, equally distributed in promoters of all the tissues analyzed. For example, the motif VRTCAGCM, revealed in the brain set, occurs

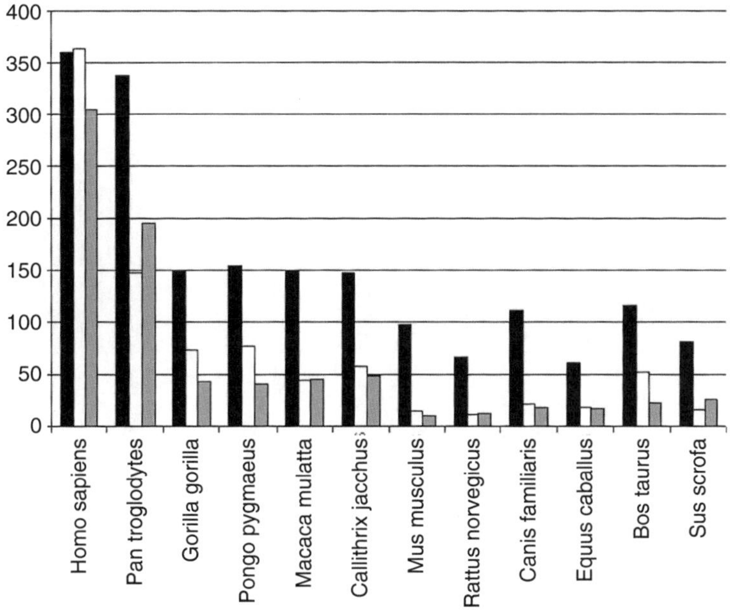

Fig. 19.3 Number of the motifs, revealed in promoters of genes, expressed in the brain set (*black bar*), excretory set (*white bar*), and lungs set (*gray bar*)

Table 19.1 Characteristics of the most significant motifs, revealed in the promoters of genes, expressed in different tissues of *Homo sapiens*

Motif	Occurrence in promoters	Occurrence in random sequences	*log*(*P*(*n*, *N*))
Brain set			
AGRRRGAA	0.37	0.05	−39.61
TTTTYATT	0.30	0.03	−34.23
VTYCCCAG	0.28	0.02	−31.00
CTTYYTCT	0.24	0.02	−30.64
CCYCWBCC	0.26	0.05	−26.50
YCCCAGSN	0.21	0.05	−25.28
Excretory set			
TTTWTWTT	0.40	0.04	−31.56
GCAGMGVC	0.40	0.02	−30.89
GGCKGBRG	0.39	0.05	−28.28
GAAMCAAW	0.29	0.08	−27.58
AAAMAMYC	0.27	0.06	−26.75
SCWGGAGY	0.29	0.05	−25.90
Lungs set			
CCYBYCYC	0.46	0.10	−33.94
TTYTTWTT	0.34	0.03	−33.36
TBSMCAGG	0.28	0.03	−22.42
HCYCARCC	0.37	0.07	−21.01
MWKSCCAG	0.39	0.09	−19.10
GGNGSCKG	0.30	0.08	−18.77

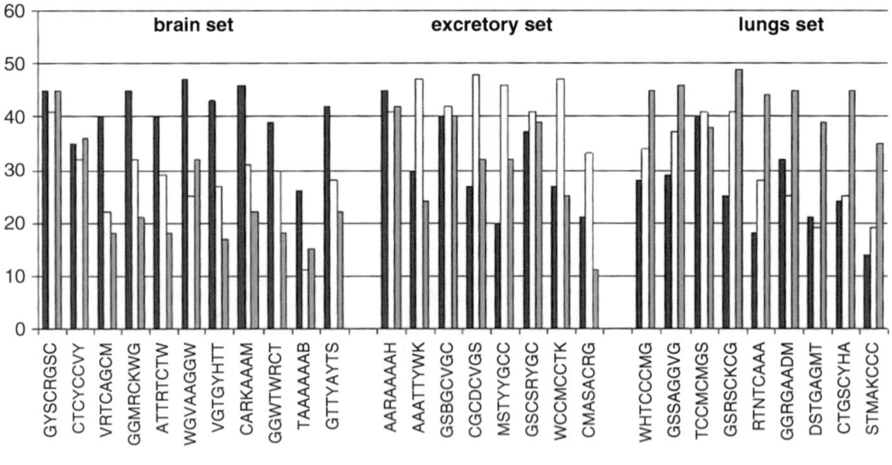

Fig. 19.4 The occurrence rate of the motifs in promoters of *H sapiens* genes, expressed in brain (*black bar*), excretory system (*white bar*), and lungs (*gray bar*)

in 40% of brain-specific sequences, but it was found only in 20% of excretory set or lung set promoters. At the same time, the motif GSBGCVGS, revealed in the excretory set, equally occurs in the sets of all tissues analyzed. We suppose that

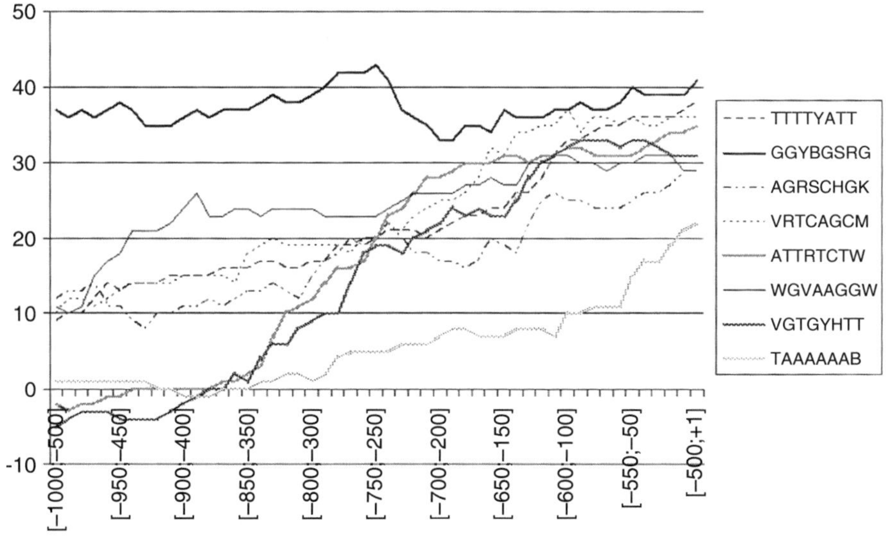

Fig. 19.5 Distribution of the brain-specific motifs of *H. sapiens* along the promoter sequences. Axis X – regions of promoter, relative to the transcription start site, axis Y – the percent of promoters, containing a motif, relative to the random level. Zero level is the occurrence rate of the motif in the set of the random sequences

these common motifs could correspond to the basic properties of promoter regions of miRNA genes, independent from the tissue specificity of gene.

Then we estimated the distribution of the motifs along the promoter sequences, relative to the transcription start site position. Figure 19.5 demonstrates the occurrence rate of the motifs in the 500-bp window, which slides along the promoter sequences with the step 10 bp. The analysis of the graphs demonstrates that most of the motifs are located in the region [−500; +1] relative to the start of transcription. At the same time, some other motifs, e.g., GGYBGSRG, are equally distributed along the promoter sequences.

The comparison of the motifs with the known binding sites of transcription factors and position weight matrixes from TRANSFAC (Matys et al. 2006), TRRD (Kolchanov et al. 2002) and Jaspar (Portales-Casamar et al. 2010) databases indicates that the majority of them have similarities with binding sites of transcription factors, which participates in development, regulation of the cell cycle, and apoptosis (Table 19.2). For all species were found motifs, presented in all three tissues. Some of them correspond to the ubiquitous binding sites like TATA-box or CCAAT-box. At the same time, part of the motifs corresponds to tissue-specific binding sites. For instance, the motifs for binding sites of TATA binding protein or SP1, important for the early development of an organism were found in all three sets of promoters. On the contrary, the motifs of CREB binding site, required for brain development and functioning were found in the set of brain-specific promoters. A lot of the motifs were not interpreted significantly. These unclassified motifs could

Table 19.2 Classification of the motifs using TRRD, TRANSFAC, and Jaspar

Brain set	Excretory set	Lungs set
SP1	SP1	SP1
SREBP1	SREBP1	SREBP1
TBP	TBP	TBP
NKX	NKX	NKX
NFk_B	NFk_B	NFk_B
IRF1	IRF1	IRF1
AP2	AP2	E2F
EGR	CP2	CP2
Max	GR	Max
CREB	GATA1-3	GATA1-3
MYT1	STAT6	STAT6
MYC	PAX	PAX
SRF	E2F	SRF

correspond to still unknown species-specific transcription factor binding sites or to some structural features of promoters, like short polyA-polyT runs, which are known to induce DNA curvature or serve as "easily melting" sites.

Finally, we estimated species-specificity of the motifs obtained in the promoters of different mammalian species. For this purpose, we calculated the average oligonucleotide similarity $H_{Oli(Homo,X)}$ between human promoters and the promoters of other mammalians using the ARGO program as follows. Let's compare the set of human promoters and the set of chimpanzee promoters. At first, we calculate the oligonucleotide similarity of the first human promoter and all promoters of chimpanzee, to find the maximal value of similarity function. Then we repeat this procedure for all other human promoters and finally estimate the oligonucleotide similarity $F_{Oli(Homo;Pan)} = \sum_{i=1}^{N} \max F_i/N$ of the sets analyzed. Finally, the value of the average oligonucleotide similarity is: $H_{Oli(Homo;Pan)} = \frac{F_{Oli(Homo;Pan)} - F_{Oli(Homo;Random)}}{F_{Oli(Homo;Homo)} - F_{Oli(Homo;Random)}}$, where $F_{Oli(Homo,Random)}$ is the oligonucleotide similarity between the set of human promoters and the set of random sequences, generated with the same mononucleotide content. $F_{Oli(Homo,Homo)}$ is the oligonucleotide similarity of the set of human promoters with itself.

To estimate the dependence of the average oligonucleotide similarity H_{Oli} on the homology of the sequences analyzed, we calculated the average homology H_{Align} of the sets using the pairwise alignment by the same way.

Figure 19.6 demonstrates the results of comparison of the human promoters and promoters of other species. The highest values of the similarities to human are typical for other primates. Promoters of all tissues demonstrate the similar values of average homology and oligonucleotide similarity. It is of interest that oligonucleotide similarity of the set of brain-specific dog promoters is slightly higher then the oligonucleotide similarity in promoters of other non-primate species and it is equal to the value of macaque.

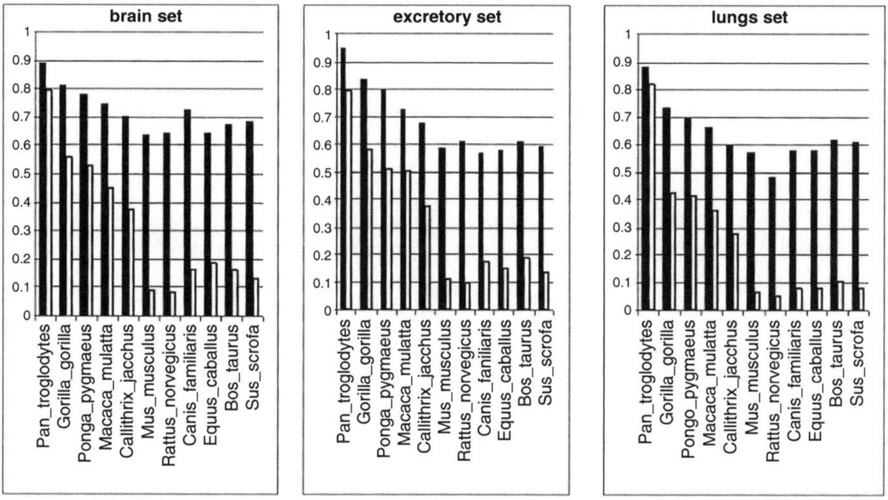

Fig. 19.6 Comparison of the oligonucleotide similarity H_{Oli} (*black bar*), the average homology H_{Align} (*white bar*) of *H. sapiens* promoters, and the promoters of other species

Comparison of the oligonucleotide similarity and average homology demonstrates that for phylogenetically close to human species like other primates, these values differ no more than per 30% in all tissues. For chimpanzee, this difference is less than 15%. For tissue-specific promoters of evolutionary distant from human species, like non-primates, the average homology values are relatively low, but oligonucleotide similarity values are rather high. The value of difference varies from 40% to 55%. We suggest that it could be explained by evolutionary conservation of the tissue-specific regulatory signals located in upstream regions of genes of phylogenetically distant species, despite of upstream region low similarities.

19.4 Conclusion

The potential regulatory signals in promoters of miRNA genes, expressed in different tissues of mammalians were revealed; these may be used for the further experimental analysis. The vast majority of the motifs obtained, correspond to the transcription factor binding sites, involved in the regulation of the organism development, tissue differentiation, cell cycle regulation, and apoptosis. It was shown that in tissue-specific promoters of evolutionarily distant species of mammalians, similarity at the level of regulatory signals is significantly higher than the average homology of sequences.

Acknowledgments The work was supported by the Russian Foundation for Basic Research (grants no. 09-04-01641-a, 11-04-12167 and 11-04-01888-a), Integration projects of the Siberian Branch of the Russian Academy of Sciences no. 26, 113, 119 and Programs of the Russian

Academy of Sciences no. 22 (project no. 8) and no. 23 (project no. 29), the Ministry of Science and Education of the Russian Federation (Contracts no. P857, no. P721).

References

Altuvia Y, Landgraf P, Lithwick G, Elefant N, Pfeffer S, Aravin A, Brownstein MJ, Tuschl T, Margalit H (2005) Clustering and conservation patterns of human microRNAs. Nucleic Acids Res 33:2697–2706

Alvarez-Garcia I, Miska EA (2005) MicroRNA functions in animal development and human disease. Development 132:4653–4662

Bartel DP (2004) MicroRNAs: genomics, biogenesis, mechanism, and function. Cell 116:281–297

Bartel DP (2009) MicroRNAs: target recognition and regulatory functions. Cell 136:215–233

Berezikov E, Guryev V, van de Belt J, Wienholds E, Plasterk RH, Cuppen E (2005) Phylogenetic shadowing and computational identification of human microRNA genes. Cell 120:21–24

Berezikov E, Thuemmler F, van Laake LW, Kondova I, Bontrop R, Cuppen E, Plasterk RH (2006) Diversity of microRNAs in human and chimpanzee brain. Nat Genet 38:1375–1377

Bernstein E, Caudy AA, Hammond SM, Hannon GJ (2001) Role for a bidentate ribonuclease in the initiation step of RNA interference. Nature 409:363–366

Bohnsack MT, Czaplinski K, Gorlich D (2004) Exportin 5 is a RanGTP-dependent dsRNA-binding protein that mediates nuclear export of pre-miRNAs. RNA 10:185–191

Borchert GM, Lanier W, Davidson BL (2006) RNA polymerase III transcribes human microRNAs. Nat Struct Mol Biol 13:1097–1101

Bracht J, Hunter S, Eachus R, Weeks P, Pasquinelli AE (2004) Trans-splicing and polyadenylation of let-7 microRNA primary transcripts. RNA 10:1586–1594

Bueno MJ, Pérez de Castro I, Gómez de Cedrón M, Santos J, Calin GA, Cigudosa JC, Croce CM, Fernández-Piqueras J, Malumbres M (2008) Genetic and epigenetic silencing of microRNA-203 enhances ABL1 and BCR-ABL1 oncogene expression. Cancer Cell 13:496–506

Cai X, Hagedorn CH, Cullen BR (2004) Human microRNAs are processed from capped, polyadenylated transcripts that can also function as mRNAs. RNA 10:1957–1966

Calin GA, Dumitru CD, Shimizu M, Bichi R, Zupo S, Noch E, Aldler H, Rattan S, Keating M, Rai K et al (2002) Frequent deletions and down-regulation of micro- RNA genes miR15 and miR16 at 13q14 in chronic lymphocytic leukemia. Proc Natl Acad Sci USA 99:15524–15529

Calin GA, Sevignani C, Dumitru CD, Hyslop T, Noch E, Yendamuri S, Shimizu M, Rattan S, Bullrich F, Negrini M, Croce CM (2004) Human microRNA genes are frequently located at fragile sites and genomic regions involved in cancers. Proc Natl Acad Sci USA 101:2999–3004

Calin GA, Ferracin M, Cimmino A, Di Leva G, Shimizu M, Wojcik SE, Iorio MV, Visone R, Sever NI, Fabbri M et al (2005) A microRNA signature associated with prognosis and progression in chronic lymphocytic leukemia. N Engl J Med 353:1793–1801

Cao X, Yeo G, Muotri AR, Kuwabara T, Gage FH (2006) Noncoding RNAs in the mammalian central nervous system. Annu Rev Neurosci 29:77–103

Caudy AA, Myers M, Hannon GJ, Hammond SM (2002) Fragile X-related protein and VIG associate with the RNA interference machinery. Genes Dev 16:2491–2496

Chang TC, Yu D, Lee YS, Wentzel EA, Arking DE, West KM, Dang CV, Thomas-Tikhonenko A, Mendell JT (2008) Widespread microRNA repression by Myc contributes to tumorigenesis. Nat Genet 40:43–50

Chen JF, Mandel EM, Thomson JM, Wu Q, Callis TE, Hammond SM, Conlon FL, Wang DZ (2006) The role of microRNA-1 and microRNA-133 in skeletal muscle proliferation and differentiation. Nat Genet 38:228–233

Croce CM, Calin GA (2005) MiRNAs, cancer, and stem cell division. Cell 122:6–7

Cullen BR (2004) Transcription and processing of human microRNA precursors. Mol Cell 16:861–865

Davis TH, Cuellar TL, Koch SM, Barker AJ, Harfe BD, McManus MT, Ullian EM (2008) Conditional loss of Dicer disrupts cellular and tissue morphogenesis in the cortex and hippocampus. J Neurosci 28:4322–4330

Dogini DB, Ribeiro PA, Rocha C, Pereira TC, Lopes-Cendes I (2008) MicroRNA expression profile in murine central nervous system development. J Mol Neurosci 35:331–337

Enright AJ, John B, Gaul U, Tuschl T, Sander C, Marks DS (2003) MicroRNA targets in *Drosophila*. Genome Biol 5:R1

Filipowicz W, Bhattacharyya SN, Sonenberg N (2008) Mechanisms of post-transcriptional regulation by microRNAs: are the answers in sight? Nat Rev Genet 9:102–114

Fiore R, Khudayberdiev S, Christensen M, Siegel G, Flavell SW, Kim TK, Greenberg ME, Schratt G (2009) Mef2-mediated transcription of the miR379–410 cluster regulates activity-dependent dendritogenesis by fine-tuning Pumilio2 protein levels. EMBO J 28:697–710

Foshay KM, Gallicano GI (2009) MiR-17 family miRNAs are expressed during early mammalian development and regulate stem cell differentiation. Dev Biol 326:431–443

Giraldez AJ, Mishima Y, Rihel J, Grocock RJ, Van Dongen S, Inoue K, Enright AJ, Schier AF (2006) Zebrafish MiR-430 promotes deadenylation and clearance of maternal mRNAs. Science 312:75–79

Grishok A, Pasquinelli AE, Conte D, Li N, Parrish S, Ha I, Baillie DL, Fire A, Ruvkun G, Mello CC (2001) Genes and mechanisms related to RNA interference regulate expression of the small temporal RNAs that control *C. elegans* developmental timing. Cell 106:23–34

He L, Thomson JM, Hemann MT, Hernando-Monge E, Mu D, Goodson S, Powers S, Cordon-Cardo C, Lowe SW, Hannon GJ, Hammond SM (2005) A microRNA polycistron as a potential human oncogene. Nature 435:828–833

He L, He X, Lowe SW, Hannon GJ (2007) MicroRNAs join the p53 network – another piece in the tumoursuppression puzzle. Nat Rev Cancer 7:819–822

Huang TH, Fan B, Rothschild MF, Hu ZL, Li K, Zhao SH (2007) MiRFinder: an improved approach and software implementation for genome-wide fast microRNA precursor scans. BMC Bioinform 8:341

Hutvágner G, McLachlan J, Pasquinelli AE, Bálint E, Tuschl T, Zamore PD (2001) A cellular function for the RNAinterference enzyme Dicer in the maturation of the let-7 small temporal RNA. Science 293:834–838

Ibáñez-Ventoso C, Vora M, Driscoll M (2008) Sequence relationships among *C. elegans*, *D. melanogaster* and human microRNAs highlight the extensive conservation of microRNAs in biology. PLoS ONE 3:e2818

Jiang Q, Wang Y, Hao Y, Juan L, Teng M, Zhang X, Li M, Wang G, Liu Y (2008) miR2Disease: a manually curated database for microRNA deregulation in human disease. Nucleic Acids Res 37:D98–D104

John B, Enright AJ, Aravin A, Tuschl T, Sander C, Marks DS (2004) Human microRNA targets. PLoS Biol 2:e363

Johnnidis JB, Harris MH, Wheeler RT, Stehling-Sun S, Lam MH, Kirak O, Brummelkamp TR, Fleming MD, Camargo FD (2008) Regulation of progenitor cell proliferation and granulocyte function by microRNA-223. Nature 451:1125–1129

Kawahara Y, Zinshteyn B, Chendrimada TP, Shiekhattar R, Nishikura K (2007) RNA editing of the microRNA-151 precursor blocks cleavage by the Dicer–TRBP complex. EMBO Rep 8:763–769

Kawahara Y, Megraw M, Kreider E, Iizasa H, Valente L, Hatzigeorgiou AG, Nishikura K (2008) Frequency and fate of microRNA editing in human brain. Nucleic Acids Res 36:5270–5280

Kawaji H, Severin J, Lizio M, Waterhouse A, Katayama S, Irvine KM, Hume DA, Forrest AR, Suzuki H, Carninci P, Hayashizaki Y, Daub CO (2009) The FANTOM web resource: from mammalian transcriptional landscape to its dynamic regulation. Genome Biol 4:R40

Ketting RF, Fischer SE, Bernstein E, Sijen T, Hannon GJ, Plasterk RH (2001) Dicer functions in RNA interference and in synthesis of small RNA involved in developmental timing in *C. elegans*. Genes Dev 15:2654–2659

Kim VN (2004) MicroRNA precursors in motion: exportin-5 mediates their nuclear export. Trends Cell Biol 14:156–159

Kim VN (2005) MicroRNA biogenesis: coordinated cropping and dicing. Nat Rev Mol Cell Biol 6:376–385

Kim HK, Lee YS, Sivaprasad U, Malhotra A, Dutta A (2006) Muscle-specific microRNA miR-206 promotes muscle differentiation. J Cell Biol 174:677–687

Kim DH, Saetrom P, Jr Snove O, Rossi JJ (2008) MicroRNA-directed transcriptional gene silencing in mammalian cells. Proc Natl Acad Sci USA 105:16230–16235

Knight SW, Bass BL (2001) A role for the RNase III enzyme DCR-1 in RNA interference and germ line development in *Caenorhabditis elegans*. Science 293:2269–2271

Kolchanov NA, Ignatieva EV, Ananko EA, Podkolodnaya OA, Stepanenko IL, Merkulova TI, Pozdnyakov MA, Podkolodny NL, Naumochkin AN, Romashchenko AG (2002) Transcription regulatory regions database (TRRD): its status in 2002. Nucleic Acids Res 30:312–317

Kosik KS (2006) The neuronal microRNA system. Nat Rev Neurosci 7:911–920

Krichevsky AM, Sonntag KC, Isacson O, Kosik KS (2006) Specific microRNAs modulate embryonic stem cell-derived neurogenesis. Stem Cells 24:857–864

Lagos-Quintana M, Rauhut R, Yalcin A, Meyer J, LendeckelW TT (2002) Identification of tissue-specific microRNAs from mouse. Curr Biol 12:735–739

Lai EC, Tomancak P, Williams RW, Rubin GM (2003) Computational identification of *Drosophila* microRNA genes. Genome Biol 4:R42

Landgraf P, Rusu M, Sheridan R, Sewer A, Iovino N, Aravin A, Pfeffer S, Rice A, Kamphorst AO, Landthaler M, Lin C, Socci ND, Hermida L, Fulci V, Chiaretti S, Foà R, Schliwka J, Fuchs U, Novosel A, Müller RU, Schermer B, Bissels U, Inman J, Phan Q, Chien M, Weir DB, Choksi R, De Vita G, Frezzetti D, Trompeter HI, Hornung V, Teng G, Hartmann G, Palkovits M, Di Lauro R, Wernet P, Macino G, Rogler CE, Nagle JW, Ju J, Papavasiliou FN, Benzing T, Lichter P, Tam W, Brownstein MJ, Bosio A, Borkhardt A, Russo JJ, Sander C, Zavolan M, Tuschl T (2007) A mammalian microRNA expression atlas based on small RNA library sequencing. Cell 129:1401–1414

Lee YS, Dutta A (2009) MicroRNAs in cancer. Annu Rev Pathol 4:199–227

Lee Y, Jeon K, Lee JT, Kim S, Kim VN (2002) MicroRNA maturation: stepwise processing and subcellular localization. EMBO J 21:4663–4670

Lee Y, Ahn C, Han J, Choi H, Kim J, Yim J, Lee J, Provost P, Rådmark O, Kim S, Kim VN (2003) The nuclear RNase III Drosha initiates microRNA processing. Nature 425:415–419

Lee Y, Kim M, Han J, Yeom KH, Lee S, Baek SH, Kim VN (2004) MicroRNA genes are transcribed by RNA polymerase II. EMBO J 23:4051–4060

Lewis BP, Shih IH, Jones-Rhoades MW, Bartel DP, Burge CB (2003) Prediction of mammalian microRNA targets. Cell 115:787–798

Li SC, Pan CY, Lin WC (2006) Bioinformatic discovery of microRNA precursors from human ESTs and introns. BMC Genomics 7:164

Lu C, Tej SS, Luo S, Haudenschild CD, Meyers BC, Green PJ (2005) Elucidation of the small RNA component of the transcriptome. Science 309:1567–1569

Lujambio A, Ropero S, Ballestar E, Fraga MF, Cerrato C, Setién F, Casado S, Suarez-Gauthier A, Sanchez-Cespedes M, Git A, Spiteri I, Das PP, Caldas C, Miska E, Esteller M (2007) Genetic unmasking of an epigenetically silenced microRNA in human cancer cells. Cancer Res 67:1424–1429

Lund E, Güttinger S, Calado A, Dahlberg JE, Kutay U (2004) Nuclear export of microRNA precursors. Science 303:95–98

Margulies M, Egholm M, Altman WE, Attiya S, Bader JS, Bemben LA, Berka J, Braverman MS, Chen YJ, Chen Z, Dewell SB, Du L, Fierro JM, Gomes XV, Godwin BC, He W, Helgesen S, Ho CH, Irzyk GP, Jando SC, Alenquer ML, Jarvie TP, Jirage KB, Kim JB, Knight JR, Lanza

JR, Leamon JH, Lefkowitz SM, Lei M, Li J, Lohman KL, Lu H, Makhijani VB, McDade KE, McKenna MP, Myers EW, Nickerson E, Nobile JR, Plant R, Puc BP, Ronan MT, Roth GT, Sarkis GJ, Simons JF, Simpson JW, Srinivasan M, Tartaro KR, Tomasz A, Vogt KA, Volkmer GA, Wang SH, Wang Y, Weiner MP, Yu P, Begley RF, Rothberg JM (2005) Genome sequencing in microfabricated high-density picolitre reactors. Nature 437:376–380

Matys V, Kel-Margoulis OV, Fricke E, Liebich I, Land S, Barre-Dirrie A, Reuter I, Chekmenev D, Krull M, Hornischer K, Voss N, Stegmaier P, Lewicki-Potapov B, Saxel H, Kel AE, Wingender E (2006) TRANSFAC and its module TRANSCompel: transcriptional gene regulation in eukaryotes. Nucleic Acids Res 34:D108–D110

McManus MT (2003) MicroRNAs and cancer. Semin Cancer Biol 13:253–258

Miska EA, Alvarez-Saavedra E, Townsend M, Yoshii A, Sestan N, Rakic P, Constantine-Paton M, Horvitz HR (2004) Microarray analysis of microRNA expression in the developing mammalian brain. Genome Biol 5:R68

Nam JW, Shin KR, Han J, Lee Y, Kim VN, Zhang BT (2005) Human microRNA prediction through a probabilistic co-learning model of sequence and structure. Nucleic Acids Res 33:3570–3581

Pasquinelli AE, Hunter S, Bracht J (2005) MicroRNAs: a developing story. Curr Opin Genet Dev 15:200–205

Portales-Casamar E, Thongjuea S, Kwon AT, Arenillas D, Zhao X, Valen E, Yusuf D, Lenhard B, Wasserman WW, Sandelin A (2010) JASPAR 2010: the greatly expanded open-access database of transcription factor binding profiles. Nucleic Acids Res 38:D105–D110

Pulipparacharuvil S, Renthal W, Hale CF, Taniguchi M, Xiao G, Kumar A, Russo SJ, Sikder D, Dewey CM, Davis MM, Greengard P, Nairn AC, Nestler EJ, Cowan CW (2008) Cocaine regulates MEF2 to control synaptic and behavioral plasticity. Neuron 59:621–633

Rao PK, Kumar RM, Farkhondeh M, Baskerville S, Lodish HF (2006) Myogenic factors that regulate expression of muscle-specific microRNAs. Proc Natl Acad Sci USA 103:8721–8726

Rodriguez A, Griffiths-Jones S, Ashurst JL, Bradley A (2004) Identification of mammalian microRNA host genes and transcription units. Genome Res 14:1902–1910

Saito Y, Liang G, Egger G, Friedman JM, Chuang JC, Coetzee GA, Jones PA (2006) Specific activation of microRNA-127 with downregulation of the proto-oncogene BCL6 by chromatin-modifying drugs in human cancer cells. Cancer Cell 9:435–443

Schratt GM, Tuebing F, Nigh EA, Kane CG, Sabatini ME, Kiebler M, Greenberg ME (2006) A brain-specific microRNA regulates dendritic spine development. Nature 439:283–289

Sempere LF, Freemantle S, Pitha-Rowe I, Moss E, Dmitrovsky E, Ambros V (2004) Expression profiling of mammalian microRNAs uncovers a subset of brain-expressed microRNAs with possible roles in murine and human neuronal differentiation. Genome Biol 5:R13

Shcherbata HR, Hatfield S, Ward EJ, Reynolds S, Fischer KA, Ruohola-Baker H (2006) The microRNA pathway plays a regulatory role in stem cell division. Cell Cycle 5:172–175

Stadler BM, Ruohola-Baker H (2008) Small RNAs: keeping stem cells in line. Cell 132:563–566

Stark KL, Xu B, Bagchi A, Lai WS, Liu H, Hsu R, Wan X, Pavlidis P, Mills AA, Karayiorgou M, Gogos JA (2008) Altered brain microRNA biogenesis contributes to phenotypic deficits in a 22q11-deletion mouse model. Nat Genet 40:751–760

Stormo GD (2000) Gene-finding approaches for eukaryotes. Genome Res 10:394–397

Vishnevsky OV, Kolchanov NA (2005) ARGO: a web system for the detection of degenerate motifs and large-scale recognition of eukaryotic promoters. Nucleic Acids Res 33:w417–w422

Wheeler G, Ntounia-Fousara S, Granda B, Rathjen T, Dalmay T (2006) Identification of new central nervous system specific mouse microRNAs. FEBS Lett 580:2195–2200

Wightman B, Ha I, Ruvkun G (1993) Posttranscriptional regulation of the heterochronic gene lin-14 mediates temporal pattern formation in *C. elegans*. Cell 75:855–862

Yang W, Chendrimada TP, Wang Q, Higuchi M, Seeburg PH, Shiekhattar R, Nishikura K (2006) Modulation of microRNA processing and expression through RNA editing by ADAR deaminases. Nat Struct Mol Biol 13:13–21

Yi R, Doehle BP, Qin Y, Macara IG, Cullen BR (2005) Overexpression of exportin 5 enhances RNA interference mediated by short hairpin RNAs and microRNAs. RNA 11:220–226

Yi R, Poy MN, Stoffel M, Fuchs E (2008) A skin microRNA promotes differentiation by repressing 'stemness'. Nature 452:225–229

Zeng Y (2009) Regulation of the mammalian nervous system by microRNAs. Mol Pharmacol 75:259–264

Index

P. Pontarotti (ed.), *Evolutionary Biology – Concepts, Biodiversity, Macroevolution and Genome Evolution*, DOI 10.1007/978-3-642-20763-1,

£149.95